MICROTECHNOLOGY AND MEMS

Springer
Berlin
Heidelberg
New York
Hong Kong
London
Milan
Paris
Tokyo

Physics and Astronomy ONLINE LIBRARY

http://www.springer.de

MICROTECHNOLOGY AND MEMS

Series Editor: H. Baltes D. Liepmann

The series Microtechnology and MEMS comprises text books, monographs, and state-of-the-art reports in the very active field of microsystems and microtechnology. Written by leading physicists and engineers, the books describe the basic science, device design, and applications. They will appeal to researchers, engineers, and advanced students.

Mechanical Microsensors
By M. Elwenspoek and R. Wiegerink

CMOS Cantilever Sensor Systems
Atomic Force Microscopy and Gas Sensing Applications
By D. Lange, O. Brand, and H. Baltes

Micromachines as Tools for Nanotechnology
Editor: H. Fujita

Modelling of Microfabrication Systems
By R. Nassar and W. Dai

Laser Diode Microsystems
By H. Zappe

Silicon Microchannel Heat Sinks
Theories and Phenomena
By L. Zhang, K.E. Goodson, and T.W. Kenny

Hans Zappe

Laser Diode Microsystems

With 167 Figures

Springer

Professor Dr. H. Zappe
Laboratory for Micro-optics
Institute of Microsystem Technology
University of Freiburg
79110 Freiburg, Germany
E-mail: zappe@imtek.uni-freiburg.de

Series Editors:

Professor Dr. H. Baltes
ETH Zürich, Physical Electronics Laboratory
ETH Hoenggerberg, HPT-H6, 8093 Zürich, Switzerland

Professor Dr. Hiroyuki Fujita
University of Tokyo, Institute of Industrial Science
4-6-1 Komaba, Meguro-ku, Tokyo 153-8505, Japan

Professor Dr. Dorian Liepmann
University of California, Department of Bioengineering
466 Evans Hall, #1762, Berkeley, CA 94720-1762, USA

ISSN 1439-6599

ISBN 3-540-40454-6 Springer-Verlag Berlin Heidelberg New York

Library of Congress Cataloging in Publication Data.
Zappe, Hans P.
Laser diode microsystems / Hans Zappe.
p.cm. – (Microtechnology and MEMS)
Includes bibliographical references and index.
ISBN 3-540-40454-6 (alk. paper)
1. Microelectronics. 2. Solid-state lasers. 3. Diodes, Semiconductor. I. Title. II. Series.
TK7874.Z37 2003 621.36'6–dc21 2003054229

Springer-Verlag Berlin Heidelberg New York
a member of BertelsmannSpringer Science+Business Media GmbH

http://www.springer.de

Typesetting: camera-ready copy by the author
Cover concept: eStudio Calamar Steinen
Cover production: *design & production* GmbH, Heidelberg

Printed on acid-free paper SPIN: 10857158 57/3141/tr - 5 4 3 2 1 0

For Max and Fran

Preface

The functionality of microsystems is increasing daily. Whereas a decade ago a purely mechanical silicon micro-machined device was justifiably considered a technical marvel, microsystems may now include electrical, magnetic, chemical, optical or biological functions. The increasingly interdisciplinary approach required for advanced work in this area is one of the reasons that microsystem technology has become one of the most exciting fields in which to work.

Optics in microsystems has seen considerable growth due in part to significant demand from industry, particularly that involved in telecommunications. Optical microsystems are in use and under development for transmission, switching and reception of optical data streams, for complex optical sensor systems, for optical data storage and a host of medical applications. As many of these applications mature, light-based microsystems will become increasingly industrially viable and of widening technological importance.

Optical microsystems require a source of photons. Due to their miniscule size, high efficiency and broad range of available wavelengths and output powers, semiconductor laser diodes (as well as, frequently, light-emitting diodes) are most often the photon source of choice for inclusion in a compact, high-performance optical microsystem. With an estimated almost \$3 billion market share in 2003, the microscopic laser diode has eclipsed all macroscopic types of lasers and is by its very nature ideal for employment in microsystems.

This text covers the physics, technology and application of laser diodes for use in optical microsystems. It is intended for students studying and working in microsystem or MEMS engineering as well as engineers and scientists active in any of the broad spectrum of disciplines in which microsystems are becoming increasingly relevant. Our tour through this field will encompass a discussion of basic laser diode physics as well as the structure, fabrication techniques and operating characteristics of a variety of diode lasers. An examination of the most important micro-optical components as well as the assembly techniques required for the construction of a complete optical microsystem will give the reader a hint of the breadth of optical technology available at these physical dimensions.

Acknowledgments

It is with pleasure that I am able to thank many individuals who have contributed their time and intellectual effort to the successful completion of this book.

A number of people have dedicated significant energy to reading and correcting large portions of the manuscript. I am very grateful to my colleagues Wolfgang Mönch and Oliver Paul here at the Institute for Microsystem Technology at the University of Freiburg, Markus Rossi of Heptagon Oy, Zurich, Joachim Wagner of the Fraunhofer Institute for Applied Solid State Physics, Freiburg, and Olav Solgaard of Stanford University, California, for their thorough reading and extensive review of portions of the text. I am likewise indebted to Henry Baltes of the ETH Zurich for his reading of and commenting on the text as well as his assistance in seeing the book to publication. Special thanks are due to my father, Hans H. Zappe, of San Jose, California, for his thorough analysis of the first chapter and his insightful comments into a number of physical phenomena.

I have the truly great pleasure of working with a devoted team in the Micro-optics Laboratory at the Institute for Microsystem Technology and many of my co-workers have contributed in some form to the text. I am indebted to Bernd Aatz and Carsten Glasenapp for the extensive measurement of laser characteristics which we use as examples in Chapter 5 and for not being given to exasperation when I wanted yet something else measured anew; Dennis Hohlfeld for providing the VCL cavity calculations and the photograph of his thermally-tunable filter shown in Fig. 6.36; Mario Hug and Florian Krogmann for micro-optical lens and component data; Alexander Hodapp for the refractive micro-lens of Fig. 6.12; Andreas Mohr for the Fresnel lens of Fig. 6.15; and David Kallweit for the calculated fiber mode profile of Fig. 6.29. I am most grateful to Nadja Katthagen for keeping things in the lab running when my brain was, as it usually is, elsewhere.

The linewidth and noise characteristics presented in Chapter 5 were done at CSEM, Zurich, and I am particularly grateful to Fabrice Monti di Sopra and Martin Hess for enabling or performing these measurements. The VCL wafer map of Fig. 5.38 was likewise a product of my CSEM days and a warm thanks to all my colleagues from those times is in order. I extend my gratitude to Rainer Michalzik of the University of Ulm, and Dieter Wiedenmann of U-L-M photonics GmbH for taking the photograph presented in Fig. 5.4 especially for this text. Many thanks are due to Markus Rossi of Heptagon Oy for supplying the photograph of the pattern generator in Fig. 6.25.

I am indebted to Josef Rosenzweig and Michael Schlechtweg of the Fraunhofer Institute for Applied Solid State Physics, Freiburg, as well as Konrad Czotscher of Lucent Technologies, Nuremberg, for the very nice eye diagram of Fig. 5.18. Grateful thanks to Eiji Higurashi of the NTT Microsystem Laboratories, Kanagawa, for the diagrams and photo of the integrated optical blood-flow sensor of Fig. 1.1 and Fig. 1.2. and to Ulrike Wallrabe of the Forschungszentrum Karlsruhe, for supplying photos and diagrams of the hybrid optical displacement sensor of Fig. 1.3 and Fig. 1.4. I am likewise grateful to Anton Ambrosy of the Alcatel Corporate Research Center, Stuttgart, for the illustrations of the transceiver structure shown in Fig. 1.5 and Fig. 1.6.

Many of the topics addressed have been the subject of courses for students in Microsystem Technology at the University of Freiburg and in short courses organized by the Fondation Suisse pour la Recherche en Microtechnique (FSRM) throughout Europe. I am grateful to all past students and course participants who provided suggestions, criticisms, comments and general feedback on the material in an effort to make the presentation clearer and more concise.

Freiburg, July 2003 *Hans Zappe*

Contents

List of Symbols

Notes:

- SI units are given except when traditional usage dictates otherwise.
- Vector quantities are given in bold-face.
  Example: distinguish electric field $\boldsymbol{E}$ from energy E
- Some quantities need to be determined from context.
  Example: temperature T and transmittance T

Symbol	Description	Unit
A	Area	m^2
a	Aperture size	m
α	Loss	cm^{-1}
A'_e	Modified electronic transition probability	$eV^2\ cm^3/s$
α_0	Background loss	m^{-1}
A_{21}	Einstein A coefficient	s^{-1}
α_a	Absorption coefficient at the laser wavelength	m^{-1}
A_e	Electronic transition probability	$eV\ cm^3/s$
α_{lef}	Linewidth enhancement factor	–
α_m	Mirror loss	m^{-1}
$\boldsymbol{B}$	Magnetic flux density	Wb/m^2
β	z-directed propagation constant	m^{-1}
β'	Difference between $\Delta\beta$ and κ in a Bragg mirror	m^{-1}
β_0	Resonance condition in a Bragg mirror	m^{-1}
β_0	Unmodulated propagation constant	m^{-1}
B_{12}, B_{21}	Einstein B coefficient	m^3/Js^2
B_α	Photon absorption coefficient	cm^6/s^2
B_s	Electronic transition probability	cm^6/s^2
BW	Bandwidth	Hz
χ	Susceptibility	–
c	Speed of light in vacuum	m/s
C_a	Photon absorption constant	$cm^{-1}eV^{-1/2}$
D	Density of an optical grey filter	–
$\boldsymbol{D}$	Electric flux density	C/m^2
D	Lens aperture size	m

Symbol	Description	Unit
d	Distance from aperture to measurement point	m
d	Spot size at the focus of an optical system	m
$\Delta\alpha$	Change in absorption due to modulation	m^{-1}
$\Delta\beta$	Change in propagation constant due to modulation	m^{-1}
$\Delta\beta$	De-tuning from resonance in a Bragg mirror	m^{-1}
Δf	Deviation of excitation from resonance frequency	–
$\Delta\phi$	Phase shift	rad
$\Delta\lambda$	Linewidth, expressed as wavelength	m
$\Delta\lambda_0$	Mode spacing (in Sect. 5.3)	m
$\Delta\lambda_{FP}$	Free spectral range, expressed as wavelength	m
D_n	Electron diffusion coefficient	cm^2/s
Δn	Index step in a fiber	–
$\Delta\nu$	Linewidth, expressed as frequency	Hz
$\Delta\nu_{FP}$	Free spectral range, expressed as frequency	Hz
D_p	Hole diffusion coefficient	cm^2/s
D_T	Total fiber dispersion	$s\ km^{-1}\ nm^{-1}$
$\Delta\omega$	Damping coefficient	rad/s
$\boldsymbol{E}$	Electric field	V/m
E	Energy	J
ε	Dielectric constant	–
ε_0	Permittivity of vacuum	F/m
E_a	Acceptor energy level	eV
E_c	Conduction band energy	eV
E_d	Donor energy level	eV
E_e	Electron energy	eV
E_F	Fermi energy	eV
E_{Fc}	Quasi-Fermi level for the conduction band	eV
E_{Fv}	Quasi-Fermi level for the valence band	eV
E_g	Energy gap	eV
E_h	Hole energy	eV
$E_{h\nu}$	Energy of a photon emitted or absorbed in a quantum well	eV
ε_m	Permittivity in a material	F/m
E_v	Valence band energy	eV
F	Finesse	–
f	Focal length	m
ϕ	Phase	rad
$f(E)$	Fermi-Dirac function	–
$\Phi(\nu)$	Optical field (photon energy density per volume)	Js/m^3
$f/\#$	"f-number" of a lens	–

Symbol	Description	Unit
Φ_0	Optical field intensity	J/m^3Hz
ϕ_{bi}	pn junction built-in potential	V
$f_c(E)$	Occupation probability for electrons in the conduction band	–
Φ_ν	Optical transition intensity	J/m^3
ϕ_n	Potential on the n-side of a pn junction	V
ϕ_p	Potential on the p-side of a pn junction	V
Φ_{sat}	Saturation optical field	Js/m^3
F_{sp}	Spontaneous emission factor	–
$f_v(E)$	Occupation probability for holes in the valence band	–
Γ	Confinement factor	–
G	Photon generation rate	cm^{-1}
γ	Gain	cm^{-1}
$g(\nu)$	Lineshape function	s
γ_d	Differential gain	m^2
γ_{th}	Threshold gain	m^{-1}
$\boldsymbol{H}$	Magnetic field	A/m
η	Efficiency	–
h	Height above the optical axis	m
h	Planck's constant	Js
h	Waveguide core half-width	m
$H(\omega)$	Transfer function	–
η_d	Differential efficiency	–
η_e	Emission efficiency	–
η_i	Internal quantum efficiency	–
η_{wp}	Wall-plug efficiency	–
I	Current	A
i	Index	–
i	Index for mode number in a laser	–
I_d	Detector current	A
I_N	Noise current	A
I_{th}	Threshold current	A
I_{th0}	Threshold current at T = 0°C	A
J	Current density	A/m^2
J_0	Saturation current density of a pn junction	A/m^2
J_n	Electron current density	A/m^2
J_p	Hole current density	A/m^2
J_{th}	Threshold current density	A/m^2
κ	Grating coupling coefficient	m^{-1}
$\boldsymbol{k}$	Propagation constant	m^{-1}

Symbol	Description	Unit
k_0	Propagation constant in free space	m^{-1}
k_B	Boltzmann constant	J/K
k_I	Imaginary part of the propagation constant	m^{-1}
k_R	Real part of the propagation constant	m^{-1}
k_z	z-directed propagation constant	m^{-1}
Λ	Grating period	m
L	Laser cavity, modulator or detector length	m
λ	Wavelength	m
λ_0	Wavelength in vacuum	m
L_c	Coherence length	m
L_c	Coherence length	m
m	Electron or hole mass	kg
m	Grating index	–
m_0	Electron rest mass	kg
$m_c{}^*$	Conduction band effective mass	kg
$m_e{}^*$	Electron effective mass	kg
$m_h{}^*$	Hole effective mass	kg
μ_n	Electron mobility	cm^2/Vs
μ_p	Hole mobility	cm^2/Vs
$m_v{}^*$	Valence band effective mass	kg
N	Number of zones in a Fresnel lens	–
N	Waveguide effective index	–
n	Electron concentration	cm^{-3}
ν	Frequency	Hz
n	Refractive index	–
n'	Excess electron concentration	cm^{-3}
N'_p	Time-varying photon density in a laser	cm^{-3}
ν_0	Laser oscillation frequency	Hz
n_0	Refractive index outside lens	–
n_0	Unmodulated refractive index	–
N_A	Acceptor doping concentration	cm^{-3}
NA	Numerical aperture	–
N_c	Effective density of states in the conduction band	cm^{-3}
n_c	Waveguide cap refractive index	–
N_D	Donor doping concentration	cm^{-3}
N_e	Electron density in a laser	cm^{-3}
n_e	Electron concentration	cm^{-3}
n_g	Waveguide core refractive index	–
N_i	Carrier concentration at energy level i	cm^{-3}

Symbol	Description	Unit
n_i	Imaginary part of the complex refractive index	–
n_i	Intrinsic carrier concentration	cm^{-3}
n_i	Refractive index of incident medium	–
n_L	Refractive index of a lens	–
ν_{mod}	Modulation frequency	Hz
N_p	Photon density in a laser	cm^{-3}
N_{p0}	Steady-state photon density in a laser	cm^{-3}
n_r	Optical refractive index (only in Eq. 3.42 and Eq. 3.46)	–
n_r	Real part of the complex refractive index	–
n_s	Waveguide substrate refractive index	–
n_t	Refractive index of transmitted medium	–
N_v	Effective density of states in the valence band	cm^{-3}
$\boldsymbol{P}$	Polarization	C/m^2
P	Power	W
p	Hole concentration	cm^{-3}
p'	Excess hole concentration	cm^{-3}
$P(r)$	Radial power distribution of a Gaussian beam	W
P_0	Power incident onto an optical interface	W
P_r	Power reflected at an optical interface	W
P_t	Power transmitted through an optical interface	W
q	Fundamental electronic charge	C
θ_c	Critical angle	°
θ_i	Incident angle	°
θ_{max}	Maximum input angle for a lens or fiber	°
θ_t	Transmitted angle	°
R	Radius of curvature of a lens	m
R	Reflectance	–
R	Resistance	Ω
R	Responsivity	A/W
ρ	Charge density	C/m^3
r	Linear electro-optic coefficient	m/V
r	Radial parameter	m
r	Reflectivity	–
ρ	Resistivity	Ωm
r_0	1/e width (radius) of a Gaussian beam	m
R_α	Electron transition rate due to photon absorption	$s^{-1}eV^{-1}cm^{-3}$
r_{Bragg}	Reflectivity of a Bragg mirror	–
ρ_c	Density of states for electrons	$eV^{-1}cm^{-3}$
R_e	Spontaneous emission rate	$s^{-1}eV^{-1}cm^{-3}$

Symbol	Description	Unit
r_L, r_R	Reflectivity of the (left, right) mirror of an etalon	–
R_{laser}	Laser junction series resistance	Ω
R_{nr}	Non-radiative recombination rate	$cm^{-3}s^{-1}$
r_{QW}	Density of states in a quantum well	$eV^{-1}cm^{-2}$
R_s	Stimulated emission rate	$s^{-1}eV^{-1}cm^{-3}$
R_{sp}	Spontaneous recombination rate	$cm^{-3}s^{-1}$
ρ_v	Density of states for holes	$eV^{-1}cm^{-3}$
S	Intensity	W/m^2
$\mathbf{S}$	Poynting vector	W/m^2
σ	Conductivity	$\Omega^{-1}m^{-1}$
s	Quadratic electro-optic coefficient	m^2/V^2
S'	Modulated intensity	W/m^2
$S(\nu)$	Power spectral density	W^2/Hz
S/N	Signal to noise ratio	–
s_i	Distance between image and lens	m
S_ν	Optical energy intensity	W/m^2
s_o	Distance between object and lens	m
σ_{th}	Thermal conductivity	$W\ m^{-1}K^{-1}$
T	Temperature	K
T	Transmittance	–
t	Layer thickness	m
τ	Lifetime	s
t	Time	s
t	Transmittivity	–
T_0	Characteristic temperature of a laser	°C
t_L, t_R	Transmittivity of the (left, right) mirror of an etalon	–
τ_{nr}	Non-radiative transition lifetime	s
τ_p	Photon lifetime	s
τ_r	Radiative transition lifetime	s
τ_{spon}	Spontaneous carrier lifetime	s
V	Volume	m^3
v	Velocity	m/s
V_A	Voltage applied across a pn junction	V
v_g	Group velocity	m/s
V_π	Half-wave voltage	V
W	Volume energy density	J/m^3
ω	Angular frequency	rad/s

Symbol	Description	Unit
ω_0	Resonance frequency	rad/s
ω_{3dB}	Modulation bandwidth	rad/s
W_E	Electric volume energy density	J/m^3
W_H	Magnetic volume energy density	J/m^3
W_i	Electronic transition rate per atom	s^{-1}
ω_R	Laser resonant frequency or relaxation oscillation frequency	rad/s
Ξ_N	Spectral noise density	W^2/Hz
Z	Impedance	Ω
ζ	Damping of the laser resonance	rad/s
Z_0	Impedance of free space	Ω

1 Introduction

The text before you addresses the physics and technology of laser diodes with a focus on their use in optical microsystems. Before beginning the technical discussion, it may be of edifying value to consider the laser diode in its historical and applications context. We thus begin with a brief laser diode history and an equally brief overview of some of the microsystem applications in which the semiconductor laser is currently playing an essential role.

1.1 Laser Diodes: A Very Brief History

The semiconductor laser is about as old as the author of this book. Both the laser diode and the author spent their early years in relative obscurity and came of age at about the same time. There the parallels end: whereas the laser diode became a shining star, known to a sizable fraction of the inhabitants of the planet, the author is still wallowing in deserved obscurity. We therefore surmise that a consideration of the history of the former is of considerably greater interest than that of the latter.

On May 16, 1960, Theodor Maiman, working at the Hughes Laboratory in Culver City, California, operated the first optical maser, a device which subsequently became known as a laser[1]. Using a flashlamp to pump a ruby crystal with polished end-facets, Maiman measured the emitted spectrum and concluded that he had indeed experienced laser emission. The editors of *Physical Review Letters* rejected his submitted publication since they feared it was "just another maser paper" (Townes 2000), but Maiman published his results in *Nature* on August 6, 1960 (Maiman 1960) and the laser era began.

The development of lasers and masers took place at a time when research could still be undertaken for its own sake, but very concrete applications were nonetheless envisaged for these devices. Advances in laser technology in the 1960s were strongly driven by the need to enhance the bandwidth of communications networks

---

[1] Both maser and laser are acronyms: maser = "**m**icrowave **a**mplification by **s**timulated **e**mission of **r**adiation" and laser = "**l**ight **a**mplification by **s**timulated **e**mission of **r**adiation". The laser was the product of a considerable body of work on masers in 1950s (Schawlow and Townes 1958), for which Townes, Basov and Prokhorov shared the Nobel Prize in 1964. Masers are still used today in relatively specialized areas, such as for time and frequency standards.

(yes, even before the internet) and the history of the semiconductor laser diode is strongly tied in with that of fiber optics and optical telecommunications (Hecht 1999).

Once the first ruby laser was successfully operated, the demonstration of a semiconductor-based laser was pursued with renewed vigor. The use of semiconductors for optical amplification and thus possibly as a laser material was probably first suggested by the mathematician John von Neumann in 1953 (Dupuis 1987). By the time that Maimain demonstrated his ruby laser, a number of groups in the USA were working on the experimental demonstration of lasing in gallium arsenide (GaAs), a compound semiconductor. Using pn diodes to generate high carrier densities in this material, four groups succeeded in 1962 and published almost simultaneously: one group from IBM (Nathan et al. 1962), two from General Electric (Hall et al. 1962; Holonyak et al. 1962) and one group at MIT Lincoln Laboratory (Quist et al. 1962).

These first semiconductor lasers were anything but user-friendly: operated at a temperature of 77 K or below, pulsed with currents on the order of 100 A for 10 μs with very long pauses between pulses, they were more a laboratory curiosity than a promising industrial light source. As a result, researchers began what was to be a ten year effort to reduce the operating currents, achieve continuous wave (CW) operation and get this to happen at room temperature or above (Casey and Panish 1978a, Chap. 1.2). A key technological advance was suggested almost simultaneously by Herbert Kroemer at the Varian Central Research Laboratory in Palo Alto, California (Kroemer 1963) and Zhores Alferov at the Ioffe Physico-Technical Institute in what was then Leningrad, in 1963. Kroemer and Alferov proposed to layer semiconductor materials of differing bandgap to confine the injected electrons of the pn junction spatially, thereby enhancing the efficiency of stimulated emission. The resulting semiconductor heterostructure forms the basis for all modern semiconductor lasers and the work on heterojunctions earned Alferov and Kroemer the Nobel prize in 2000 (along with Jack Kilby for the invention of the integrated circuit).

Heterostructures were more easily proposed than realized and it required considerable development in epitaxy technology before the first useful layers could be fabricated. The first successes were reported by groups from IBM (Rupprecht, Woodall and Pettit 1967) and from the Ioffe Institute (Alferov et al. 1968), using liquid-phase epitaxy and the GaAs/AlGaAs material system. It took another two years before the first single heterojunction lasers were demonstrated, more or less simultaneously at RCA (Kressel and Nelson 1969), Bell Labs (Hayashi, Panish and Foy 1969) and the Ioffe Institute in Russia (Alferov et al. 1970). These lasers still required low ambient temperatures and could still only be operated pulsed.

The final step was the move to a double heterostructure, one in which the charge carriers and the optical field are truly confined to a thin, virtually two-dimensional layer. Using such a layer configuration, the group at Bell Labs achieved room temperature, CW lasing on June 1, 1970 (Hayashi et al. 1970). Due to the vagaries of Cold War publication delays, translation difficulties, security concerns and abject jealousies, the demonstration of essentially the same type of device from the Ioffe Institute only became known in the west some time later, although it is now generally agreed that the Russian group had indeed succeeded a short time before (Alferov et al. 1971). Not long thereafter, a team at Nippon Electric in Japan successfully demonstrated a double heterostructure laser of their own (Sakuma et al. 1971).

By 1970, a semiconductor laser with generally acceptable operating parameters could thus be fabricated and continued refinement of the double heterostructure resulted in further improvement of the operating characteristics (Panish 1975). Despite the impressive volume of very exciting physics which had been learned in the process of laser development, most of the industrial researchers in the field (or at least their CFOs) had a real, revenue-generating application in mind. Many workers in the field thought of the laser as one of the key components in future optical telecommunications systems; for this application, the several minutes of lasing to which the engineer was treated before the laser diode gave up the ghost was decidedly too little. The next engineering goal was then to improve semiconductor laser lifetime.

Laser diodes with 1 000 hour lifetimes were demonstrated by 1973 (Hartman et al. 1973) and 10^6 hours (a bit more than 100 years) was expected through accelerated lifetime testing by 1977 (Hartman, Schumaker and Dixon 1977). These lifetimes, although demonstrated in the laboratory, showed sufficient promise for deployment of semiconductor lasers in the field, where device failures would adversely affect the stability of a communications network and be expensive to replace. The achievable characteristics seemed to be of sufficient merit to allow the formation of the first company to commercialize the device; Laser Diode Laboratories was spun out of RCA in 1969 but did not begin selling product until 1975. Their first lasers had prices between $250 and $350 (Hecht 1999, p. 158).

The increased use of optical fiber for laser transmission now instigated a push to longer wavelengths. The original GaAs-based lasers of the early 1970s emitted at a wavelength around 800 nm but the development of optical fibers soon showed that operation at 1 300 nm or 1 550 nm would be essential for the success of long-range optical communications. At 1 300 nm, optical dispersion in a fiber is minimized whereas at 1 550 nm absorption is at a minimum, providing impetus for the move to longer wavelengths, an evolution which would require the development of new types of semiconductors. The material and epitaxy problems were severe but laser diode wavelengths crept to longer and longer values: the first InGaAsP-based laser, emitting between 1.1 and 1.2 μm, was demonstrated by Bell Labs in 1976 (Hsieh, Rossi and Donnelly 1976) and lifetimes of 1 500 hours were reported a year later (Shen, Hsieh and Lind 1977). The 1.55 μm absorption window was reached not long thereafter and this wavelength range is still used in laser diode-based optical communications systems today.

It was at about this time that the semiconductor laser reached consumers, at least a few of them. In April 1977, the first commercial optical fiber-based telecommunications system was put into operation in Long Beach, California. By the mid

1980s, fiber trunk lines had been installed between major metropolitan areas and in December 1988, TAT-8, the first submarine fiber-optic transatlantic cable, using 1.3 μm lasers, was inaugurated; all of a sudden, users wondered why that annoying half-second delay which characterized transatlantic telephone conversations had disappeared. By 1998, transmission bandwidth of a single fiber was measured in Terabits/second (Chraplyvy and Tkach 1998) and by 2001 in tens of Terabits/second (Bigo et al. 2001). The application of the results of such hero experiments to commercial networks is experiencing a slight hiatus at the time of writing but the public's undying need to push ever more terabytes of data about the planet will likely provide impetus for even faster optical networks.

The author proposes, however, that a different event may have had greater implications for the optical microsystem engineer who today happily uses semiconductor laser diodes. On October 1, 1982, Sony introduced a piece of equipment with the designation CDP-101 onto the Japanese market. At a price of ¥168 000 (about $625 at the time), this consumer electronic device used the light from a semiconductor laser to read digitally-coded audio data from a 12-inch, aluminum-coated polycarbonate disc. Whereas the first purchasers of this compact disc player only had a choice between piano music of Frédéric Chopin, played by Claudio Arrau, Billy Joel's album "52nd Street" and the "The Visitors" by Abba, by 1983, following introduction onto the US market, 800 000 compact discs were sold (Betschon 2002). Five years later, CDs outsold LPs for the first time[2] and it is estimated that in the twenty years since their introduction, 500 million CD players and 110 billion CDs were produced and sold. While most consumers were only vaguely aware of it, the laser had made it to the living room and, even today, few users realize that there is a laser lurking on their bookshelf.

Among the keys to the rapid success of the CD as a data storage medium was the use of a non-contact optical reader head, which uses a laser diode for illumination. Although conceptually based on the highly-developed telecommunications lasers, CD lasers are medium-tech and were originally 780 nm gain-guided GaAs-based laser diodes. The relevant difference between these two types of lasers is volume: in 2002, an estimated 356 million semiconductor lasers were sold for optical data storage applications, at an average price of $2.17 each (Steele 2002). This number represents 86% of all diode lasers sold such that our need to not only transmit but also store ever more terabytes of data has resulted in the laser diode becoming an off-the-shelf commodity item. This latter fact is of particular relevance for the microsystem engineer: the semiconductor laser is an optical component that delivers considerable performance at low cost and is available mass-produced from a variety of vendors. In the following pages, we will learn how it works, what it looks like, what its typical operating characteristics are and how we can incorporate it into an optical microsystem.

[2] For younger readers, the LP refers to a 12" vinyl disc on which an analog signal was encoded by a wobbly spiral groove and read by dragging a mechanical stylus through it. Your parents thought it the epitome of high tech.

1.2 Laser Diodes in Microsystems

Laser diodes are now used in more than communications networks or optical data storage devices; the availability of these light sources with a variety of characteristics and at low cost have made them of particular value for optical microsystems. Microsystems refer to assemblies of components, whether electrical, mechanical, optical or otherwise with device size scales in the micrometer range and system size scales below several millimeters. Closely related to MEMS (Micro-Electro-Mechanical Systems) and often strongly based on MEMS technology, microsystems provide high functionality in systems with very compact dimensions, in which the attendant advantages not only in size but also robustness, power-consumption, manufacturability and ultimately cost open new application horizons. Optical microsystems use light and the source of light is frequently the laser diode hence our present interest in this component.

To give the reader a taste of what is possible using laser diodes in microsystems, we would like to present a few examples of state-of-the-art optical microsystem applications, as developed in industrial and academic laboratories. This selection is intended to be representative; the body of work in this area is large and growing.

1.2.1 Non-Invasive Blood-Flow Monitoring

There is considerable academic, medical and social demand for the development of compact sensors for continuous monitoring of physiological parameters; of particular interest are quasi-autonomous devices which may be worn by a patient allowing unobtrusive, uninterrupted measurement of certain vital statistics. In addition, non-invasive techniques are to be preferred, such that optical measurement approaches are often the most suitable.

An excellent example of what microsystem technology can achieve in this regard is an optical microsystem for monitoring blood flow. Developed at the NTT Telecommunications Energy Laboratories in Kanagawa, Japan, the sensor uses the Doppler effect to continuously and non-invasively measure circulation in a patient (Higurashi, Sawada and Ito 2002). Scattering of an incident light beam by moving blood results in a small wavelength shift from which flow velocity may be determined.

As seen in the schematic drawing and photograph of Fig. 1.1, the blood-flow sensor consists of a DFB laser diode mounted along with a photodetector on a silicon-based substrate; the substrate also includes a waveguide and provides mechanical, thermal and electric contact to the optical devices. The laser wavelength was chosen to be 1310 nm, as the human dermis transmits well at that wavelength with little absorption due to water. The waveguide edge was fabricated with a convex shape, allowing horizontal focussing without the need for an external lens and a silicon cap was used to suppress optical cross-talk between laser and detector.

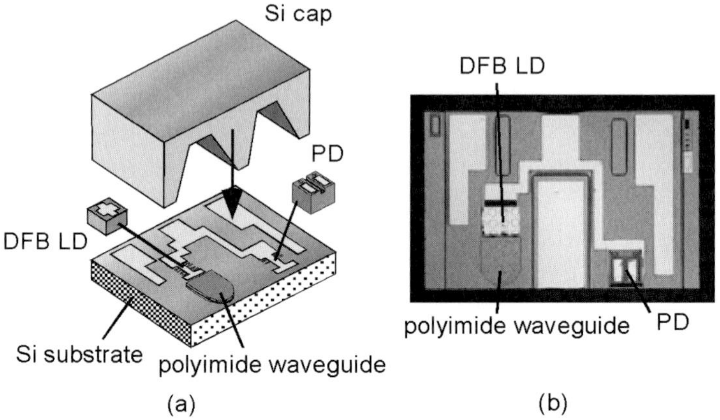

Fig. 1.1. Non-invasive optical blood-flow measurement system; a) schematic view of the laser and detector arrangement; b) photograph of the assembled microsystem. Diagram courtesy of NTT

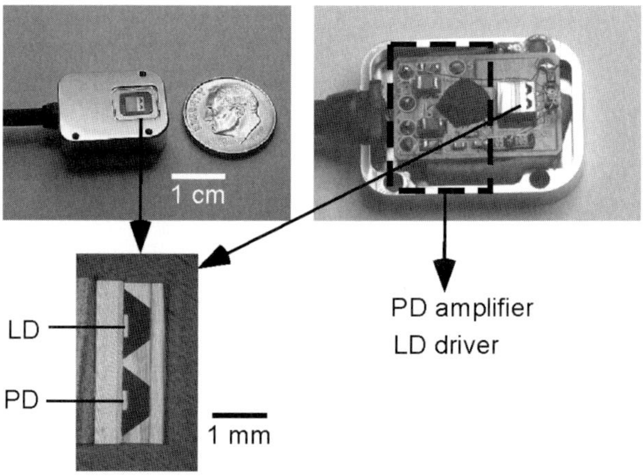

Fig. 1.2. Photographs of the packaged blood flow sensor showing the drive and amplifier circuitry as well as the mounted chip. The detail *lower left* shows the emitting waveguide and detector surface. Photos courtesy of NTT

This hybridly assembled optical sensor, with dimensions 1.5 x 2 x 3 mm^3, was subsequently packaged together with a laser diode driver and photodiode amplifier as shown in Fig. 1.2. Package dimensions are 6.5 x 15 x 20 mm^3, allowing easy access of the completed sensor to fingertips or other body parts for blood flow measurements. Demonstration of the performance of the sensor showed that measurement of heartbeat and determination of occlusions, leading to reduced blood flow, were easily done while manually holding the sensor onto the skin. The small size and low power consumption essential for successful application of such a sensor may only be achieved through use of a semiconductor laser diode as the light source.

1.2.2 Optical Distance Measurement

Determination of distance to an object, as well as monitoring its displacement, movement or vibration, is likewise a measurement which is ideally performed optically. Distance measurements using light are contact-free, may be performed under adverse conditions (such as in a vacuum or explosive atmosphere) and are typically of high accuracy.

Microsystem technology provides the means to reduce the size and power consumption while increasing the robustness of an optical distance sensor. At the Institute for Microstructure Technology at the Forschungszentrum Karlsruhe in Karlsruhe, Germany, a micro-optical distance sensor employing a semiconductor laser diode and a position-sensitive detector (PSD) has been developed and demonstrated (Wallrabe et al. 2002). Using a triangulation approach, the light emitted from the laser and reflected from the measurement object is directed onto the PSD, where the position of the reflected light beam allows determination of the distance between sensor and object.

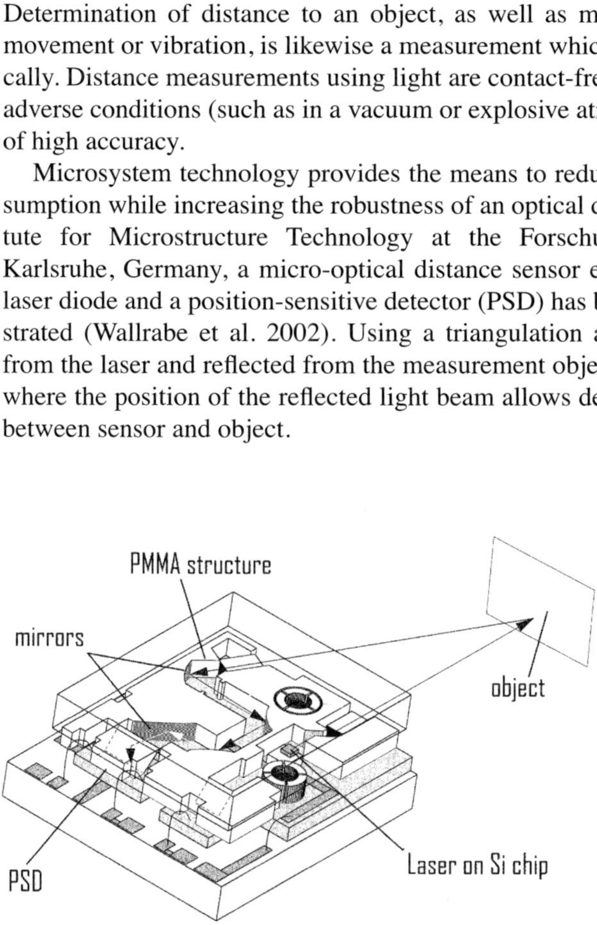

Fig. 1.3. Schematic view of the distance sensor. The lower component (silicon) serves as an electrical and thermal mount for the laser diode and PSD whereas the upper chip (PMMA on ceramic) contains the mirrors and lenses required for the optical functions. Diagram courtesy of the Forschungszentrum Karlsruhe

The schematic diagram of Fig. 1.3 shows the structure of the microsystem. A lower substrate, fabricated from silicon, acts as a mount for the laser diode, a monitor photodiode and the PSD; electrical and thermal contact is provided for these components. The upper chip is fabricated in PMMA, an optically transparent plastic, and defines the optical path using mirrors and lenses where reflective coatings are employed to generate mirror surfaces; this component is mounted on the silicon substrate using flip-chip techniques.

The two components, shown in the photograph of Fig. 1.4, are mounted together and aligned using glass micro-spheres and appropriate alignment features. The entire device has dimensions 2.5 x 6.7 x 7.2 mm^3 and is packaged on a standard TO8 mount. Using the light from the laser diode emitting at $\lambda = 685$ nm as the measurement signal, sensors with a measurement range of 1 mm at 2 μm resolution as well as 10 mm measurement range at 40 μm resolution were demonstrated.

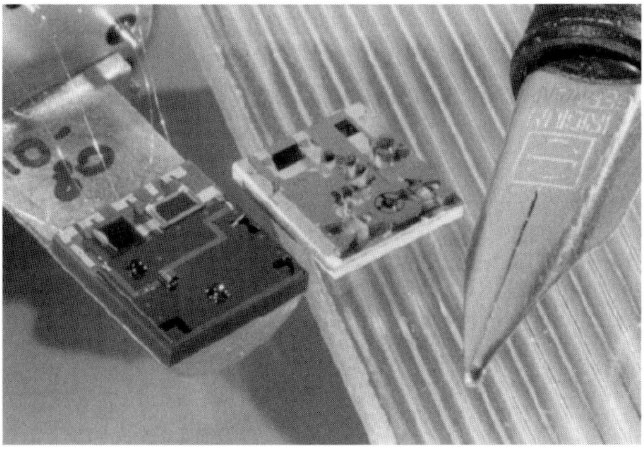

Fig. 1.4. Photograph of the distance sensor before alignment of the Si and PMMA components; the laser diode is mounted on the left side of the silicon chip, shown on the left. Photo courtesy of the Forschungszentrum Karlsruhe

1.2.3 Optoelectronic Transceiver

One of the most active research areas for optical microsystems has been the development of components and subsystems for optical communications; although the industry has recently seen a sobering in its outlook, it is nevertheless likely that this application will represent the single largest use for optoelectronics and optical microsystems for the foreseeable future. Of primary industrial interest is the development of systems which bring optical data communications closer to the consumer, preferably with a fiber connection in every home. Key to the acceptance of such concepts by consumers will be price, such that the availability of technologies which allow the fabrication of modules with complex optical functionality at low cost will be essential if the field is to develop.

A step in this direction is the development of complete transceiver modules, which allow reception and transmission of optical data on a single optical fiber. At the Alcatel Corporate Research Center in Stuttgart, Germany, a silicon-based transceiver platform has been developed which allows hybrid integration of all the necessary components (Blume et al. 2000). The module is based on silica waveguides deposited by flame hydrolysis on the silicon platform and subsequently patterned by etching. As can be seen in the schematic diagram of Fig. 1.5, a directional coupler directs the signal received from the optical fiber at $\lambda = 1.55\ \mu m$ to a photodiode; a $\lambda = 1.3\ \mu m$ InP-based laser diode is then used to transmit data back onto the fiber and a monitor diode is used to sample the laser emission. As seen in Fig. 1.6, the optoelectronic components are mounted on the silicon substrate using flip-chip techniques coupled with micro-machined alignment features; alignment tolerances between laser and waveguide of better than $0.5\ \mu m$ were demonstrated. These technologies ease assembly of the entire optical microsystem and thereby potentially reduce the fabrication costs.

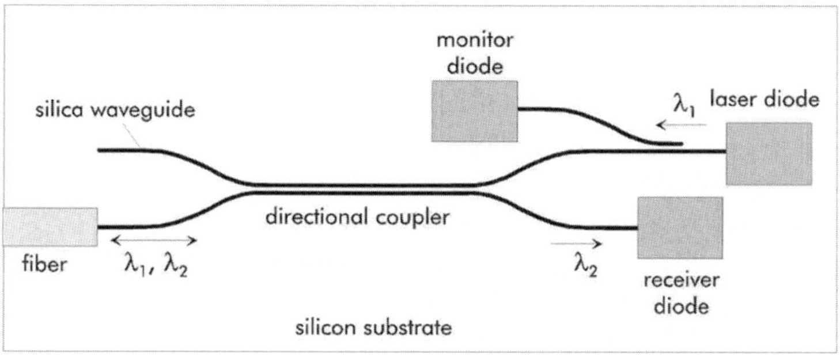

Fig. 1.5. Schematic top view of the optical transceiver hybridly assembled on a silicon substrate. Light at λ_2 is received from the fiber and detected at the receiver diode; light at λ_1 is transmitted by the laser diode back onto the fiber. Diagram courtesy of Alcatel SEL

Using etched U-grooves to mount the fibers, system performance allowed detection sensitivities of –33 dBm with 2 dBm emission power in the fiber and less than 2 dB optical loss due to fiber coupling. These transceiver modules were fabricated in medium quantities on a pilot production line, demonstrating the industrial viability of the assembly techniques; full testing, including thermal cycling and high-humidity tests, allowed determination of process yield as well as long-term reliability. Such results are very encouraging for the acceptance and further development of laser diode-based optical microsystems, also for applications outside the telecommunications area.

1.3 Thematic Outline

The purpose of this text is to introduce semiconductor laser diodes to those who would like to incorporate them into optical microsystems. Along the way, the

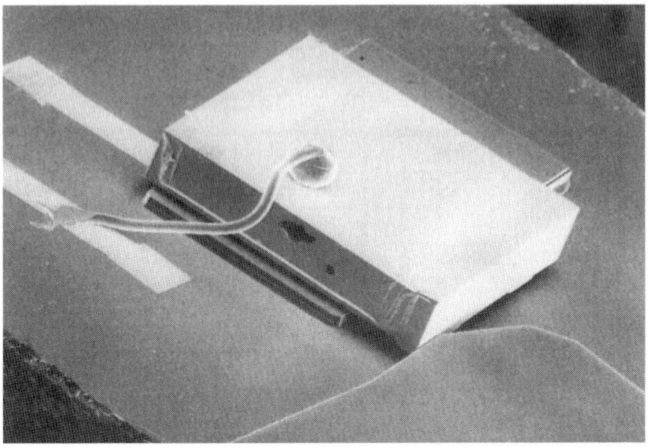

Fig. 1.6. SEM photograph of a flip-chip mounted laser diode on the silicon substrate of a hybridly-assembled optoelectronic transceiver; the alignment posts are clearly seen. Photograph courtesy of Alcatel SEL

reader should develop an appreciation for how lasers work in general, how semiconductor lasers look and operate and finally how laser diodes may be combined with other micro-optical components to yield complete microsystems.

We thus begin in Chapter 2 with an overview of the physics of lasing, intended for readers with little or no background in quantum electronics. We present the wave and quantum views of light, introduce the concept of an optical resonant cavity and thereby derive the essential physics required for understanding how a laser operates. In Chapter 3 we turn to semiconductors, first reviewing their essential electrical and optical properties and from there describing how these materials may be made to lase. We conclude with a discussion of semiconductor-based optical waveguides, as these structures are a constitutive part of most laser diodes.

With these basics in hand (or in brain), the reader is then prepared for the exposé of the variety of laser diode structures presented in Chapter 4. This chapter outlines a number of the most important types of laser diodes and includes a discussion of their structure, operating principle and fabrication considerations. Each laser type is also evaluated with respect to its utility in an optical microsystem, with factors such as power consumption, size and beam profile explicitly considered. Chapter 5 follows closely with an in-depth analysis of laser diode behavior and characteristics. Designed to give the reader a detailed view of which and how laser diode measurements should be performed, the chapter concludes with an summary of typically expected characteristics, providing guidelines for what might be reasonably expected in a measurement and what might not.

The final two chapters then move in the direction of optical microsystems. Since optical systems require more components than just laser light sources, Chapter 6 gives an overview of micro-optical components essential in a variety of systems: lenses, diffractive optics, detectors, modulators and fibers are discussed with a view to combining them with laser diodes in a system. The assembly of optical microsystems, including laser diodes and the components discussed in Chapter 6, is the topic of Chapter 7. With a discussion of the various hybrid and monolithic assembly techniques available, this final chapter takes a look at the relevant fabrication and assembly technologies that allow manufacture of complete optical microsystems.

The literature references in the text serve a dual purpose, namely to provide the reader with a resource to delve further into a given topic and occasionally to allow placing a discussion into its historical context. For the former, we have striven to provide either the most modern or most comprehensive reference (preferably both) and for the latter, the reader will notice that the reference is occasionally older than he or she is. Due to the appallingly short half-life of internet addresses and, at the time of writing, photonics startup companies, we have not burdened the text with reference to these.

The emphasis on optical microsystem applications implies that a few aspects of laser diodes are treated only parenthetically or not at all. In particular, power laser diodes (those emitting Watts to hundreds of Watts), while of considerable industrial and academic interest, are at present not really suitable for use in compact microsystems and are thus not prominent in the text which we now invite the reader to sit back and enjoy.

2 Basic Laser Physics

Lasers emit light. Before we can employ this fact in the design of an optical microsystem using semiconductor lasers, it is essential that we understand light and how lasers go about emitting their particular version of it. This first chapter thus provides an overview of laser and light physics.

We begin with two descriptions of light, the first based on its wave properties and the second on its quantum aspects. The former is advantageous for understanding phenomena such as propagation and interference, whereas the latter provides a useful picture for light emission from materials and optical amplification. Based on a mixture of these two views, we will then be able to describe optical gain, the function of resonant cavities and ultimately, laser oscillation. Whereas many classic texts discuss laser physics in greater depth (Siegmann 1986; Silvfast 1996), the coverage here is designed to be sufficient for an understanding of the material which follows.

2.1 The Handwaving Laser

For those readers who might not have encountered the internals of a laser before, we begin with a brief, heuristic explanation of how a laser works and what it needs to do so. This "handwaving" explanation, so called because the explainer generally uses an excess of bodily motion in lieu of equations on the blackboard to explain physical phenomena, should provide an indication of why the topics to be addressed subsequently are relevant.

Consider the diagrams of Fig. 2.1, which show schematically the three physical processes we require for the fabrication of a laser. In the beginning, we need light: through the injection of energy into an optical medium, photons are generated by the transitions of electrons from higher to lower atomic energy levels in the material. The medium may be a semiconductor, a doped glass, a gas or any one of a variety of materials which emit light when properly stimulated. The energy source, known as a pump, may be a voltage, a current or another light source such as a flash bulb or another laser.

The generated light must then be amplified. Pumping adds further energy to the system such that the photon intensity is multiplied and thus increases. Photons generated in the material stimulate further electronic transitions and thus the generation of more photons. We then have a growing light beam in the laser medium.

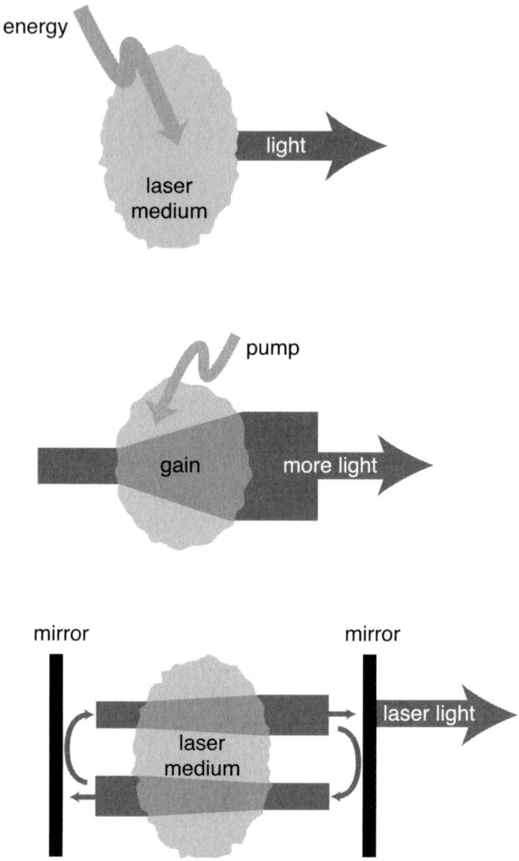

Fig. 2.1. Lasing in three easy steps: injection of energy to generate light *top*; pumping to provide optical amplification *middle*; and a resonator to provide feedback *bottom*

Finally, the amplifying, light emitting optical medium is placed into a resonator, shown in the figure as a cavity formed by two parallel mirrors. The resonator has three functions: it forces the light to traverse the laser medium repeatedly, thus growing exponentially in intensity. In addition, it generates a preferred laser direction, since the components of the generated optical field which oscillate between the mirrors (i.e., move horizontally in the figure) are amplified more strongly than those which move in other directions. Finally, the resonator also selects a very narrow band of wavelengths for resonance and thus lasing.

At least one of the mirrors is semi-transparent, allowing a small portion of the resonant light to be emitted. This emitted light is then the "laser beam", characterized by its high intensity, well-defined beam direction and narrow emission spectrum.

In summary, a laser requires a source of photons, a photon amplifier and a resonant cavity. In the following sections, we will consider these components in greater detail, with the intention of providing an overview of the physics of lasers.

2.2 Wave Properties of Light

The wave picture of light originated with Christiaan Huygens (1629–1695) (Hecht 1987) and remains the most useful description for the understanding of propagation, interference and coherence, phenomena which form an essential part of laser physics. In the wave picture, light is described as an electromagnetic wave propagating in a space, either vacuum or filled with material (Jackson 1975; Kraus and Carver 1973; Sadiku 2001). The following is a somewhat dense review of the essentials of the wave description of electromagnetic radiation.

2.2.1 The Wave Equation

The wave properties of light are predicated on the use of harmonic functions for the mathematical representation of the light wave, where these result from a solution of the wave equation.

Field Quantities

Electromagnetic waves are oscillating electric and magnetic fields, $\boldsymbol{E}$ [V/m] and $\boldsymbol{H}$ [A/m], respectively. These fields are related to the electric and magnetic flux densities, $\boldsymbol{D}$ [C/m^2] and $\boldsymbol{B}$ [Wb/m^2] as well as the polarization $\boldsymbol{P}$ [C/m^2] and magnetization $\boldsymbol{M}$ [Wb/m^2] as

$$\boldsymbol{D} = \varepsilon_m \boldsymbol{E} = \varepsilon_0 \boldsymbol{E} + \boldsymbol{P} \tag{2.1}$$

and

$$\boldsymbol{B} = \mu_m \boldsymbol{H} = \mu_0 \boldsymbol{H} + \boldsymbol{M} . \tag{2.2}$$

In general, the field variables $\boldsymbol{E}, \boldsymbol{D}, \boldsymbol{H}$ and $\boldsymbol{B}$ as well as $\boldsymbol{P}$ and $\boldsymbol{M}$ are vector quantities, having both a magnitude and direction, as indicated in this text by the boldface symbol.

The constants ε_m [F/m] and μ_m [H/m] represent the permittivity and permeability, respectively, of the material in which the field propagates. These in turn are often defined as

$$\varepsilon_m = \varepsilon \cdot \varepsilon_0 \tag{2.3}$$

and

$$\mu_m = \mu \cdot \mu_0 \tag{2.4}$$

where $\varepsilon_0 = 8.85\text{x}10^{-12}$ F/m and $\mu_0 = 4\pi\text{x}10^{-7}$ H/m are the permittivity and permeability in free space. The parameters ε and μ are dimensionless numbers defining how different the material is from vacuum. ε is known as the dielectric constant and, whereas ε = 1 for vacuum, typically ε > 1 for any material.

Maxwell's Equations

The spatial and time dependence of the electric and magnetic fields are related by Maxwell's equations, in honor of James Maxwell (1831–1879), and summarized in their differential form as follows:

$$\nabla \cdot \varepsilon_m \boldsymbol{E} = \rho \tag{2.5}$$

$$\nabla \cdot \mu_m \boldsymbol{H} = 0 \tag{2.6}$$

$$\nabla \times \boldsymbol{H} = \boldsymbol{J} + \varepsilon_m \frac{\partial \boldsymbol{E}}{\partial t} \tag{2.7}$$

$$\nabla \times \boldsymbol{E} = -\mu_m \frac{\partial \boldsymbol{H}}{\partial t} . \tag{2.8}$$

In the previous expressions, ρ [C/m^3] represents the charge density and $\boldsymbol{J}$ [A/m^2] the current density.

We recall the definitions of the vector operators in the Cartesian coordinate system, namely divergence

$$\nabla \cdot \boldsymbol{A} = \frac{\partial \boldsymbol{A}}{\partial x} + \frac{\partial \boldsymbol{A}}{\partial y} + \frac{\partial \boldsymbol{A}}{\partial z} \tag{2.9}$$

and curl

$$\nabla \times \boldsymbol{A} = \begin{vmatrix} \boldsymbol{u}_x & \boldsymbol{u}_y & \boldsymbol{u}_z \\ \frac{\partial}{\partial x} & \frac{\partial}{\partial y} & \frac{\partial}{\partial z} \\ \boldsymbol{A}_x & \boldsymbol{A}_y & \boldsymbol{A}_z \end{vmatrix} \tag{2.10}$$

for a vector $\boldsymbol{A}$ with the Cartesian components defined by the unit vectors $\boldsymbol{u}_x$, $\boldsymbol{u}_y$ and $\boldsymbol{u}_z$.

Wave Equation

Combining Maxwell's equations allows us to derive the wave equation (Zappe 1995), which in turn relates the spatial to the time dependence of either the electric or magnetic field. Written in terms of **E** or **H** the wave equation may take either of the forms

$$\nabla^2 \boldsymbol{E} = \varepsilon_m \mu_m \cdot \frac{\partial^2 \boldsymbol{E}}{\partial t^2} = \frac{1}{v^2} \cdot \frac{\partial^2 \boldsymbol{E}}{\partial t^2} \tag{2.11}$$

or

$$\nabla^2 \boldsymbol{H} = \varepsilon_m \mu_m \cdot \frac{\partial^2 \boldsymbol{H}}{\partial t^2} = \frac{1}{v^2} \cdot \frac{\partial^2 \boldsymbol{H}}{\partial t^2} . \tag{2.12}$$

In the previous expressions, v [m/s] represents the velocity of the wave and is given by $v = \sqrt{1/(\varepsilon_m \mu_m)}$. For the case in which the wave is propagating in vacuum, it is easy to show that $v \equiv c = \sqrt{1/(\varepsilon_0 \mu_0)} = 3 \times 10^8$ m/s.

2.2.2 Solutions to the Wave Equation

Three-Dimensional Solution

The wave equations are solved by harmonic functions, most generally sine and cosine functions. Using the spatial vector **r** and the time variable t, it may be shown by substitution that

$$\boldsymbol{E}_r = E_0[\cos(\pm(\boldsymbol{k}_r \cdot \boldsymbol{r}) \pm \omega t) + \sin(\pm(\boldsymbol{k}_r \cdot \boldsymbol{r}) \pm \omega t)]\boldsymbol{u}_r \tag{2.13}$$

represents a general solution to the wave equation for the electric field, Eq. 2.11. The electric field $\boldsymbol{E}_r$ is thus a vector quantity, whose direction is defined by the unit vector $\boldsymbol{u}_r$ to be in the **r** direction and with magnitude E_0 [V/m].

We have introduced the spatial frequency, or propagation constant, $\boldsymbol{k}_r$ [m^{-1}] and the temporal frequency ω [rad/s] in Eq. 2.13. These two parameters, which will be discussed more thoroughly below, are characteristic of the wave and define its propagation behavior.

The harmonic solution to the wave equation, Eq. 2.13, may also be usefully written in exponential form, such that

$$\boldsymbol{E}_r = E_0[\exp j(\pm(\boldsymbol{k}_r \cdot \boldsymbol{r}) \pm \omega t)]\boldsymbol{u}_r , \tag{2.14}$$

where we recall that $j = \sqrt{-1}$. Exponential notation will prove to be a useful means to express field and intensity quantities. Through Euler's rule, named for Leonhard Euler (1707–1783),

$$e^{j\phi} = \cos\phi + j\sin\phi , \tag{2.15}$$

we are reminded that this exponential form represents a sum of sinusoids.

One-Dimensional Solution

The general, three-dimensional forms of the wave equation, Eq. 2.11 or Eq. 2.12, may be simplified by considering the case in which the electric field $\boldsymbol{E}_x$, for example, is polarized in $\boldsymbol{x}$ direction[1], as given by the unit vector $\boldsymbol{u}_x$, and propagates solely in the z direction. For this special case, the wave equation for the electric field becomes

$$\frac{\partial^2 \boldsymbol{E}_x}{\partial z^2} = \frac{1}{v^2}\frac{\partial^2 \boldsymbol{E}_x}{\partial t^2} \tag{2.16}$$

which then leads to a simpler solution, either as an explicit sum of harmonic functions

$$\boldsymbol{E}_x = E_0[\cos(\pm(\boldsymbol{k}_z \cdot z) \pm \omega t) + \sin(\pm(\boldsymbol{k}_z \cdot z) \pm \omega t)]\boldsymbol{u}_x \tag{2.17}$$

or equivalently in exponential form

$$\boldsymbol{E}_x = E_0[\exp j(\pm(\boldsymbol{k}_z \cdot z) \pm \omega t)]\boldsymbol{u}_x \,. \tag{2.18}$$

It should thus be clear from either the three-dimensional or one-dimensional solutions that the electric fields which constitute an electromagnetic wave vary sinusoidally in both in time and space.

2.2.3 Spatial and Temporal Frequencies

For the solutions to the wave equation and thus the mathematical description of a propagating electromagnetic wave, Eq. 2.13 or Eq. 2.14, we introduced two parameters: the spatial frequency, or propagation constant, k and the temporal or angular frequency, ω. Whereas the former is in general a vector quantity, the latter is a scalar; we will write k as a scalar here.

Definition of k and ω

The propagation constant k has units [m^{-1}] and is defined as

$$k = \frac{2\pi}{\lambda}, \tag{2.19}$$

where λ [m] represents the wavelength of the electromagnetic wave in the material in which it propagates. For propagation in vacuum, the wavelength is typically written as λ_0 such that for a material with optical refractive index n, the wavelength in the material is given by

[1] Polarization will be discussed further in Sect. 2.2.4 below.

$$\lambda = \frac{\lambda_0}{n}. \tag{2.20}$$

The propagation constant in vacuum k_0 is then related to the propagation constant in a material k as

$$k = \frac{2\pi}{\lambda} = nk_0 = \frac{2\pi n}{\lambda_0}. \tag{2.21}$$

Likewise, the angular frequency ω with units [rad/s] is defined by

$$\omega = 2\pi\nu \tag{2.22}$$

for frequency ν [Hz = s^{-1}]. The frequency of a wave is related to the wavelength and velocity v as

$$v = \lambda\nu \tag{2.23}$$

where in vacuum we have

$$c = \lambda_0\nu. \tag{2.24}$$

What Do k and ω Mean?

Most readers will likely be able to imagine the physical relevance of temporal frequency, but spatial frequency is often a more difficult concept to grasp. We can help in the understanding by comparing the role of these two parameters in the arguments of the solutions to the wave equation.

Thinking back to Eq. 2.17, we see that the argument $(\pm kz \pm \omega t)$ represents a phase, which we may define as ϕ. Taking the first time derivative of the phase, we then obtain

$$\frac{d\phi}{dt} = \omega. \tag{2.25}$$

Thus the angular frequency ω represents the change in phase of the wave with respect to time.

Alternatively, taking the first spatial derivative of the phase, considering here the simplified one-dimensional solution, results in

$$\frac{\partial\phi}{\partial z} = k \tag{2.26}$$

implying that k represents the change in phase of the wave with respect to space or position.

These results may become clearer upon considering the plots of Fig. 2.2. On the left we see a “snapshot” in space, i.e. we stand at one point in space and see how the wave varies in time. As expected, the time variation is sinusoidal with a period

$2\pi = \omega t$. The "snapshot" on the right is in time: for a fixed time, we see how the wave varies in space. The variation is again sinusoidal, this time with a period $2\pi = kz$. Thus ω and k represent the temporal and spatial frequency and are proportional to the reciprocal of the period.

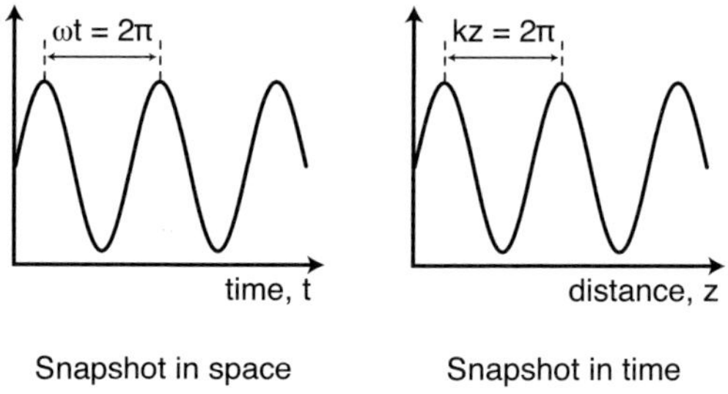

Fig. 2.2. A harmonic function frozen in space *left* and time *right*

The reason we belabor this point is that the propagation constant k is one of the key parameters for the description of the transmission of light, particularly through media. For a mathematical description of the effect of absorption or, crucially for lasers, the effect of optical amplification, the propagation constant is essential. We will see k quite often in the next few chapters.

Sign Conventions

A final consideration in the analysis of the harmonic solutions to the wave equation are the signs of the arguments. The ± for each term in the phase, $(\pm kz \pm \omega t)$, allows for the general solution to move to negative or positive times or positions. As shown schematically in Fig. 2.3, the + sign implies that waves move to the left on the time or space axis, whereas the – sign implies movement to the right.

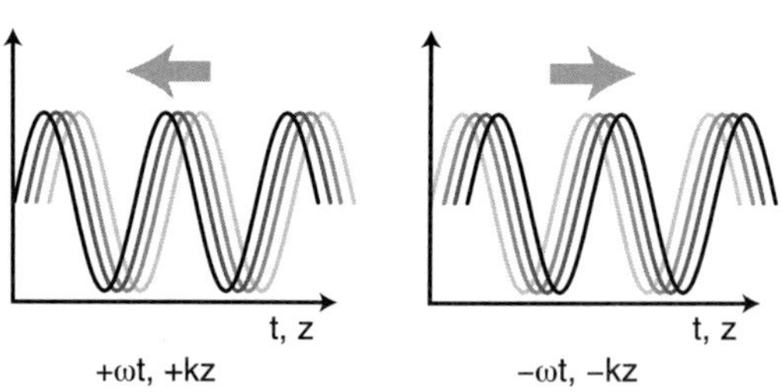

Fig. 2.3. Movement of a harmonic wave with positive and negative phase terms

To simplify our lives, we will employ the sign convention $(+kz + \omega t)$ such that the harmonic solutions to the wave equation will have the simplified form

$$\boldsymbol{E}_r = E_0[\exp j(\boldsymbol{k}_r \cdot \boldsymbol{r} + \omega t)]\boldsymbol{u}_r = E_0 e^{j(\boldsymbol{k}_r \cdot \boldsymbol{r})} e^{j\omega t} \boldsymbol{u}_r \ . \tag{2.27}$$

This definition is arbitrary and many authors use alternative sign conventions.

Example 2.1

Assignment: *What are the typical values for frequency, angular frequency and propagation constant for the electromagnetic waves emitted from a HeNe laser, a gas laser often found in optics laboratories?*

Solution: *We know that the red HeNe vacuum wavelength is* $\lambda_0 = 633\ nm$. *From this value we can calculate:*

Frequency: $\nu = c/\lambda_0 \approx 474THz$

Angular frequency: $\omega = 2\pi\nu \approx 3\times10^{15}\ rad/s$

Propagation constant: $k = 2\pi/\lambda_0 \approx 9.9\times10^{6} m^{-1} = 9.9\mu m^{-1}$

These represent typical values for the visible wavelength range.

2.2.4 Polarization

The harmonic electromagnetic waves which derive from the solution of the wave equation for electric or magnetic fields are examples of transverse electromagnetic (TEM) waves. TEM implies that the electric and magnetic fields are orthogonal to each other and both are orthogonal to the propagation direction.

The polarization of a TEM wave is defined by the direction of the electric field[2]. For the simplest case, linear polarization as shown schematically in Fig. 2.4, the electric field has a constant direction and magnitude orthogonal to the propagation direction. If the Cartesian axes are chosen so that the electric field extends along the, for example, $\boldsymbol{x}$ direction, we may say that the wave is "x-polarized" and express this with the notation $\boldsymbol{E}_x$.

In general, however, the length (i.e., magnitude) and direction of the electric field vector may change as the wave propagates. This gives rise to elliptical polarization, shown schematically in Fig. 2.5, for which the electric field vector rotates about the propagation axis and concomitantly varies in magnitude. The tip of the electric field vector traces an ellipse when viewed in the propagation direction, hence the nomenclature employed.

Elliptical polarization represents the most general case from which other polarization cases may be derived. If, for example, the electric field vector rotates about the propagation direction but does not change in magnitude as it does so, the elec-

[2] The polarization referred to in this section is not the same as the polarization $\boldsymbol{P}$ defined in Eq. 2.1; see footnote on page 28.

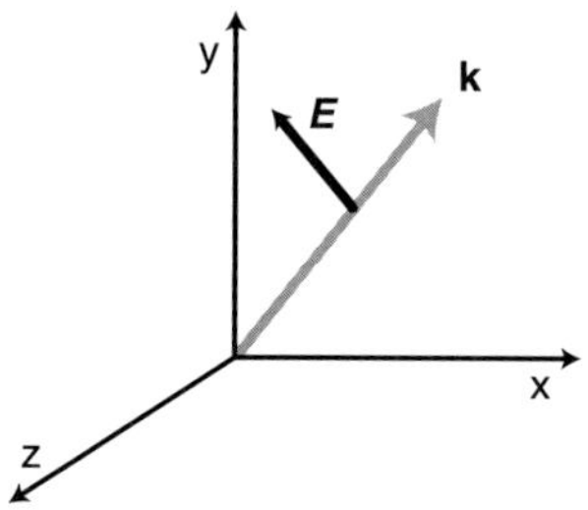

Fig. 2.4. Schematic depiction of a linearly polarized TEM wave

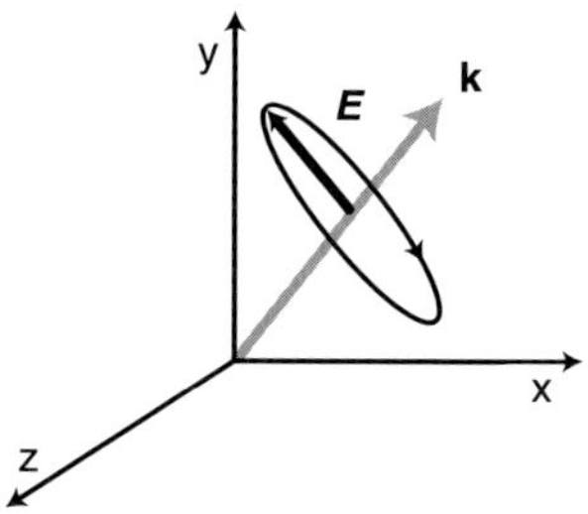

Fig. 2.5. Schematic depiction of an elliptically polarized TEM wave

tric field vector traces a circle, hence circular polarization. For the case in which one axis of the ellipse is reduced to zero, we return to the linear polarization case.

Polarization may in general be defined as a superposition of two orthogonal linearly polarized electric field vectors subject to a phase offset; useful graphical and mathematical constructions have been developed to help visualize and define polarization direction (Kraus and Carver 1973; Born and Wolf 1980, pp. 31–32).

2.2.5 Energy

Poynting Vector and Intensity

The power flow of an electromagnetic wave is given by the Poynting vector, named after John Henry Poynting (1852–1914). The Poynting vector $\boldsymbol{S}$ [$\mathrm{W/m^2}$] is defined as

$$\boldsymbol{S} = \boldsymbol{E} \times \boldsymbol{H}\,, \tag{2.28}$$

where the vector (or cross) product implies that the direction of $\boldsymbol{S}$ is orthogonal to both $\boldsymbol{E}$ and $\boldsymbol{H}$.

Also of interest is the time-average of power flow, since measurements of power always require a minimum measurement time which is usually longer than the period of wave oscillation. The time average is given by

$$\langle S\rangle = \frac{1}{T}\int_0^T S(t')dt' = \frac{1}{2}|\boldsymbol{E}||\boldsymbol{H}| , \tag{2.29}$$

where T [s] represents the temporal period of the wave.

The ratio of the magnitudes of the electric and magnetic fields gives the impedance, either of the medium in which the wave propagates, Z,

$$Z = \frac{|\boldsymbol{E}|}{|\boldsymbol{H}|} = \sqrt{\frac{\mu_m}{\varepsilon_m}} \tag{2.30}$$

or of vacuum, if the wave is propagating in free space, Z_0

$$Z_0 = \frac{|\boldsymbol{E}|}{|\boldsymbol{H}|} = \sqrt{\frac{\mu_0}{\varepsilon_0}} = 377\Omega , \tag{2.31}$$

the latter representing the impedance of free space. Recalling the definitions of the permittivity and permeability (Eq. 2.3 and Eq. 2.4), the expression for the impedance of a medium may be expanded as

$$Z = \sqrt{\frac{\mu_m}{\varepsilon_m}} = \sqrt{\frac{\mu\mu_0}{\varepsilon\varepsilon_0}} = \sqrt{\frac{\mu}{\varepsilon}}Z_0 = \frac{1}{n}Z_0 \tag{2.32}$$

for optical refractive index n; we will discuss this parameter in greater detail in Sect. 2.2.8.

Using the impedance, we can write simple expressions for the magnitude of the Poynting vector:

$$S = Z|\boldsymbol{H}|^2 = \frac{1}{Z}|\boldsymbol{E}|^2 . \tag{2.33}$$

Including the effect of the time average, these expressions become

$$\langle S\rangle = \frac{1}{2}Z|\boldsymbol{H}|^2 = \frac{1}{2Z}|\boldsymbol{E}|^2 \tag{2.34}$$

and by recalling our definition of Z, we are left with

$$\langle S\rangle = \frac{Z_0}{2n}|\boldsymbol{H}|^2 = \frac{n}{2Z_0}|\boldsymbol{E}|^2 . \tag{2.35}$$

The Poynting vector is a measure of power flow and thus has units of power per area [W/m^2]; this quantity is often loosely denoted the intensity of a wave. Intensity is insofar of importance in that most physical measurements of optical field quantities are actually measurements of the time average of intensity, as given by Eq. 2.34 or Eq. 2.35. It is useful to keep in mind that measured signals are thus usually proportional to $\boldsymbol{E}^2$ or $\boldsymbol{H}^2$ rather than $\boldsymbol{E}$ or $\boldsymbol{H}$.

Energy Density

Whereas the Poynting vector defines the power flow of a wave, a further parameter of interest is the energy density per unit volume, W [J/m^3]. For the electric and magnetic fields, we may write, respectively,

$$W_E = \frac{1}{2}\varepsilon_m|\boldsymbol{E}|^2 \tag{2.36}$$

$$W_H = \frac{1}{2}\mu_m|\boldsymbol{H}|^2 , \tag{2.37}$$

where $W_E = W_H$.

The total energy density in a system, W_T, is given by the sum of the electric and magnetic components, so that we may write

$$W_T = W_E + W_H . \tag{2.38}$$

Since the electric and magnetic components are equal for TEM waves, the total energy density amounts to twice W_E or W_H.

Example 2.2

Assignment: *Compare the peak optical electric fields and energy densities of light from a 100 W, 10 cm diameter light bulb, measured on the surface of the bulb, with that from a 1 mW laser pointer emitting from an aperture 1 mm in diameter, measured at the aperture.*

Solution: *The comparison is tricky since the light bulb has a broad spectrum which extends far into the infrared wavelength range whereas the laser emits at virtually a single wavelength. Let's be generous and assume that 5% of the power is emitted at visible wavelength so that we have*

$$\langle S_{Light\ bulb}\rangle = 159\ W/m^2$$

Peak electric field in free space: $|\boldsymbol{E}| = \sqrt{2Z_0\langle S\rangle} \approx 346\ V/m$

Electric energy density: $W_E = \frac{1}{2}\varepsilon_0|\boldsymbol{E}|^2 \approx 0.53\ \mu J/m^3$

Total energy density: $W_T = 2W_E \approx 1.06\ \mu J/m^3$

For the laser, we obtain

$$\langle S_{Laser}\rangle = 10^3\ W/m^2$$

Peak electric field in free space: $|\boldsymbol{E}| = \sqrt{2Z_0\langle S\rangle} \approx 868\ V/m$

Electric energy density: $W_E = \frac{1}{2}\varepsilon_0|\boldsymbol{E}|^2 \approx 3.34\ \mu J/m^3$

Total energy density: $W_T = 2W_E \approx 6.68\,\mu J/m^3$

We see that, although the laser has a factor 10^5 less total power than the light bulb, the low efficiency of the latter coupled with the small emission area of the former implies that the laser power density is actually higher.

2.2.6 Interference

When two electromagnetic waves meet in space, they interact with each other and the resultant field is the sum of the two original waves. Since the wave equation, Eq. 2.11, is linear, its solutions add so that the electric (or magnetic) field at any point is given by the sum of all fields which meet at the point. This phenomenon is known as interference and is central to understanding a wide variety of optical and laser effects.

We may understand interference by considering the simple case of two electric fields, $\boldsymbol{E}_1$ and $\boldsymbol{E}_2$, which meet in space,

$$\boldsymbol{E}_1 = \boldsymbol{E}_{10}\sin(\omega t + k_1 z) \tag{2.39}$$

$$\boldsymbol{E}_2 = \boldsymbol{E}_{20}\sin(\omega t + k_2 z)\,. \tag{2.40}$$

$\boldsymbol{E}_1$ and $\boldsymbol{E}_2$ have different magnitudes, the same temporal frequency ω but different propagation constants k_1 and k_2; such fields may result, for example, if the light from a single laser source is split into two parts, which meet after propagating through two different materials. The total field at that point, $\boldsymbol{E}_T$,

$$\boldsymbol{E}_T = \boldsymbol{E}_1 + \boldsymbol{E}_2 \tag{2.41}$$

may then easily be calculated using a bit of trigonometry. Recalling the identity

$$\sin(\alpha + \beta) = \sin\alpha\cos\beta + \cos\alpha\sin\beta \tag{2.42}$$

we then have

$$\begin{aligned}\boldsymbol{E}_T = {} & \boldsymbol{E}_{10}(\sin\omega t\cos k_1 z + \cos\omega t\sin k_1 z) \\ & + \boldsymbol{E}_{20}(\sin\omega t\cos k_2 z + \cos\omega t\sin k_2 z)\end{aligned} \tag{2.43}$$

$$\begin{aligned}\boldsymbol{E}_T = {} & (\boldsymbol{E}_{10}\cos k_1 z + \boldsymbol{E}_{20}\cos k_2 z)\sin\omega t \\ & + (\boldsymbol{E}_{10}\sin k_1 z + \boldsymbol{E}_{20}\sin k_2 z)\cos\omega t\end{aligned} \tag{2.44}$$

$$\boldsymbol{E}_T = \boldsymbol{E}_0\cos kz\sin\omega t + \boldsymbol{E}_0\sin kz\cos\omega t \tag{2.45}$$

yielding finally

$$\boldsymbol{E}_T = \boldsymbol{E}_0\sin(kz + \omega t)\,. \tag{2.46}$$

Thus the sum of the two input fields is also a sinusoid, albeit with distinct values for magnitude $\boldsymbol{E}_0$ and propagation constant k. We may evaluate the magnitude by recalling that we just defined

$$\boldsymbol{E}_0 \cos kz = \boldsymbol{E}_{10} \cos k_1 z + \boldsymbol{E}_{20} \cos k_2 z \tag{2.47}$$

$$\boldsymbol{E}_0 \sin kz = \boldsymbol{E}_{10} \sin k_1 z + \boldsymbol{E}_{20} \sin k_2 z\,. \tag{2.48}$$

By squaring and adding the left and right sides of Eq. 2.47 and Eq. 2.48, and using a further trigonometric identity, we obtain

$$\boldsymbol{E}_T^2 = \boldsymbol{E}_{10}^2 + \boldsymbol{E}_{20}^2 + 2\boldsymbol{E}_{10}\boldsymbol{E}_{20}\cos(\Delta\phi) \tag{2.49}$$

where we recall that $\boldsymbol{E}^2$ is proportional to the intensity of the wave (recall Eq. 2.33). In Eq. 2.49 we have defined a phase shift $\Delta\phi$ as

$$\Delta\phi = (k_1 z - k_2 z) \tag{2.50}$$

which essentially represents the optical path difference that two waves from a common source will have taken or, more generally, just the phase difference between the two waves which interfere.

We see from Eq. 2.49 that the intensity of the field which results from the interference of two waves is *not* given by the sum of the intensities of the input fields. Instead, the intensity varies strongly with the relative phase of the two fields. In particular, if $\Delta\phi = 0, 2\pi, 4\pi, \ldots$, we have constructive interference in which the intensity is a maximum given by

$$\boldsymbol{E}_T^2 = \boldsymbol{E}_{10}^2 + \boldsymbol{E}_{20}^2 + 2\boldsymbol{E}_{10}\boldsymbol{E}_{20} \tag{2.51}$$

and for $\Delta\phi = \pi, 3\pi, 5\pi, \ldots$, destructive interference results, yielding a minimum in the intensity, namely

$$\boldsymbol{E}_T^2 = \boldsymbol{E}_{10}^2 + \boldsymbol{E}_{20}^2 - 2\boldsymbol{E}_{10}\boldsymbol{E}_{20}\,. \tag{2.52}$$

For the special case where $\boldsymbol{E}_{10} = \boldsymbol{E}_{20} = \boldsymbol{E}$, the maximum constructive interference yields $\boldsymbol{E}_T = 4\boldsymbol{E}^2$ and destructive interference has $\boldsymbol{E}_T = 0$.

Thus interference can result in a net field intensity of zero, even for non-zero input fields, if the two input waves are exactly out of phase. Constructive interference is employed to build up the resonance in a laser cavity, which we will see in Sect. 2.5.1 and are also the basis for diffractive optics, which we will consider in Sect. 6.2.

2.2.7 Coherence

An essential aspect of laser light is its coherence, a characteristic which distinguishes the light waves emitted from a laser from those emitted by most other light sources. Two electromagnetic waves are termed to be coherent if they have the same frequency, a fixed phase relationship to one another and if this frequency equivalence and phase relationship is stable over a given time. Two or more coherent waves may give rise to the interference phenomena just discussed, essential for the understanding of both laser operation as well as many of the applications of laser light.

Coherence is in practice never absolute but rather defined by a coherence length, L_C [m], or coherence time, t_C [s], where the two are related as

$$L_C = ct_C \tag{2.53}$$

for c equal to the speed of light. These two parameters represent complementary descriptions of temporal coherence.

The coherence time may be understood as follows: imagine two waves which have the same frequency and are initially *in phase*, the latter term implying a relative phase shift of 0 or $2i\pi$ for integer i. Since the relative phase, in practice, does not remain constant over time, this parameter shifts as the wave propagates and at some point the waves are *out of phase*, implying a phase shift of $i\pi$ for integer i. This happens after a propagation time t_C, the coherence time; alternatively, the wave has then propagated a distance equivalent to the coherence length, L_C.

The coherence time or length is defined by the linewidth of an optical field, a parameter which will be defined in detail in Sect. 2.3.4 below. Linewidth defines the spread of wavelengths $\Delta\lambda$ (or frequencies) present in a light wave of center wavelength λ_0, and always has a value > 0, even for lasers. The coherence length is related to the linewidth as

$$L_C = \frac{\lambda_0^2}{\Delta\lambda} \tag{2.54}$$

from which we see that a narrower linewidth (i.e., a smaller spread of wavelengths in an optical signal) implies a larger coherence length. A large coherence length implies that two waves may, for example, experience a long path length difference and still interfere with each other.

Example 2.3

Assignment: *Compare the coherence lengths of an incandescent light bulb, an LED, a Hg arc lamp and a laser.*

Solution: *We need to estimate the linewidth of these light sources. Typical values are*

Light bulb with peak at $\lambda_0 = 2\ \mu m$: $\Delta\lambda = 3000$ nm (a rough estimate based on a black body spectrum with $T = 3{,}000$ C)

LED at $\lambda_0 = 690$ nm: $\Delta\lambda = 1.6$ nm

Hg arc lamp at $\lambda_0 = 546$ nm: $\Delta\lambda = 0.5$ nm

Semiconductor laser at $\lambda_0 = 780$ *nm:* $\Delta\lambda = 0.02$ *pm*

Using Eq. 2.54, we can easily calculate the coherence length. Since coherence is never absolute (i.e., L_C *is never infinite since* $\Delta\lambda$ *is never 0), we can also assign a measure of the level of coherence of these light sources:*

Light bulb: $L_C = 1.3$ *µm ("incoherent")*

LED: $L_C = 298$ *µm ("semi-coherent")*

Hg arc lamp: $L_C = 596$ *µm ("semi-coherent")*

Laser: $L_C = 30.4$ *m ("coherent")*

2.2.8 Light–Matter Interaction

In Sect. 2.2.2, we solved the wave equation derived from Maxwell's equations but neglected to consider the effect of a material in which the light wave propagates in detail. We will now remedy this situation by expressly including material effects in a more complete form of the wave equation. The interaction between an electromagnetic wave and a material takes place through polarization: the wave induces a polarization of atoms and electrons in the material, which in turn generates a field which interacts with the field of the incident wave[3]. The permittivity is a measure for the strength of this interaction.

Polarization and Susceptibility

Recalling that the polarization of the material may be included in the definition of the electric flux density,

$$\boldsymbol{D} = \varepsilon_m \boldsymbol{E} = \varepsilon_0 \boldsymbol{E} + \boldsymbol{P} \tag{2.55}$$

we may write the polarization as

$$\boldsymbol{P} = (\varepsilon_m - \varepsilon_0)\boldsymbol{E} = (\varepsilon\varepsilon_0 - \varepsilon_0)\boldsymbol{E} = \varepsilon_0(\varepsilon - 1)\boldsymbol{E}\,. \tag{2.56}$$

The factor $(\varepsilon - 1)$ is a dimensionless number indicating how different the material is from vacuum. In vacuum, $\varepsilon = 1$ such that polarization is zero since there is nothing to polarize; as permittivity increases for $\varepsilon > 1$, the effect of the material becomes stronger.

The term $(\varepsilon - 1)$ is often defined to be the susceptibility χ,

$$\chi = (\varepsilon - 1) \tag{2.57}$$

[3] The polarization referred to here is that of an electric field due to a redistribution of charges in a material. This is not exactly the same as the polarization of a TEM wave discussed in Sect. 2.2.4. To avoid confusion, $\boldsymbol{P}$ is sometimes referred to as "polarizibility".

which is also a dimensionless quantity whose value ranges from 0 (for vacuum) to positive values for materials. Based on this definition, we may write the polarization as

$$\boldsymbol{P} = \varepsilon_0 \chi \boldsymbol{E} . \tag{2.58}$$

In addition, the square root of the dielectric constant, ε, is defined as the refractive index, n,

$$n = \sqrt{\varepsilon} \tag{2.59}$$

which may then further be related to susceptibility as

$$n = \sqrt{1+\chi} . \tag{2.60}$$

Wave Equation in a Material

From the definition of electric flux density, Eq. 2.55, we may re-write two of Maxwell's equations, Eq. 2.5 and Eq. 2.7 (Gauss' and Ampère's laws, respectively, named for Johann Gauss (1777-1855) and André Marie Ampère (1775-1836)), explicitly as a function of polarization as

$$\nabla \cdot \boldsymbol{E} = -\frac{1}{\varepsilon_0} \nabla \cdot \boldsymbol{P} \tag{2.61}$$

$$\nabla \times \boldsymbol{H} = \varepsilon_0 \frac{\partial \boldsymbol{E}}{\partial t} + \frac{\partial \boldsymbol{P}}{\partial t} \tag{2.62}$$

where we have assumed that both charge density, ρ, and current density, $\boldsymbol{J}$, in the material are zero and that $\mu = 1$. In these latter two expressions, the effect of non-unity ε, or equivalently non-zero χ, is absorbed into the polarization term $\boldsymbol{P}$.

By employing the definitions above, we may then use these modified forms of Maxwell's equations to derive a revised form of the wave equation which explicitly includes the effect of the material on the wave through either susceptibility

$$\nabla^2 \boldsymbol{E} = \frac{1}{v^2} \cdot \frac{\partial^2 \boldsymbol{E}}{\partial t^2} = \frac{1}{c^2}[1+\chi] \cdot \frac{\partial^2 \boldsymbol{E}}{\partial t^2} \tag{2.63}$$

or dielectric constant

$$\nabla^2 \boldsymbol{E} = \frac{\varepsilon}{c^2} \cdot \frac{\partial^2 \boldsymbol{E}}{\partial t^2} \tag{2.64}$$

or refractive index

$$\nabla^2 \boldsymbol{E} = \frac{n^2}{c^2} \cdot \frac{\partial^2 \boldsymbol{E}}{\partial t^2} . \tag{2.65}$$

ε, χ and n

All three of the representations of the wave equation above are equivalent, since dielectric constant, susceptibility and refractive index are all simply related by

$$\varepsilon = \chi + 1 = n^2 . \tag{2.66}$$

These parameters are fully equivalent means for expressing the effect of a material on an electromagnetic wave and which one uses is a matter of practicality and, often, taste.

In general, these parameters are not only material-dependent but also wavelength dependent: thus the refractive index of a single material is quite different for electromagnetic waves in the "electronic" (MHz) frequency range and the "optical" (THz) frequency range. In addition, the values are in general complex where the real and imaginary parts define different aspects of wave propagation, as we will see in Sect. 2.3.4 below.

General Propagation Constant

Solving one of the wave equations Eq. 2.63 - Eq. 2.65 will give rise to a wave solution of the same form as we saw previously,

$$\boldsymbol{E}_r = E_0[\exp j(\pm \boldsymbol{k}_r \cdot \boldsymbol{r} \pm \omega t)]\boldsymbol{u}_r \tag{2.67}$$

save that the value for propagation constant k may now be expressed as a function of either dielectric constant, susceptibility or refractive index. Thus Eq. 2.21 becomes

$$k = \frac{2\pi}{\lambda_0} n = \frac{2\pi}{\lambda_0} \sqrt{\varepsilon} = \frac{2\pi}{\lambda_0} \sqrt{1+\chi} . \tag{2.68}$$

As we mentioned above, χ, ε and n are in general complex quantities; the reasons for this fact will be discussed in detail in the next chapter. If we separate n, for example, into its real and imaginary parts

$$n = n_R + jn_I , \tag{2.69}$$

Equation 2.68 then implies that the propagation constant will likewise be complex and may be written in the form

$$k = k_R + jk_I . \tag{2.70}$$

Our expression for the electric field, Eq. 2.67, then takes the following form:

$$\boldsymbol{E} = E_0[e^{\pm jk_R z} e^{\mp k_I z} e^{\pm j\omega t}] . \tag{2.71}$$

Equation 2.71 represents the most general form of the electric field solution for a propagating wave. The argument of the exponential has been separated into three terms and the field then written as a product of three exponentials. Of these, the first ($\pm jk_R z$) represents the propagating part of the wave, recalling that $\exp[jkz]$ is, through Euler's relation, a sum of harmonic (sine and cosine) functions.The second term ($\mp k_I z$) describes the change in intensity of the wave with distance, either decreasing due to absorption (for a negative argument) or increasing due to amplification (for a positive argument). The third term ($\pm j\omega t$) stands for the time dependence, again a harmonic function.

Whereas this type of description for an electromagnetic wave in a material holds in general, it is of particular interest for lasers. As we will see below, when energy is pumped into a lasing material, an optical field is amplified since the imaginary part of the refractive index becomes positive; the resulting growth (rather than the more typical attenuation) of the light wave is key for the understanding of laser behavior.

2.3 Quantum Properties of Light

The wave picture of light is useful for describing many aspects of light propagation. On the other hand, to understand the emission, absorption and amplification of light, as it takes place in a laser, the quantum description of light, as developed by Max Planck (1858–1947), is essential (Planck 1948). Whereas in the previous section we considered harmonic wave functions, in this section we consider the photon.

As the elementary quantum of radiation, the photon is typically characterized by an energy E [J] given by[4]

$$E = h\nu = \frac{h\omega}{2\pi} = \frac{hc}{\lambda} \tag{2.72}$$

where ν [Hz = s^{-1}] represents the frequency of the corresponding wave, ω [rad/s] the rotational frequency, λ [m] the wavelength and h Planck's constant where $h = 6.626\times10^{-34}$ Js. A useful rule of thumb relating the energy of a photon to its wavelength is given by

$$E[eV] = \frac{1.24}{\lambda[\mu m]} . \tag{2.73}$$

Note the units for energy (eV not J) and wavelength (μm not m).

---

[4] Note that we represent energy by an italicized E which should be distinguished from the boldface $\boldsymbol{E}$ employed for the electric field vector.

2.3.1 Photon Emission and Absorption

The atoms in many physical systems, such as isolated atoms in a gas or strongly coupled atoms in a crystalline solid, have numerous discrete energy states. These may include vibrational or rotational states (in the case of gasses) or various excited electronic states in which electrons undergo transitions to different energy levels (as is often the case in solids). Energy conservation during a transition from one state to another usually involves photons: when the atom undergoes a transition from a higher energy state to a lower one, the difference in energy is taken away by a photon. Correspondingly, if a photon with a certain energy is absorbed by an atom, this energy gives rise to a transition of the atom to a higher energy level.

Whereas gas lasers often employ a wide range of rotational and vibrational atomic states for the for lasing transitions, semiconductor lasers are based on electronic transitions: added energy from incident photons raises electrons from the valence band of the atoms of the crystal to into the conduction band whereas electrons falling to lower energy levels emit the difference in energy again as photons. Although the discussion which follows is applicable to other types of atomic systems for lasers, we should bear in mind that for semiconductor lasers, the energy levels we refer to are those for electrons in the band structure of the material.

Photon Emission

Consider the simple two-level system of Fig. 2.6; this may correspond to two allowed electron energies of, for example, the valance and conduction bands of a semiconductor. An electron which falls from E_2 to E_1 gives rise to the *emission* of a photon with energy $h\nu = E_2 - E_1$.

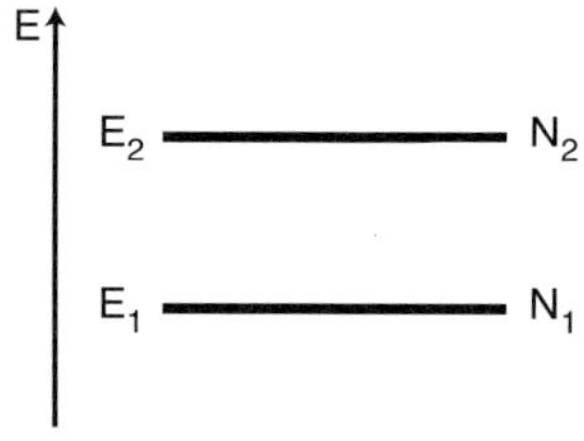

Fig. 2.6. Two discrete energy levels E_1 and E_2, with electron concentrations N_1 and N_2

The generation of a photon through this process is known as spontaneous emission and takes place at a rate

$$\left.\frac{dN_2}{dt}\right|_{spon} = -A_{21}N_2 \tag{2.74}$$

where N_2 represents the electron concentration in energy level E_2. The coefficient A_{21} [s^{-1}] is the decay rate and is also known as the "Einstein A coefficient" (Einstein 1917).

The differential equation of Eq. 2.74 may easily be solved to yield the time variation of the electron concentration in the upper energy level, namely

$$N_2(t) = N_2(0)e^{-A_{21}t} = N_2(0)e^{-\frac{t}{\tau_2}} \tag{2.75}$$

in which we have defined the spontaneous lifetime $\tau_{spon} = \tau_2$ as $\tau_2 = 1/A_{21}$. Thus a given electron population will decay exponentially with time to a lower energy level, emitting photons as it does so. Typical values for lifetime in semiconductor materials are in the ps to ns range.

Photon Absorption

The *absorption* of a photon with energy $h\nu = E_2 - E_1$ by the two-level system of Fig. 2.6 raises an electron from E_1 to E_2. We may represent the density of photons by the definition of an optical field, $\Phi(\nu)$ [$JHz^{-1}m^{-3}$ = Js/m^3] whose units imply optical energy per unit frequency per unit volume. The photon absorption rate, defined as the rate at which the electron concentration N_1 decreases with time, may then be written as

$$\left.\frac{dN_1}{dt}\right|_{abs} = -B_{12}N_1\Phi(\nu) \tag{2.76}$$

and is thus proportional to both the number of electrons at E_1 and the value of the optical field. The constant B_{12} [m^3/Js^2] is defined as an absorption rate and is also known as the "Einstein B coefficient".

Stimulated Emission

The presence of an optical field $\Phi(\nu)$ leads not only to absorption of photons by the atomic system but also to the phenomenon of stimulated emission. In this process, a photon incident on an electron at the higher energy level, E_2, will "stimulate" it to undergo a radiative transition to the lower level, thereby emitting a second photon with the same energy and phase as the incident photon. Stimulated emission causes a loss of electrons from the upper energy level (and concomitant emission of photons) at a rate

$$\left.\frac{dN_2}{dt}\right|_{stim} = -B_{21}N_2\Phi(\nu) \tag{2.77}$$

with an absorption/emission rate constant B_{21} [m^3/Js^2].

Transition Summary

As schematically shown in Fig. 2.7, three basic processes are relevant for a first understanding of how a multi-level system may interact with photons. Absorption transfers the energy from an incident photon to an electron raised to a higher energy level; spontaneous emission generates a photon when an electron falls from a higher energy level to a lower one of its own accord; and stimulated emission results in the emission of a photon due to an electronic transition but this transition is instigated by an incident photon, resulting in two photons of identical energy and phase. The combination of these three processes will allow us to explain the behavior of a laser on a quantum scale.

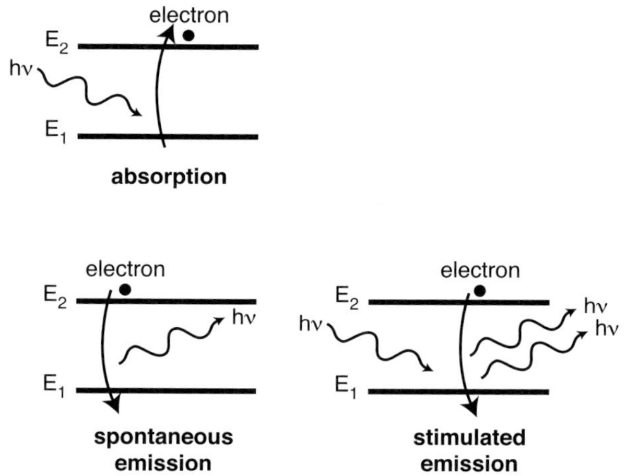

Fig. 2.7. The three relevant electronic transitions in a two-level system: absorption, spontaneous emission and stimulated emission

2.3.2 Thermal Equilibrium

If no energy is added to (or subtracted from) the atomic system in question, it is at thermal equilibrium. In this case, the number of transitions from E_1 to E_2 must equal the number of transitions from E_2 to E_1. We have one absorption process (Eq. 2.76) and two emission processes (Eq. 2.74 and Eq. 2.77) where at thermal equilibrium

$$\frac{dN_1}{dt} = \frac{dN_2}{dt} \tag{2.78}$$

which in turn implies that

$$A_{21}N_2 + B_{21}N_2\Phi(\nu) = B_{12}N_1\Phi(\nu) . \tag{2.79}$$

We will return to this last equation in a moment, but we shall first consider a few aspects of the energy distribution of electrons and the form of the optical field at thermal equilibrium.

Energy Distribution of Electrons

The energy distribution of atoms and systems of these in general is described, at thermal equilibrium, by a Maxwell-Boltzmann distribution, named in honor of Maxwell and Ludwig Boltzmann (1844–1906). Electrons are fermions and thus have an energy distribution described by the Fermi-Dirac function (see Eq. 3.3 in Sect. 3.1.2); for electron energies considerably above the Fermi energy, we may however also approximate the energy distribution of electrons by a Maxwell-Boltzmann distribution.

The number of particles (atoms or electrons in our case) in an energy interval dE at an energy E, $N(E)dE$, may be expressed by (Silvast 1996, Chap. 6.2)

$$N(E)dE = \frac{N}{k_B T} e^{-\frac{E}{k_B T}} dE \tag{2.80}$$

for total number of particles N in a system at temperature T. This expression, of the same form as the Maxwell-Boltzmann distribution, tells us that we have many particles with low energy and exponentially fewer at higher energies. In addition, as the temperature of the system increases, we will have more particles at higher energies. We may see this clearly in Fig. 2.8, in which the particle energy distribution for two system temperatures (roughly those of ice and steam) are shown.

Using the distribution of Eq. 2.80, we can write a simple relation for the relative numbers of particles at two different energies, for example at the two energies E_1 and E_2 of our simple two-level system. We can derive

$$\frac{N_2}{N_1} = \exp\left[-\frac{(E_2 - E_1)}{k_B T}\right] \tag{2.81}$$

which states that the ratio of particle concentrations is exponentially related to the energy difference between them; this expression is valid for energies considerably above $k_B T$. In particular, as energy increases it is exponentially less likely to find particles (e.g., electrons) at that energy. For that reason, we expect the higher energy level E_2 of our system to be considerably less populated than the lower energy level E_1, such that spontaneous and stimulated emission at thermal equilibrium are very unlikely due to the small value of N_2.

Example 2.4

Assignment: *Determine the number of atoms in block of material at room temperature which have an energy high enough to generate a visible photon.*

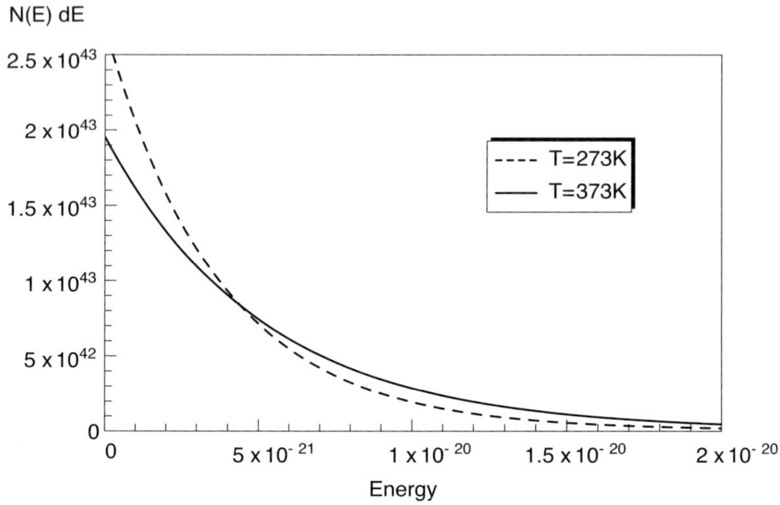

Fig. 2.8. The Maxwell-Boltzmann distribution of particle density as a function of energy for two temperatures, 273 K and 373 K

Solution: *We will take $N = 10^{23}$ atoms/cm^3. At $T = 300$ K, $k_BT \approx 26$ meV. Furthermore, let's assume that the visible wavelength range is $\lambda < 750$ nm, including the ultraviolet; visible photons thus have an energy greater than*

$$E = \frac{1.24}{0.75\mu m} \approx 1.77\ eV.$$

Thus we want to know how many particles with $E > 1.77$ eV are expected to be present in a distribution with thermal energy k_BT. We integrate the Maxwell-Boltzmann-like distribution through this energy range:

$$N_{visible} = \int_{1.77}^{\infty} \frac{N}{k_BT} e^{-\frac{E}{k_BT}} dE = -Ne^{-\frac{E}{k_BT}}\Bigg|_{1.77}^{\infty} \approx 3\times10^{-7}\ cm^{-3}.$$

We thus expect about one visible photon for every $1/3\times10^{-7} \approx 3\times10^{6}$ cm^3 of material.
If we repeat the calculation for $T = 1000$°C, we obtain a photon density of about 10^{14} cm^{-3}, which help to explain why hot things (like light bulb filaments) visibly glow.

Optical Field

The energy distribution of photons in a system at thermal equilibrium, which defines our optical field $\Phi(\nu)$, may be characterized by the black body spectrum. In a material system with refractive index n and temperature T, the energy density per unit frequency and unit volume of photons with frequency ν, $\Phi(\nu)$, is given by

$$\Phi(\nu) = \frac{8\pi n^3 h\nu^3}{c^3}\left(e^{\frac{h\nu}{k_B T}} - 1\right)^{-1} . \tag{2.82}$$

We may simplify the expression somewhat by defining an optical field intensity factor Φ_0 [$\mathrm{Jm^{-3}Hz^{-1}}$]

$$\Phi_0 = \frac{8\pi h}{\lambda^3} \tag{2.83}$$

such that

$$\Phi(\nu) = \frac{\Phi_0}{e^{h\nu/k_B T} - 1} ; \tag{2.84}$$

we recall that $\lambda = \lambda_0/n$ for refractive index n. A typical value for Φ_0 for, for example, a HeNe laser emitting at $\lambda = 633$ nm, is $\Phi_0 \approx 6\times10^{-14}$ $\mathrm{J/m^3Hz}$.

This relationship, which holds for all materials with a temperature T, is characterized by a peaked concentration distribution, where the peak frequency (or wavelength or energy) is characterized solely by the temperature and moves to higher frequency (or shorter wavelength or higher energy) with increasing temperature. Examples for two temperatures are shown in Fig. 2.9, where we see that at higher temperatures, the movement of the peak to shorter wavelengths is accompanied by a significant increase in intensity.

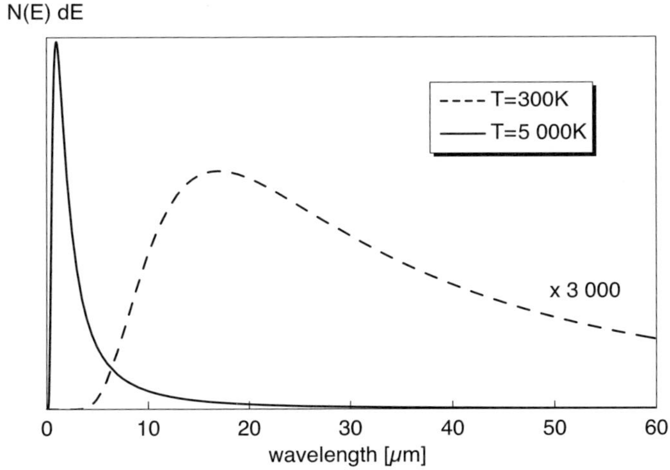

Fig. 2.9. Calculated black-body spectrum of two bodies at T = 300 K and T = 5 000 K. The characteristic for T = 300 K has been multiplied by a factor 3000

2.3.3 Transition Rates

Based on what we have learned about the characteristics of a system at thermal equilibrium, we may now derive expressions for the expected transition rates of electrons between the two energy levels. Recall that at thermal equilibrium, we know the energy distribution of electrons (Eq. 2.80) and the energy distribution of photons (Eq. 2.82). We also know from Eq. 2.79 that the upward and downward transition rates must be balanced.

Relationship Between Transition Rates

From these three equations, we may write the ratio of the number of electrons at the higher energy to the number at lower energy as

$$\frac{N_2}{N_1} = \frac{B_{12}\Phi(\nu)}{A_{21} + B_{21}\Phi(\nu)} = e^{-\frac{h\nu}{k_B T}} \tag{2.85}$$

which leads to

$$\frac{N_2}{N_1} = \frac{B_{12}\Phi(\nu)}{\left(e^{-\frac{h\nu}{k_B T}}\right)\left[A_{21} - B_{21}\Phi_0\left(\frac{1}{e^{-\frac{h\nu}{k_B T}} - 1}\right)\right]} = e^{-\frac{h\nu}{k_B T}} . \tag{2.86}$$

This relationship may be simplified as

$$\frac{N_2}{N_1} = \frac{B_{12}\Phi_0}{\left[A_{21}\left(e^{-\frac{h\nu}{k_B T}} - 1\right) - B_{21}\Phi_0\right]} = e^{-\frac{h\nu}{k_B T}} \tag{2.87}$$

leading to the expression

$$B_{12}\Phi_0 - B_{21}\Phi_0 e^{-\frac{h\nu}{k_B T}} = A_{21}\left(1 - e^{-\frac{h\nu}{k_B T}}\right) . \tag{2.88}$$

We can see by inspection that Eq. 2.88 may only be solved if the following are true:

$$B_{12} = B_{21} \equiv B \tag{2.89}$$

and

$$A_{21} = B_{12}\Phi_0 \equiv B\Phi_0 \equiv A \,. \tag{2.90}$$

What do these conditions imply? The first (Eq. 2.89) states that the probabilities of upward and downward induced transitions at thermal equilibrium are equal. Thus the probability that an incident photon is absorbed is the same as that it generates a further photon through stimulated emission, an effect known as the principle of detailed balance.

The second expression (Eq. 2.90) suggests that the rate of spontaneous emission is the same as the rate of stimulated emission, again at thermal equilibrium. Thus generated photons are as likely to come from a spontaneous emission process as a stimulated emission process.

Implications for Lasers

As we shall see in greater detail below, the generation of laser light depends critically on stimulated emission. The amplification of photons, central to laser operation, requires that stimulated emission dominates over absorption and also that stimulated emission dominates over spontaneous emission.

The first requirement implies that

$$\frac{\text{stimulated emission}}{\text{absorption}} = \frac{B_{21}N_2\Phi(\nu)}{B_{12}N_1\Phi(\nu)} = \frac{N_2}{N_1} \gg 1 \,. \tag{2.91}$$

For stimulated emission to dominate over absorption, then, considerably more electrons must be at the higher energy level than at the lower; we recall from Eq. 2.81 that this is unlikely at thermal equilibrium.

Fulfilling the second requirement above stipulates that

$$\frac{\text{stimulated emission}}{\text{spontaneous emission}} = \frac{B_{21}N_2\Phi(\nu)}{AN_2} = \frac{B\Phi(\nu)}{A} \gg 1 \,. \tag{2.92}$$

Thus a high optical field is required to assure that stimulated emission dominates over spontaneous emission; in the absence of an optical field, spontaneous emission is the only source of photons.

These last two relations are central to the understanding of why lasers are configured as they are. From Eq. 2.91, we see that a laser is operated far from thermal equilibrium; through a "pumping" mechanism, energy is injected into the system, assuring that the density of electrons at higher energies is higher than at lower energies. Secondly, Eq. 2.92 implies that the optical field intensity has to be high; this restriction in turn implies that generated photons need to be confined to the region where they are created, thus inducing an efficient generation of further photons due to the high optical field. As we will see below, a system in which stimulated emission represents the dominant transition is capable of amplifying optical signals and thus exhibits optical gain.

Example 2.5

Assignment: *Determine what optical power levels would be required to assure a dominance of stimulated emission in a GaInP-based 670 nm semiconductor laser.*

Solution: *We can make the following assumptions concerning the laser:*
Wavelength $\lambda_0 = 670$ *nm;*
Refractive index $n = 3.5$*;*
Transition linewidth $\Delta\nu = 1$ *THz.*
From Eq. 2.92 and Eq. 2.90, we see that a dominance of stimulated emission requires that

$$\frac{B\Phi(\nu)}{A} > 1$$

or

$$\Phi(\nu) > \frac{A}{B} = \Phi_0 .$$

We may calculate Φ_0 *from Eq. 2.83, keeping in mind that* $\lambda = \lambda_0/n$ *, yielding a value* $\Phi_0 \approx 2.4$ *pW/m*^{3}*Hz; the optical field* $\Phi(\nu)$ *must thus be at least this large.*
We may relate this to the optical power required by realizing that the intensity S is related to the optical field as

$$S = \frac{c}{n}\Delta\nu\Phi_0 \approx 203\ \mathrm{MW/m^2}.$$

Assuming that a typical GaInP laser waveguide has a cross-sectional area of 0.5 x 5.0 µm, this intensity value leads to a required power level of $P \approx 509\ \mu W$*. We should bear in mind that this power level is just enough to assure that stimulated emission is equal to spontaneous emission; for stimulated emission to truly dominate spontaneous emission, we need to go to higher power levels.*

2.3.4 Spectral Shape

An additional important factor in determining the characteristics of optical amplification in a material is the spectral width of the optical transition; the intensity of optical emission as a function of frequency ν (recalling that $c = \lambda\nu$) is given by the lineshape function $g(\nu)$ [Hz^{-1} = s], an example of which is shown in Fig. 2.10. The lineshape function has a peak at a certain wavelength λ_0 or frequency ν_0 and a spectral width $\Delta\lambda$ or $\Delta\nu$ defined as the full-width half-maximum.

Although we have represented our two-level system as one with two exactly defined energies, the spectral width of the optical field is not zero (i.e., all photons are not emitted at exactly one wavelength) but has a finite, non-zero linewidth

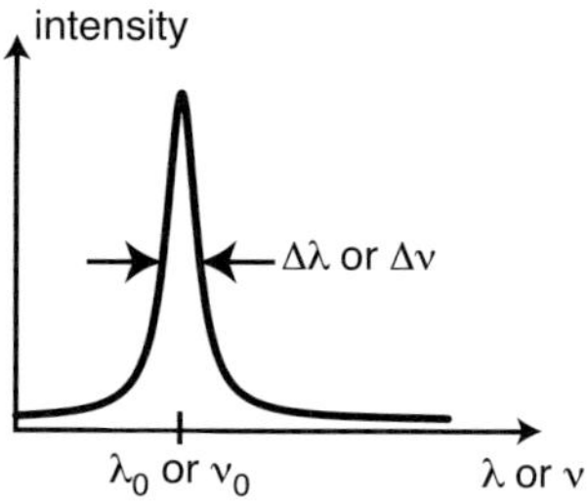

Fig. 2.10. A typical lineshape function $g(\nu)$ for a laser system with a Lorentzian intensity distribution. The width $\Delta\nu$ or $\Delta\lambda$ is defined as the full-width half-maximum

(measured in λ or ν). This width derives from the finite lifetime of the electrons at the upper energy; the electrons do not decay instantaneously but with a characteristic spontaneous lifetime τ_{spon} such that the transition probability decays exponentially with time. Such a characteristic in the time domain implies, through the Fourier transform, a finite width in the frequency domain.

Susceptibility

We saw in Sect. 2.2.8 that the effect of a material on a wave may be described by the susceptibility, χ. Using a simple material model, it can be shown (Zappe 1995) that the susceptibility of a material with a resonance frequency ω_0 and a damping coefficient $\Delta\omega$ may be expressed as a function of excitation frequency ω as

$$\chi = \frac{\chi_0'}{(\omega_0^2 - \omega^2 - j\omega\Delta\omega)} \tag{2.93}$$

for χ_0' a constant; the resonance frequency is one at which efficient energy transfer takes place between an incident photon and the atoms of the material whereas the damping coefficient defines how quickly an induced resonance is damped and is essentially a measure of the resultant spectral linewidth.

Through definition of the parameter Δf as the normalized deviation of the excitation frequency from the resonance frequency ($\Delta f = (\omega - \omega_0)/\Delta\omega$), we may rewrite the susceptibility as

$$\chi(\omega) = -j\chi_0\left(\frac{1}{1 + j\Delta f}\right) \tag{2.94}$$

which, after separating into real and imaginary components, becomes

$$\chi(\omega) = -X_0\left[\frac{\Delta f}{1 + \Delta f^2} + j\frac{1}{1 + \Delta f^2}\right] . \tag{2.95}$$

The real and imaginary parts of this complex expression for susceptibility, which vary with Δf as shown in Fig. 2.11, have distinct physical implications. The real part essentially defines the "refractive index" as we presented it in Sect. 2.2; when writing a value for real n, this parameter is generally given as $n = \sqrt{\mathrm{Re}(\chi) + 1}$.

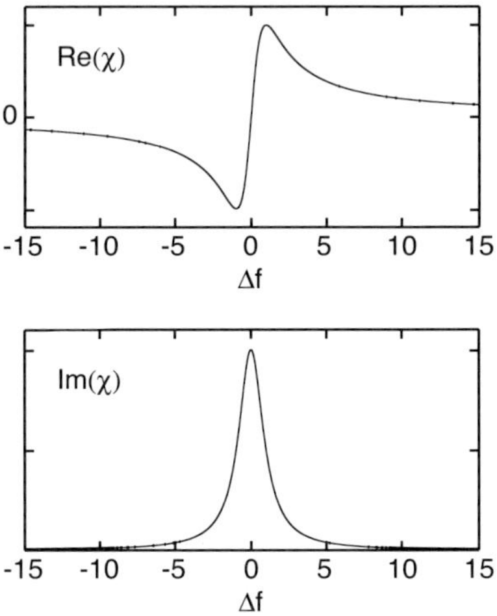

Fig. 2.11. Variation of the real *top* and imaginary *bottom* components of susceptibility with detuning Δf from the resonance frequency

The imaginary part of Eq. 2.95 defines the absorbing or amplifying response of a material to an incident wave. At thermal equilibrium, $\mathrm{Im}(\chi)$ will describe absorption only but in the non-equilibrium case of a pumped laser material the spectral width of stimulated emission leading to amplification will also be characterized by $\mathrm{Im}(\chi)$. We see from the figure that the imaginary component has a peaked characteristic, centered at the resonant frequency (i.e., $\Delta f = 0$), with a spectral width defined by $\Delta\omega$; the mathematical form of this curve is that of a Lorentzian.

Lineshape

Thus the spectral form of the optical transition responsible for absorption and amplification is defined by the lineshape function, $g(\nu)$, which typically has the Lorentzian form shown in Fig. 2.11. The lineshape function defines only the shape of the characteristic in wavelength, and is defined so that the normalization

$$\int_{-\infty}^{\infty} g(\nu)d\nu = 1 \tag{2.96}$$

is satisfied. The intensity of the transition is defined by the parameter Φ_ν [J/m^3] such that we may define the optical field $\Phi(\nu)$ as

$$\Phi(\nu) = \Phi_\nu \cdot g(\nu) . \tag{2.97}$$

We note explicitly that the optical field $\Phi(\nu)$ has dimensions energy per unit volume per unit frequency, whereas the constant Φ_ν has dimensions energy per unit volume.

Linewidth Broadening

The lineshape as predicted by $g(\nu)$ is defined by a linewidth, $\Delta\omega$ or $\Delta\nu$; this linewidth, as we mentioned above, derives from the spontaneous lifetime of electrons in the upper energy level and represents a lower limit for this parameter. Numerous physical mechanisms, however, may result in a broadening of the linewidth so that this lower limit is rarely reached (Yariv 1991, 150–155).

We may define two forms of linewidth broadening. For homogeneous broadening, the linewidth broadening effects cause all the atoms to exhibit the same broadened value of $\Delta\omega$; these effects include non-radiative transitions between the two energy levels, atomic or phonon collisions which lead to a de-phasing of the optical fields and finite electron lifetime. Inhomogeneous broadening refers to a linewidth increase due to physical mechanisms which give rise to various $\Delta\omega$ values, different for different atoms in the material. Inhomogeneous broadening may result from Doppler shifts (in gas lasers, for example) or crystalline strain (in semiconductor lasers). Which of these mechanisms is relevant and dominates the optical behavior depends strongly on the laser material and operating conditions.

2.4 Optical Amplification

The existence of stimulated emission, in which a photon incident onto a material generates a second identical photon, is one of the key processes for lasing. This photon generation mechanism leads to optical amplification and thus optical gain.

2.4.1 Absorption vs. Amplification

As we see schematically in Fig. 2.12, a light wave traversing a material interacts with it resulting, for most materials at thermal equilibrium, in a loss of intensity due to absorption of photons. This effect is illustrated by the first (upper) case in the figure, where absorption is typically characterized by an absorption coefficient, α, where $\alpha > 0$.

We recall that optical absorption is due to transitions of electrons from E_1 to E_2 and is given by

$$\left.\frac{dN_1}{dt}\right|_{abs} = -B_{12}N_1\Phi(\nu) . \tag{2.98}$$

Incident photons may also give rise to stimulated emission, involving an electronic transition from E_2 to E_1 and this occurs at the rate

$$\left.\frac{dN_2}{dt}\right|_{stim} = -B_{21}N_2\Phi(\nu) . \tag{2.99}$$

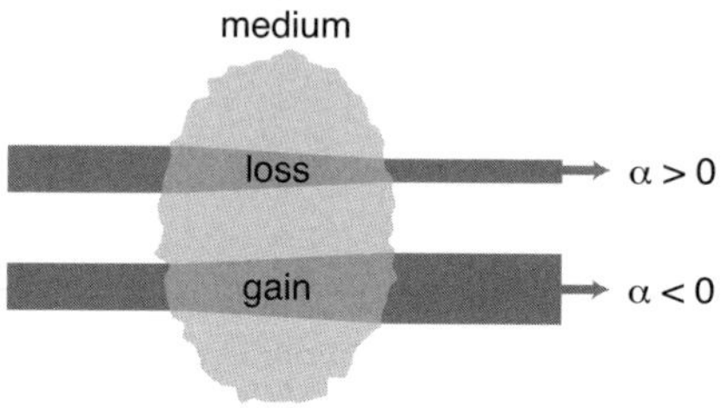

Fig. 2.12. Schematic representation of a light wave traversing an optical medium, subject either to absorption (loss) or amplification (gain)

Recalling that $B_{12} = B_{12} = B$, the relations between A and B (Eq. 2.90), the definition of $A = 1/\tau$ as well as the definition of Φ_0 (Eq. 2.83), we may express the time rate of change of electron concentration in level 1 or level 2 as

$$\frac{dN_{1,2}}{dt} = \frac{\lambda^3}{8\pi h\tau}N_{1,2}\Phi_\nu g(\nu) = W_iN_{1,2} . \tag{2.100}$$

The parameter W_i [s^{-1}], defined as

$$W_i = \frac{\lambda^3\Phi_\nu g(\nu)}{8\pi h\tau} , \tag{2.101}$$

expresses a transition rate per atom such that the product $W_iN_{1,2}$ [s^{-1}] represents the electronic transition rate (upwards or downwards) of $N_{1,2}$ atoms.

We recall that an electronic transition involves the absorption or emission of a photon. Since the rate of photon absorption is given by W_iN_1 and the rate of photon emission by W_iN_2 , we may write the *net* photon emission rate as

$$W_i(N_2 - N_1) \text{ [photons per volume per second]} \tag{2.102}$$

and thus the net power emission, due to photons with frequency ν, as

$$W_i(N_2 - N_1)h\nu \text{ [power per volume per second].} \tag{2.103}$$

From the last two expressions, we see that, for the case where $N_2 > N_1$, there is net power *emission* rather than *absorption* from the material and the second case illustrated in Fig. 2.12, namely optical amplification or gain, becomes relevant.

2.4.2 Gain

The amplification of photons in a material leads to optical gain, a parameter which we may define more clearly as follows. Let us define an optical energy intensity S_ν, [W/m^2], essentially the Poynting vector of Sect. 2.2.5, related to the optical field volume energy density Φ_ν [J/m^3] as

$$S_\nu = \frac{c}{n}\Phi_\nu = \lambda\nu\Phi_\nu \tag{2.104}$$

for wavelength λ, frequency ν and refractive index n; the optical energy intensity gives us the optical power flow per unit area.

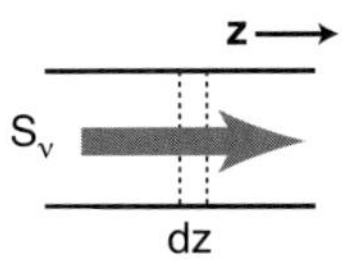

Fig. 2.13. Optical power flow in the z direction

For an optical field propagating along the z axis through a material, as shown schematically in Fig. 2.13, the incremental change in energy intensity dS_ν may be written as

$$dS_\nu = [W_i(N_2 - N_1)h\nu]dz \tag{2.105}$$

where the term in brackets essentially defines the power gain per volume. Using the definition from Fig. 2.101, this relation becomes

$$dS_\nu = \frac{\lambda^2}{8\pi\tau}(N_2 - N_1)g(\nu)S_\nu dz\ . \tag{2.106}$$

This simple differential equation has the solution

$$S_\nu = S_{\nu 0}e^{\gamma z} \tag{2.107}$$

for $S_{\nu 0}$ a constant equal to the optical intensity at $z = 0$ and the gain γ [m^{-1}] is defined as

$$\gamma = \frac{\lambda^2}{8\pi\tau}(N_2 - N_1)g(\nu)\ . \tag{2.108}$$

These last two expressions are of fundamental importance for lasers. Eq. 2.107 tells us that the optical field will grow exponentially with propagation distance in the material, so long as $\gamma > 0$. From Eq. 2.108, we see that gain is positive if $N_2 > N_1$ implying that the material is not at thermal equilibrium; in addition, we see that a wide lineshape function $g(\nu)$ and a small lifetime τ are required to obtain a large value for γ. Gain and loss represent two means for expressing the same phenomenon, and we may typically write that $\gamma = -\alpha$. such that a negative loss represents positive gain.

Example 2.6

Assignment: *Estimate the gain in the ruby laser, the first type of laser to be demonstrated*[5].

Solution: *Ruby is Al_2O_3 with added Cr, 0.05% by weight. Used as a laser material, it emits at $\lambda_0 = 694.3$ nm, has a refractive index $n = 1.76$ and a spontaneous lifetime $\tau = 3$ ms. Let's assume that $g(\nu) \approx 1/(\text{FWHM linewidth}) \approx 3$ ps (corresponding to $\Delta\lambda \approx 0.53$ nm) and take a typical value for $N_2 - N_1 = 8 \times 10^{18}\ cm^{-3}$.*

Using Eq. 2.108 and remembering that $\lambda = \lambda_0/n$, we obtain

$$\gamma = \frac{\lambda^2}{n^2} \frac{1}{8\pi\tau}(N_2 - N_1)g(\nu) \approx 0.5\ cm^{-1}.$$

While this gain value is typical for many solid-state lasers, we will see that gain in semiconductor lasers is several orders of magnitude higher.

Population Inversion

The requirement that $N_2 > N_1$ for $\gamma > 0$, if we consider the implications of Eq. 2.81, implies that the system must be far from thermal equilibrium; we recall that in thermal equilibrium, typically $N_2 \ll N_1$ so that gain γ is always negative (or loss α positive). As shown schematically in Fig. 2.14, the state of inversion, for which $N_2 > N_1$, is a non-equilibrium state only achieved through the addition of energy to the system, a process known as pumping. Pumping may be realized by optical or electrical means, whereby the latter is of greater significance for semiconductor lasers.

It should be added that, strictly, inversion is not possible for a simple two-level system which we have used for our model. Most real lasers are represented by more complicated multi-level systems, of which excellent descriptions may be found in the literature (Siegmann 1986).

Gain Saturation

As more and more energy is added to the material through pumping, the population of the upper level N_2 increases so that stimulated emission dominates. Stimulated (and spontaneous) emission reduce N_2 since the electrons have to decay to a lower energy to emit a photon. As the resultant optical field increases, N_2 thus decreases

---

[5] We will consider the gain of semiconductor lasers in Chapter 3 below.

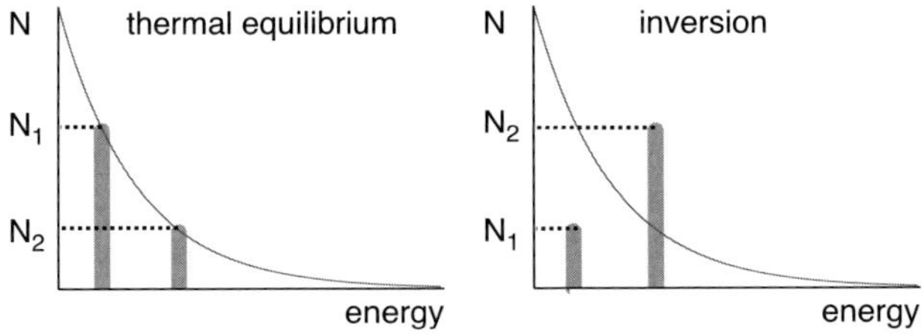

Fig. 2.14. Electron concentrations at two energy levels for a system at thermal equilibrium *left* and one with population inversion *right*. The exponential decrease in population as a function of energy of the Maxwell-Boltzmann distribution is shown for both

despite continued pumping; the higher the optical field, the quicker the decrease in N_2.

These two competing processes imply that gain does not increase arbitrarily with pumping. Recalling from Eq. 2.108 that $\gamma \propto (N_2 - N_1)$, the high optical field generated by stimulated emission then causes a reduction of gain as the optical field intensity increases. Thus the optical field Φ does not grow endlessly, but saturates at a value Φ_{sat} which is primarily a function of the transition lifetime and is typically defined as the field intensity for which the stimulated emission rate equals the spontaneous emission rate due to the lifetime of the upper energy level. As a result, the gain actually begins to decrease for increasing Φ and then saturates at a level equal to the background losses of the cavity (equal to the threshold gain, defined in Eq. 2.139 in Sect. 2.6.2 below).

A related point concerns how inversion may be achieved. It occasionally seems reasonable that a high optical field Φ might lead to gain in an optically absorbing system (i.e., change the sign of γ). From Eq. 2.108, we see that this cannot be the case: gain is only positive for an inverted electron population which in turn can only be reached through pumping. Optical pumping schemes, it should be added, do not consist of a high optical field Φ but instead use another optical field, usually at a shorter wavelength, to increase the population of the upper energy level, such that inversion is reached, gain becomes positive and the thus generated optical field Φ begins to grow.

Hole Burning

Gain saturation need not necessarily occur uniformly in a laser, either spatially or spectrally. The optical field may exhibit locally enhanced values, such that the gain is locally saturated or even reduced; this effect manifests itself as a "hole" in either the spectral or spatial distribution of the laser (Yamada and Suematsu 1981).

As shown schematically in Fig. 2.15, gain saturation at a certain frequency ν_h within the laser spectrum may result in spectral hole burning at that frequency, noticeable as a distinct dip in the output spectrum; this effect is particularly prominent in inhomogeneously broadened laser materials. Alternatively, spatially local

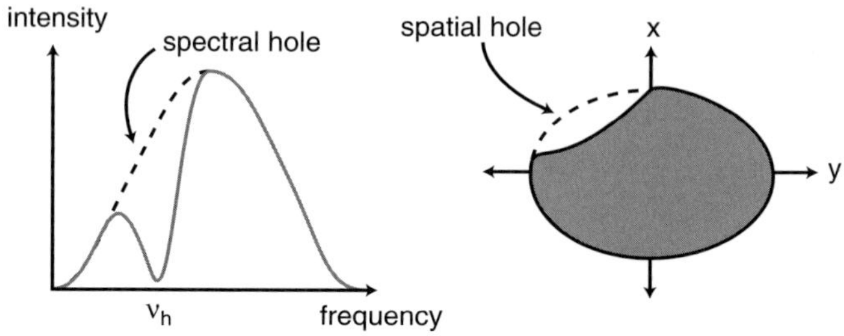

Fig. 2.15. Spectral hole burning *left* may cause a dip in the output spectrum whereas spatial hole burning *right* may result in a local decrease of laser output intensity

gain saturation or gain reduction due, perhaps, to non-uniform current injection, local heating effects or non-uniformities may give rise to spatial hole burning, as also shown in Fig. 2.15. This latter effect may result in mode instabilities and consequently unstable laser emission.

2.4.3 Laser Beam Growth

Based on our knowledge of the quantum interactions between light and discrete energy levels in laser materials as we have described in this section, we may summarize the effects to describe what happens to a light beam as it traverses a population-inverted material. Let's consider the schematic depiction of such a material as shown in Fig. 2.16. A small optical field is generated at the left and the resultant optical wave propagates in the **z** direction through a block of material subject to uniform pumping. If the pump level is sufficiently high to achieve population inversion in the material, stimulated emission dominates over spontaneous emission and the intensity of the wave grows exponentially as it propagates until the field intensity reaches its saturation level, Φ_{sat}. Once the saturation field is reached, gain remains at its saturation level and the optical field increases linearly (not exponentially) with propagation distance, as it continues to absorb energy from the material.

The point at which Φ_{sat} is reached is termed the laser *threshold*, at which point the laser begins to "lase"; the spontaneous emission is strongly reduced and most of the energy injected into the system results in an increase in optical field intensity.

It is of interest to consider how long a propagation path through the material is required before saturation (and thus lasing) is achieved. For the HeNe laser, a gas laser found in larger quantities in most optics laboratories, the "material" is a gas mixture of He and Ne, pumped by an electric discharge. The relatively low level of gain for this system ($0.15\ \text{cm}^{-1}$) implies that a propagation length of about 30 m is required before the threshold is reached. Such a laser length is impractical even for the most generous laboratory space, such that an alternative means for achieving

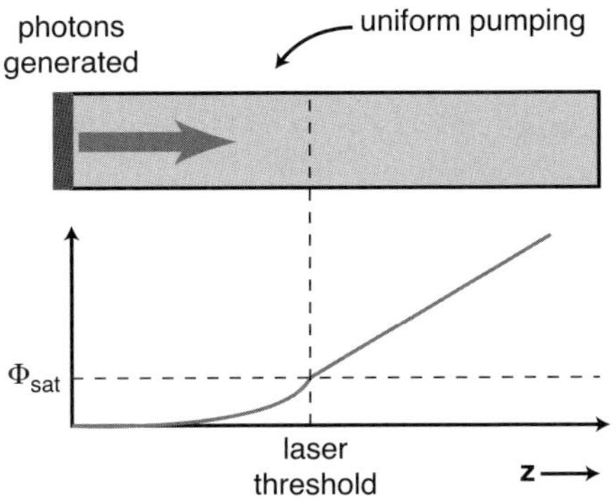

Fig. 2.16. Schematic depiction of the growth of an optical field as it propagates through a uniformly pumped block of material

this path length is required. One approach is through the use of a resonant cavity, to which we turn in the next section.

2.5 Optical Resonant Cavities

A material subject to pumping can amplify an optical signal, as we have just seen. The growth in the optical field, however, can require a considerable length of material and also takes place in all directions simultaneously; in addition, the spectral width of the generated optical field can be relatively large. These factors suggest that amplification through stimulated emission is a necessary but not sufficient condition for the fabrication of a laser.

The use of an optical resonant cavity is the next step toward lasing. The resonator allows the field to traverse the same stretch of material multiple times, reducing the overall laser length. The orientation of the cavity gives rise to an optical resonance in a particular physical direction, resulting in the generation of a directed optical beam from the homogeneously distributed optical field. Finally, the cavity selects certain frequency components of the optical field, known as modes, which have a considerably smaller spread in wavelength than the field generated in the material.

2.5.1 Fabry–Perot Etalon

One of the simplest yet most widespread forms of optical cavity is the Fabry–Perot etalon, developed in 1899 by Charles Fabry (1867–1945) and Alfred Perot (1863–1925) for spectroscopic applications (Fabry and Perot 1899). The etalon, as depicted in Fig. 2.17. consists of two parallel mirrors surrounding the gain medium; the mirrors may be planar (the case for semiconductor lasers) or concave (as found in most macroscopic lasers). Light generated by the gain medium is reflected back and forth within the cavity, increasing in intensity and ultimately leading to laser emission in the direction normal to the plane of the mirrors. Light is emitted through one or both of the mirrors, since one or both have reflectance $R < 1$.

A variety of other resonant cavity configurations have been developed and used as laser resonators (Siegmann 1986, chap. 19); we will see examples of some applicable to semiconductor lasers when we discuss laser structures in Chapter 4. For non-semiconductor lasers, examples of other configurations include confocal spherical mirror resonators, ring etalons and external cavities based on gratings. We will analyze the planar Fabry–Perot here as it represents the cavity used in the simplest form of semiconductor laser.

Field Analysis

Let us analyze the plane-parallel mirror cavity shown in Fig. 2.17. The two mirrors, with reflectivity and transmittivity r_L, t_L and r_R, t_R for the left and right mirrors, respectively, are separated by a distance L in which a material with refractive index n is found. We will assume for the moment that this material does not amplify the optical field in the cavity and will defer the discussion concerning a resonator with gain until Sect. 2.6.

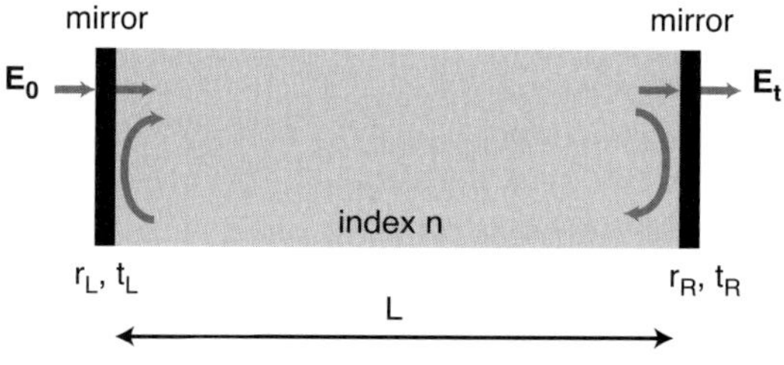

Fig. 2.17. A Fabry–Perot cavity formed by two planar parallel mirrors

Key to understanding the function of the Fabry–Perot cavity is the phase matching between electric field components which have undergone one or more reflections at the mirrors (Lipson, Lipson and Tannhauser 1995). Recalling that the electric field of the optical field $\boldsymbol{E}$ propagates with a propagation constant k, we realize that the phase shift that the wave undergoes after one round-trip through the cavity is given by

$$\phi = 2kL = \frac{4\pi n L}{\lambda_0} . \tag{2.109}$$

For relative phase shifts of $\phi = 2i\pi$ with integer i, the waves are in phase.

Let us assume an initial optical electric field value $\boldsymbol{E}_0$ which enters the cavity from the left. Due to the addition of the waves subject to multiple reflections, and thus multiple round trips through the cavity, the transmitted optical electric field which emerges at the right, $\boldsymbol{E}_t$, is given by

$$\boldsymbol{E}_t = \boldsymbol{E}_0 t_L t_R + \boldsymbol{E}_0 (t_L r_R r_L t_R) e^{j\phi} + \boldsymbol{E}_0 (r_R r_L)^2 t_R e^{j2\phi} + \dots \quad (2.110)$$

Simplifying matters by setting

$$r_L = r_R = r \quad (2.111)$$

and

$$t_L = t_R = t \quad (2.112)$$

and relating reflectance and transmittance to reflectivity and transmittivity as $R = r^2$ and $T = t^2$, Eq. 2.110 may be written as

$$\boldsymbol{E}_t = \boldsymbol{E}_0 T (1 + R e^{j\phi} + R^2 e^{j2\phi} + \dots) \quad (2.113)$$

which, by summing the series takes the relatively simple form

$$\frac{\boldsymbol{E}_t}{\boldsymbol{E}_0} = \frac{T}{1 - R e^{j\phi}} . \quad (2.114)$$

The ratio of output to input electric fields is thus a function of not only the reflectance and transmittance of the mirrors, but also the phase shift due to traversal of the cavity.

As the ratio of intensities (rather than fields) corresponds to the quantity typically measured, we may calculate this from the relation

$$\frac{W_t}{W_0} = \frac{\boldsymbol{E}_t \boldsymbol{E}_t^*}{\boldsymbol{E}_0 \boldsymbol{E}_0^*} = \frac{T^2}{(1 - R e^{j\phi})(1 - R e^{-j\phi})} \quad (2.115)$$

where $\boldsymbol{E}^*_{t,0}$ represents the complex conjugate. Expanding using Euler's relation, this yields

$$\frac{W_t}{W_0} = \frac{T^2}{(1 - R\cos\phi - jR\sin\phi)(1 - R\cos\phi + jR\sin\phi)} \quad (2.116)$$

or

$$\frac{W_t}{W_0} = \frac{T^2}{1 + R^2 - 2R\cos\phi} . \quad (2.117)$$

Recalling that $T = 1 - R$ and using the well-known trigonometric identity $1 - \cos\alpha = 2\sin^2(\alpha/2)$, we may express the transmission intensity through the cavity as a function of mirror reflectance as

$$\frac{W_t}{W_0} = \frac{(1-R)^2}{(1-R)^2 + 4R\sin^2\left(\frac{\phi}{2}\right)}. \tag{2.118}$$

2.5.2 Resonator Characteristics

A number of parameters are useful for the characterization of a resonant cavity, such as the Fabry–Perot etalon. These define the width of the resonance and the distance between adjacent resonance peaks.

Finesse

The relationship of Eq. 2.118 predicts a sharply peaked transmission characteristic which varies with phase, as seen in Fig. 2.18. The transmission has its maximum value (unity) at $\phi = 2i\pi$ for $i = 0, 1, 2...$ and goes through a minimum for $\phi = i\pi$ where $i = 1, 3, 5...$ at which point the intensity ratio is reduced to

$$\frac{W_t}{W_0} = \frac{(1-R)^2}{(1+R)^2}. \tag{2.119}$$

Since we recall from Eq. 2.109 that the phase is related to the ratio of wavelength to Fabry–Perot cavity length, the phase matching condition for maximal transmission (i.e., resonance), $\phi = 2i\pi$, is satisfied for a cavity length

$$L = \frac{i\lambda}{2} = \frac{i\lambda_0}{2n} \tag{2.120}$$

or alternatively a wavelength

$$\lambda_0 = \frac{2nL}{i}. \tag{2.121}$$

Thus the resonance condition for a cavity of fixed length is satisfied for certain resonance wavelengths.

The width of the resonance peak is typically characterized by the cavity finesse, F, defined as

$$F = \frac{\pi\sqrt{R}}{1-R} \tag{2.122}$$

such that the cavity transmission characteristic (Eq. 2.118) takes the form

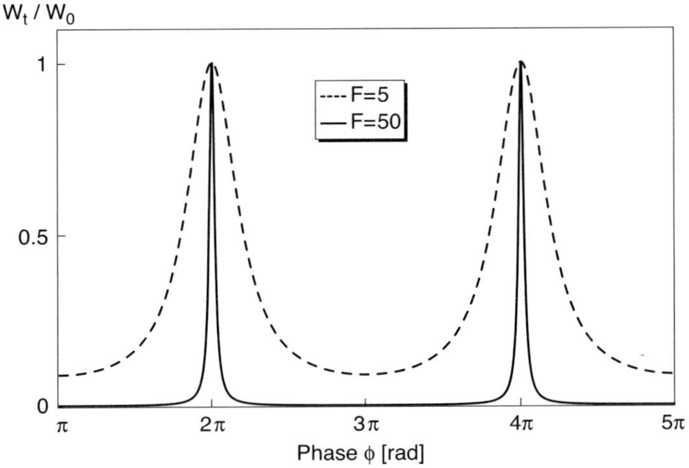

Fig. 2.18. Transmission characteristic of a Fabry–Perot etalon as a function of round-trip phase for finesse $F = 5$ and $F = 50$

$$\frac{W_t}{W_0} = \frac{1}{1 + \left(\frac{2F}{\pi}\right)^2 \sin^2\left(\frac{\phi}{2}\right)} . \tag{2.123}$$

The finesse defines the "sharpness" of the peak: high values for F imply narrow resonances, as illustrated in Fig. 2.18. As the finesse is only a function of mirror reflectance R, a high-quality cavity with large F is achieved through the use of highly reflecting mirrors. Values of $F \approx 200$ are routine and systems with $F \approx 10\,000$ are available. For the plots in Fig. 2.18, $F = 5$ is reached using mirrors with $R = 0.54$ and $F = 50$ with $R = 0.94$.

Free Spectral Range and Spectral Width

From the plot of Fig. 2.18 and the resonance condition for wavelength (Eq. 2.121), we see that the transmission resonance repeats periodically in wavelength or, alternatively by recalling that $\lambda\nu = c$, in frequency. The resonance frequencies are thus given by

$$\nu = \frac{ic}{2Ln} \tag{2.124}$$

so that the spacing between resonance peaks, in frequency, is simply

$$\Delta\nu_{FP} = \frac{c}{2Ln} . \tag{2.125}$$

$\Delta\nu_{FP}$ [Hz] is termed the free spectral range (*FSR*) of the resonator, as illustrated in Fig. 2.19, and should not be confused with the linewidth, denoted $\Delta\nu$.

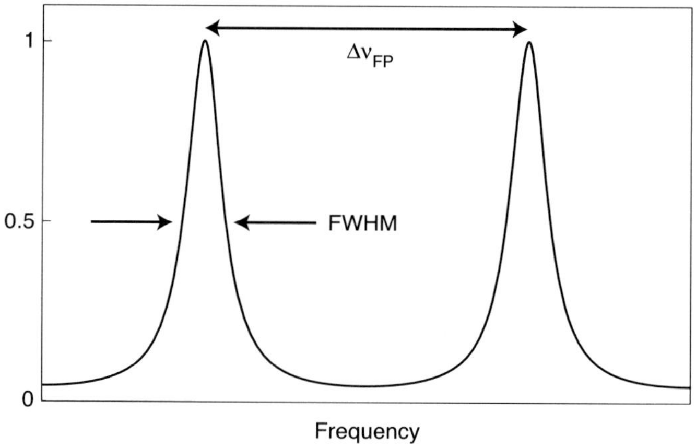

Fig. 2.19. Definition of the free spectral range ($\Delta\nu_{FP}$) and spectral width (*FWHM*) of the etalon transmission characteristic

The *FSR* expressed in wavelength, $\Delta\lambda_{FP}$ [m], may be related to the value expressed in frequency through the relation

$$\Delta\lambda_{FP} = \left(\frac{\lambda^2}{c}\right)\Delta\nu_{FP}\,. \tag{2.126}$$

The last two parameters are equivalent for the description of a resonant cavity.

Also of interest is the spectral width of the transmission resonance. The width is defined as the point at which the transmission has decreased to half of its maximum value (full-width half-maximum or *FWHM*). We may approximate the value of spectral width from Eq. 2.123 by realizing that, at resonance, $\sin(\phi) \approx \phi$, so that

$$\frac{W_t}{W_0} = \frac{1}{2} \text{ for } \frac{F\phi}{\pi} = 1 \tag{2.127}$$

from which we may define

$$FWHM = 2\phi = \frac{2\pi}{F} = \frac{\Delta\nu_{FP}}{F} \tag{2.128}$$

since a phase shift of 2π represents the spacing between resonance peaks, which we just defined as *FSR*. As for the free spectral range, spectral width may be expressed in either wavelength or frequency, the two being related by

$$FWHM_{\lambda} = \left(\frac{\lambda^2}{c}\right) FWHM_{\nu} \,. \tag{2.129}$$

Example 2.7

Assignment: *What is the finesse, FSR in frequency and wavelength as well as the spectral width of the resonance for an 850 nm GaAs-based Fabry–Perot laser cavity with L = 200 µm, given that the uncoated semiconductor mirrors have R = 0.31 and the resonator n = 3.45?*

Solution: *We may calculate the finesse from*

$$F = \frac{\pi\sqrt{R}}{1-R} = 2.54$$

and the free spectral range, in frequency units (GHz) as

$$\Delta\nu_{FP} = \frac{c}{2Ln} = 217\ GHz.$$

This last value may be converted to the FSR expressed in wavelength units (nm) by

$$\Delta\lambda_{FP} = \left(\frac{\lambda^2}{c}\right) \Delta\nu_{FP} = 0.52\ nm.$$

Finally, the spectral width is found from

$$FWHM = \frac{\Delta\nu_{FP}}{F} = 86\ GHz,$$

which is equivalent to approximately 0.21 nm in wavelength units. We see thus that the spacing between resonances for the GaAs Fabry–Perot laser is only about 2.5 times larger than the FWHM of the resonance peak.

2.6 Laser Oscillation

In the last few sections, we have introduced all the physical mechanisms necessary for lasing. As outlined in Sect. 2.1, a laser needs a source of photons, optical amplification or gain and a resonant cavity. We have seen that a physical system with two energy levels can generate photons, that energy pumped into such a system can lead to population inversion and thus gain and finally how a resonant cavity may be configured using a pair of mirrors. A final step is now required: combining these phenomena and fabricating a laser.

2.6.1 The Etalon with Gain

Let's combine a pumped, optically active medium with a resonator as shown schematically in Fig. 2.20. As we will see in considerably greater detail in the following chapters, the medium could be a piece of a light-emitting semiconductor, such as GaAs, the pumped energy could be delivered through an injected current and the cavity could be formed by the cleaved crystalline facets of the semiconductor.

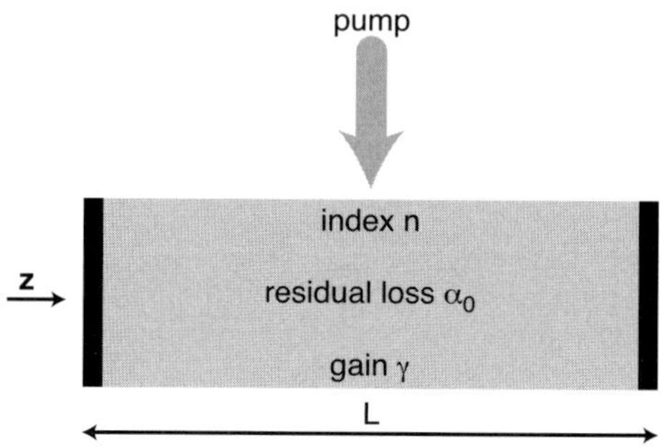

Fig. 2.20. An optical cavity consisting of a pumped active gain medium formed between two parallel mirrors

Through pumping, the semiconductor begins to emit photons, generated by recombination, and higher levels of pumping lead to gain. The effect may be summarized in a change of the propagation constant of the generated optical wave propagating in the material. Based on Eq. 2.71 (Sect. 2.2.8), we may express the propagation *intensity* of a wave W (recalling that $W \propto E^2$) in the form

$$W = W_0 \cdot e^{j2k_R z} \cdot e^{-2k_I z} \cdot e^{j2\omega t} . \tag{2.130}$$

The complex propagation constant is often written in the form

$$k = k_R + jk_I = \beta + j\frac{\alpha}{2} \tag{2.131}$$

for which

$$\beta = \frac{2\pi n}{\lambda_0} \tag{2.132}$$

for real refractive index n and the parameter α is the absorption coefficient or loss.

In absence of pumping, the real refractive index has a value n_0 and the loss a value α_0, known as the residual background loss. However, the high density of electrons injected during pumping will change both of these parameters, such that

$$\beta \rightarrow \beta_0 + \Delta\beta \tag{2.133}$$

and

$$\alpha \rightarrow \alpha_0 - \gamma \tag{2.134}$$

where the gain γ is positive, thereby reducing the total loss in the cavity as it gets larger with pumping. Thus the propagation equation, Eq. 2.130, in the presence of pumping, takes the form

$$W = W_0 \cdot e^{j2(\beta_0 + \Delta\beta)z} \cdot e^{-(\alpha_0 - \gamma)z} \cdot e^{j2\omega t} . \tag{2.135}$$

From this last expression, we see that two things happen to the propagating wave when the optical medium is pumped. From the first exponential of the three, we see that the change in real refractive index changes the propagation speed of the wave. This is not essential for lasing, but the fact that it happens will affect the wavelength at which the laser ultimately lases (see the discussion in Sect. 5.3.4).

The second exponential shows the effect of gain. When γ grows to the point where $\gamma > \alpha_0$, the argument of the exponential becomes positive and the intensity of the wave grows with propagation. At this pump level, then, all residual losses have been compensated and all further energy goes into increasing the intensity of the optical field. This is the laser threshold, the point at which the laser lases.

2.6.2 Threshold

Equation 2.135 showed us how the amplitude and phase of an optical wave propagating in a pumped medium grows and changes with time. By including the effect of a resonant Fabry–Perot cavity, we may then derive expressions for the threshold gain and oscillation wavelength of the laser.

The transmission of the cavity was given for electric fields by the expression in Eq. 2.114 from which the resonance condition (when the denominator equals zero) may be expressed as

$$Re^{j\phi} = 1 . \tag{2.136}$$

By equating the complex phase, ϕ, to the arguments of Eq. 2.135 while ignoring the $(j\omega t)$ term, fulfilling the resonance condition for a cavity of length L then implies

$$R \cdot e^{j2(\beta_0 + \Delta\beta)L} \cdot e^{-(\alpha_0 - \gamma)L} = 1 \tag{2.137}$$

where we recall that R is the mirror reflectivity. This last relationship must be satisfied at laser resonance. We see that the phase has an imaginary (first exponential) and a real (second exponential) part: these must each equal unity and may thus be separately analyzed.

Amplitude

Looking at the real exponent of Eq. 2.137 first, we have the condition

$$R \cdot e^{-(\alpha_0 - \gamma)L} = 1 \tag{2.138}$$

which may be re-written as

$$\gamma = \alpha_0 + \frac{1}{L}\ln\left(\frac{1}{R}\right) = \gamma_{th} \,. \tag{2.139}$$

The gain expressed in Eq. 2.139 is known as the threshold gain, namely that value required to offset all optical losses in the cavity. α_0 is the background or residual loss and $(1/L)\ln(1/R)$ is the mirror loss, accounting for photons which are transmitted through the mirrors (since $R < 1$) and are emitted from the laser as useful light.

Lasing begins when the gain has reached its threshold value and we thus speak of the lasing threshold as the point at which pumping is sufficiently strong to overcome all optical losses.

Propagation

The exponential of Eq. 2.137 with the imaginary argument defines a condition on the propagation of the wave through the resonant cavity. In particular, since

$$R \cdot e^{j2(\beta_0 + \Delta\beta)L} = 1 \tag{2.140}$$

we have the condition that

$$2(\beta_0 + \Delta\beta)L = 2i\pi \tag{2.141}$$

for integer $i = 1, 2, 3\ldots$. Recalling from Eq. 2.132 that the real part of the complex propagation constant is given by

$$\beta = \beta_0 + \Delta\beta = \frac{2\pi n}{\lambda_0} \tag{2.142}$$

for vacuum wavelength λ_0 and (real) refractive index of the cavity n, the condition of Eq. 2.141 may then be expressed as a condition on wavelength, namely

$$\lambda_0 = \frac{2nL}{i} \text{ for } i = 1, 2, 3\ldots. \tag{2.143}$$

This last relationship gives us the allowed lasing wavelengths in the cavity; only the wavelengths λ_0 (for various values of index i) satisfy the phase condition for resonance and are thus amplified in the cavity. Since only discrete wavelength values are permitted, these are termed the *longitudinal modes* of the laser; each mode has its own wavelength and more than one longitudinal mode may occasionally exist in a laser cavity.

2.6.3 Lasing Summary

The two relationships Eq. 2.139 and Eq. 2.143 are key to understanding laser threshold and oscillation. The first tells us what gain is necessary to overcome all photon losses (and thus how much energy we have to inject into the system via pumping) and the second defines at what wavelength oscillation takes place and thus at which wavelength (or frequency) the laser will lase.

We may understand the interplay between these conditions by considering the schematic plot of Fig. 2.21. This diagram shows the gain spectrum, namely the variation of γ with wavelength λ, $\gamma(\lambda)$, for two different pumping levels; the gain spectrum is typically a peaked characteristic with a distinct maximum at a certain wavelength[6]. Shown as a horizontal dotted line is also the sum of background and mirror loss of the cavity, $\alpha = \alpha_0 + (1/L)\ln(1/R)$.

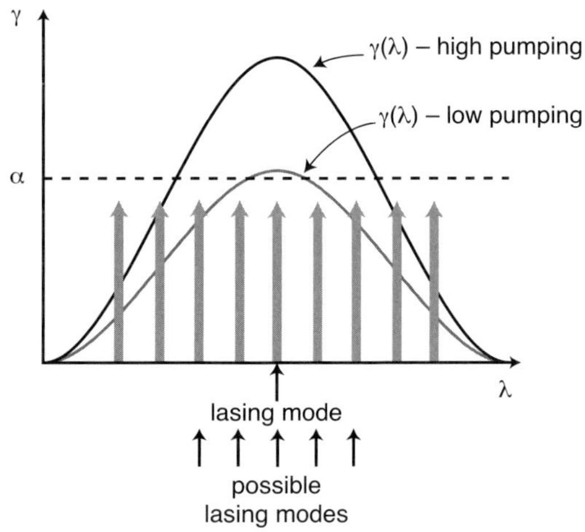

Fig. 2.21. Schematic plot of gain threshold and lasing wavelength. The gain spectra $\gamma(\lambda)$ vary with wavelength; the vertical arrows are allowed cavity wavelengths. When $\gamma > \alpha$, the laser lases at the allowed wavelength with the maximum $(\gamma - \alpha)$ value

Superimposed on the gain spectrum is a series of vertical arrows indicating the allowed laser wavelengths, namely those which satisfy the cavity resonance condi-

[6] We will consider the form of the gain spectrum for semiconductor lasers in detail in Chapter 3.

tion. Each wavelength corresponds to a different value of index i in Eq. 2.143 and thus represents a distinct laser mode.

We reach lasing as follows: through pumping, the gain spectrum increases in magnitude eventually rising above the cavity losses α at some wavelength. When $\gamma > \alpha$ at the wavelength of a laser mode allowed by the cavity (indicated by the vertical arrows in Fig. 2.21), then that mode begins to lase. As pumping increases, the gain becomes concentrated in that lasing mode which subsequently increases in intensity.

With further pumping, the spectral width of the gain spectrum above the loss level (i.e., with $\gamma > \alpha$) increases, so that more allowed cavity modes have gain above the threshold value. With sufficient pumping, these modes can also begin to lase, such that the spectrum displays one or more modes, each at a distinct wavelength. Whether a single or multiple modes actually lase is a function of numerous laser parameters, which we will discuss in the following chapters.

Example 2.8

Assignment: *Given a GaAs laser cavity of length* $L = 200\ \mu m$, *index* $n = 3.45$, *mirror reflectance* $R = 0.31$, *background losses* $10\ cm^{-1}$ *and a gain peak at 850 nm. Estimate the threshold gain, the lasing wavelength and the spacing between adjacent longitudinal modes. Compare this latter value with the spectral width due to the cavity resonance.*

Solution: *Given the background losses, cavity length and mirror reflectance, we may calculate the threshold gain from*

$$\gamma_{th} = \alpha_0 + \frac{1}{L} ln\left(\frac{1}{R}\right) = 68.6\ cm^{-1},$$

a not-un-typical value for semiconductor lasers.

The oscillation wavelength of the laser may be estimated from the phase condition of the cavity,

$$\lambda_0 = \frac{2nL}{i} \quad for\ i = 1, 2, 3\ldots;$$

since the gain peaks at 850 nm, we can find the value for integer i which leads to a cavity resonance closest to this value. We find that $i = 1623$ *yields a resonance wavelength of 850.3 nm, which would be the theoretical lasing wavelength given these structural parameters. This value for i implies that there are 1623 full wavelengths in the length of the cavity.*

We may determine the mode spacing by calculating the wavelength for $i = 1622$ ($\lambda = 850.8$ *nm) or* $i = 1624$ ($\lambda = 849.8$ *nm); from these values, we see the mode spacing is about 0.5 nm. These values compare favorably with those calculated in Example 2.7 (which used the same cavity dimensions) and from we see that the expected spectral width is about 0.21 nm.*

We have thus arrived at the point where the array of physical phenomena discussed in this chapter combine to yield a laser which emits light. Laser characteristics, including the spectral behavior and the required electrical input to reach threshold, will be considered in detail for a variety of laser types in Chapter 4 and Chapter 5.

3 Semiconductors for Lasers

The fact that semiconductor laser diodes are made from semiconductors is in large part responsible for their success. Semiconductor materials exist in great variety, resulting in a large range of accessible laser wavelengths, and the optical gain per unit volume in optically active semiconductor materials is so high that lasers of very small physical volume may be conceived. This latter fact becomes doubly relevant when we consider that semiconductor process technology, highly developed for the fabrication of electronic integrated circuits, may be applied to the manufacture of semiconductor lasers. As a result, these may be produced in large volumes, with high precision and at low cost.

The characteristics of semiconductors as they are relevant for lasers are considered here in highly concentrated form. Many physical aspects will have been considered in most advanced solid-state physics curricula (Ashcroft and Mermin 1976; Kittel 1996; Madelung 1978; Ridley 1988; Seeger 2002; Smith 1978; Ziman 1972) and we will emphasize here those facets of semiconductor physics particularly important for understanding lasers.

3.1 Electrical Properties

Semiconductor lasers convert electrical power into optical power such that the electrical characteristics of semiconductors are relevant for understanding their operation. Both the optical and electrical characteristics of semiconductors are in large part defined by the energy bands of the material. The energy band structure relates the energy E [eV or J][1] of carriers (electrons or holes) in the material to their momentum $p = \hbar k$ [Nm], where k is the wave vector [m^{-1}] and $\hbar = h/(2\pi)$ for Planck's constant $h = 6.62 \times 10^{-34}$ Js .

We recall that in free space, the relationship between E and k is given by

$$E = \frac{\hbar^2 k^2}{2m} \tag{3.1}$$

[1] SI units for energy and length are J and m, but traditionally eV and cm are used for energies in the bandgap and lengths in semiconductors. We bow to tradition in this section and the reader is cautioned to consider units carefully before starting a numerical calculation. At least in our field bridges don't collapse if we make a numerical error.

for carrier mass m [kg]. The wave vector k is the same as the wave vector defined for photons in Eq. 2.19 save that the wavelength is now electron or hole wavelength. This parabolic relationship between energy and momentum is also found in non-crystalline materials, where the optical refractive index of the material must also be included in the definition of k.

3.1.1 The Band Structure

In a semiconductor crystal, the effect of the periodic atomic lattice results in a relationship between E and k which is considerably more complex than the simple parabolic relationship given in Eq. 3.1. This relationship gives rise to a so-called band structure, unique to each kind of semiconductor material and defining the energy states of carriers in the crystal (Kittel 1963). An example of a band structure relevant for semiconductor lasers is that of GaAs, shown in simplified form in Fig. 3.1; the conduction band energy, E_c [eV], and valence band energy, E_v [eV], represent the allowed values of energy and momentum and thus the permitted states that electrons and holes may occupy. The band structure in the figure shows how the energy bands vary in different crystalline directions through the material; the [111] (or L) and [100] (or X) directions are shown.

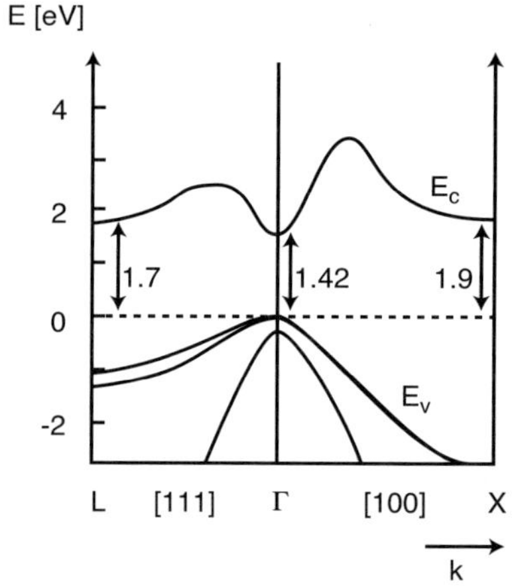

Fig. 3.1. The band structure of GaAs at 300 K showing the zone center (Γ) and two crystalline directions, L and X

Several aspects of the band structure are of interest. We note of course that E_c and E_v vary with k and that this variation differs for different crystalline directions. The conduction and valence bands approach each other most closely at the center of the zone shown, labeled Γ; the energy difference at that point, 1.42 eV for the case of GaAs, is known as the bandgap energy E_g [eV]. The bandgap is the optically and electrically most relevant part of the band structure and we will address a few important aspects of it below.

In addition, the valence band at the Γ point consists of three distinct bands, as is seen in detail in Fig. 3.2. For reasons which will become clear below, the uppermost is termed the heavy hole band, the second the light hole band and the bottom one, slightly separated from the others, the split-off band. This valence band configuration is particular to GaAs; other crystals will have a different structure.

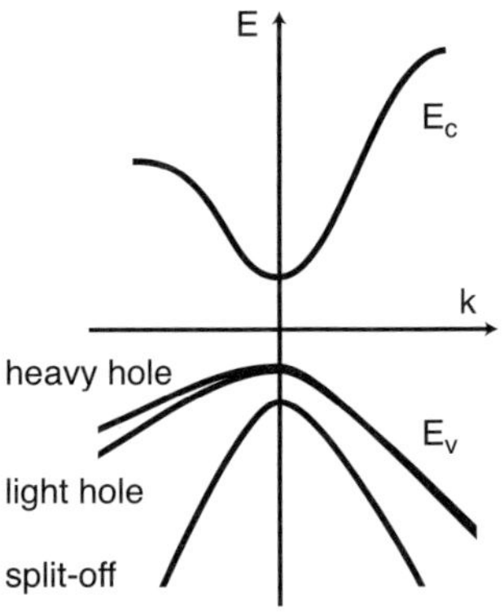

Fig. 3.2. Detail of the band structure near the bandgap (the Γ point) of GaAs; note the three valence band components: the heavy-hole, light-hole and split-off bands

Effective Mass

The band structure defines the allowed energies for electrons in a crystal. The bands therefore affect the motion of charge carriers and, for a given applied electric field, these will accelerate more slowly in the crystal than in free space. We may characterize this effect of the material on carrier dynamics by defining an effective mass, generally smaller than the physical mass, of the electrons and holes in the crystal.

Effective mass, m^* [kg], is derived from the curvature of the bands as

$$m^* = \hbar^2 \left(\frac{\partial^2 E}{\partial k^2} \right)^{-1} . \tag{3.2}$$

The stronger the curvature of the bands, then, the "lighter" the carrier; this observation explains the nomenclature heavy-hole and light-hole inn the description of the two valence bands of GaAs above. In Table 3.1, the effective mass for electrons and holes in several semiconductor materials is summarized, where values are given as a function of electron rest mass, $m_0 = 9.1 \times 10^{-31}$ kg .

Whereas the effective mass for the conduction band is straightforward to calculate, the complexity of the valence band structure may imply that a weighted aver-

Table 3.1. Effective mass for electrons and holes; m_0 = electron rest mass; the hole effective mass for GaAs represents a combination of the heavy- and light-hole bands. GaN data is for the wurtzite structure (Casey and Panish 1978b; Nakamura and Chichibu 2000)

Material	m_e* (electrons)	m_h* (holes)
InAs	0.023 m_0	0.4 m_0
GaAs	0.068 m_0	0.5 m_0
InP	0.08 m_0	0.4 m_0
GaN	0.18 m_0	0.54 m_0
Si	0.33 m_0	0.56 m_0
GaP	0.82 m_0	0.6 m_0

age of the light-hole and heavy-hole masses must be employed for determining a single value for hole effective mass, m_h*. Such an average was performed for GaAs in the table above.

Bandgap Energy

GaAs is an example of a direct-bandgap semiconductor, meaning that the conduction band minimum and the valence band maximum are at the same point on the k axis. Thus electron transitions between the bands do not require a change in momentum, implying that these transitions may be mediated by the absorption or emission of a photon only. Direct bandgap semiconductors, from which all semiconductor lasers are presently made, emit light efficiently; indirect bandgap semiconductors, such as silicon, do not.

The energy difference between conduction band minimum and valence band maximum, E_g, varies for different semiconductors; Table 3.2 gives a few example values for semiconductor materials used for lasers. As we will see in Sect. 3.3 below, attaining a certain desired lasing wavelength requires that a semiconductor of the appropriate bandgap energy is used.

Table 3.2. Bandgap energies and the corresponding wavelengths for some relevant laser materials at 300 K; silicon is included for comparison

Material	Bandgap energy [eV]	Wavelength [μm]	Type
GaN	3.39	0.366	Direct
InN	2.40	0.517	Direct
GaP	2.26	0.549	Indirect
GaAs	1.42	0.873	Direct
InP	1.35	0.919	Direct
Si	1.12	1.107	Indirect
InAs	0.36	3.444	Direct

3.1.2 The Fermi Level

The band structure defines the allowed energies for electrons and holes in a semiconductor crystal. The actual concentration of electrons and holes at these allowed energies is a function of two parameters: the density of available states for an electron or a hole and the probability that this state is occupied. We consider the occupation probability here and will discuss the density of states in Sect. 3.1.3.

Fermi Energy

Electrons are fermions, such that the probability that an electron is found at a particular energy is given by the Fermi-Dirac function (Kittel and Kroemer 1980),

$$f(E) = \frac{1}{1 + e^{(E - E_F)/(k_B T)}} \quad (3.3)$$

named in honor of Enrico Fermi (1901–1954) and Paul Dirac (1902–1984). In this expression, $f(E)$ defines the probability of finding an electron at energy E, k_B is the Boltzmann constant ($k_B = 1.38\times10^{-23}$ J/K $= 8.625\times10^{-5}$ eV/K) and T the temperature in Kelvin; E_F is defined as the Fermi energy (Huang 1987). As may be seen from the characteristic of Fig. 3.3, in which $f(E)$ is plotted for various temperatures, the Fermi-Dirac distribution is a strong function of temperature.

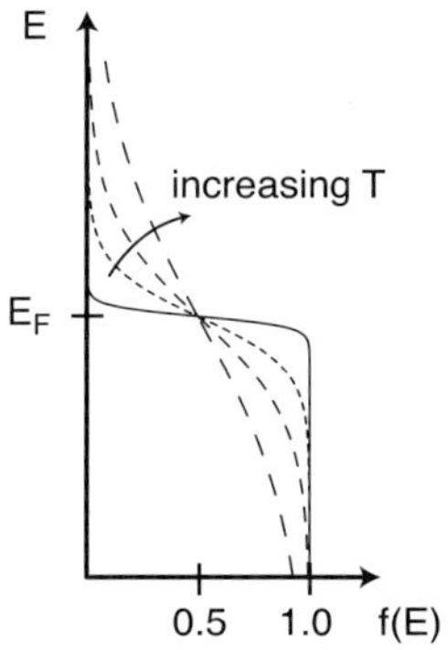

Fig. 3.3. Energy distribution of electrons as defined by the Fermi function for various temperatures

We also see from the figure that the electron distribution revolves about E_F: at $T = 0$ K, all electron states below E_F have occupation probability $f(E) = 1$ and all electron states above E_F have occupation probability $f(E) = 0$. For temperature $T > 0$ K, at which the thermal energy is given by $k_B T$, the distribution broadens so that the occupation probability for electrons with energies $E > E_F$ becomes nonzero, and with increasing temperature higher electron energies are allowed. In contrast, electrons with energy $E < E_F$ no longer have unity occupation probability at higher temperatures, so that some states below the Fermi level will remain empty.

The occupation probability of holes, which are essentially missing electrons, is given by the probability that a certain state is empty rather than filled. Thus if $f(E)$ as given by Eq. 3.3 defines the probability that an electron is found at energy E, the probability that a hole exists at that energy is then simply

$$f_{hole}(E) = 1 - f(E) \ . \tag{3.4}$$

For an undoped semiconductor, the Fermi energy lies approximately halfway between E_v and E_c. Thus the Fermi-Dirac distribution implies that, at thermal equilibrium, the valence band is filled with electrons and the conduction band is relatively empty of electrons, and thus filled with holes. We recall that this distribution of course changes strongly with temperature and note that the non-equilibrium situation, such as when electrons are pumped into the semiconductor, also results in a shift of the electron distribution; we will consider the latter case below.

Maxwell–Boltzmann Distribution

For energies considerably larger than the thermal energy, $E \gg k_B T$, the Fermi-Dirac distribution may be approximated by the Maxwell-Boltzmann distribution of Eq. 2.80, so that the energy distribution of electrons above the Fermi level varies roughly as

$$f(E) \approx e^{-\frac{(E - E_F)}{(k_B T)}} \ . \tag{3.5}$$

We recall that, in semiconductors, E_c is up to several eV larger than E_v, such that the exponential relationship of Eq. 3.5 implies

$$f(E_c) \ll f(E_v) \ . \tag{3.6}$$

As mentioned above, then, the probability of electron occupation in the conduction band at thermal equilibrium is much smaller than its occupation probability in the valence band, but is non-zero. As we have already seen in Sect. 2.3.2 the relationship between the two is an exponential one such that small variations in the Fermi energy can lead to significant changes in occupation probability for a given energy.

Example 3.1

Assignment: *Calculate for what energy above the Fermi level the Maxwell-Boltzmann and Fermi-Dirac distributions agree to better than 1% at room temperature.*

Solution: *We compare the two distributions given in Eq. 3.3 and Eq. 3.5, respectively; the Maxwell-Boltzmann distribution predicts a larger value for $f(E)$. At room temperature, $k_B T \approx 26\ meV$, such that the difference between the two distributions*

$$\left(e^{-(E-E_F)/(k_BT)} - \frac{1}{1+e^{(E-E_F)/(k_BT)}} \right) < 0.01$$

can be shown to be satisfied for $(E–E_F) > 59$ meV.
Thus for energies greater than about $2k_BT$, the Maxwell-Boltzmann distribution is an adequate approximation for the energy distribution of electrons, for which strictly the Fermi-Dirac distribution applies.

Quasi-Fermi Levels

At thermal equilibrium, a single Fermi level determines the occupation probabilities and thus ultimately the concentrations of electrons and holes. In non-equilibrium, such as when a material is pumped to achieve lasing, separate Fermi levels, termed quasi-Fermi levels, are defined for the conduction and valence bands. The occupation probability may then be determined through the positions of the two quasi-Fermi levels, defined to account for carrier concentrations in the two bands separately.

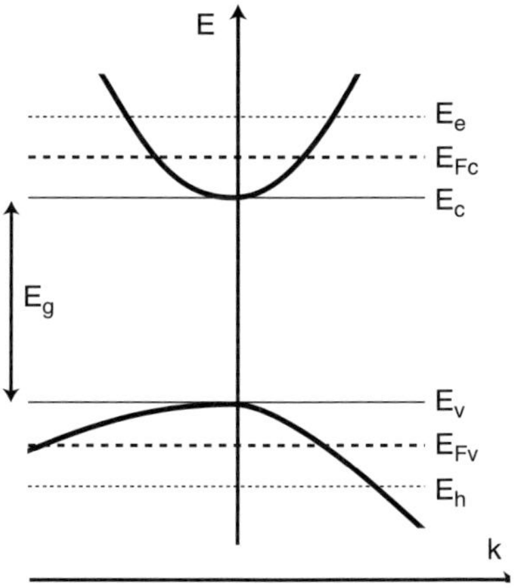

Fig. 3.4. Definition of the quasi-Fermi levels and electron and hole energies at the bandgap of a direct-bandgap semiconductor

In Fig. 3.4, the energies for electrons and holes are separately defined to reflect the non-equilibrium case. For the valence band, hole energy is defined to have a value E_h below the valence band energy E_v; the corresponding quasi-Fermi level is given by E_{Fv}. Similarly, in the conduction band, electron energy is defined as E_e above the conduction band energy E_c, and the corresponding quasi-Fermi level is E_{Fc}.

Using these definitions for the quasi-Fermi levels, the occupation probability for electrons in the conduction band, $f_c(E)$, and holes in the valence band, $f_v(E)$, may be expressed as a function of electron and hole energies by

$$f_c(E) = f(E) = \frac{1}{e^{(E_e - E_{Fc})/(k_B T)} + 1} \tag{3.7}$$

and

$$f_v(E) = 1 - f(E) = \frac{1}{e^{(E_{Fv} - E_h)/(k_B T)} + 1} . \tag{3.8}$$

The separation between E_{Fc} and E_{Fv} is a measure of the departure of the system from thermal equilibrium; the harder the pumping, the further the quasi-Fermi levels move into the respective bands. At thermal equilibrium, the quasi-Fermi levels are equal to each other and thus to the Fermi level, or $E_{Fc} = E_{Fv} = E_F$ (Fukuda 1999).

3.1.3 Carrier Concentration

The Fermi functions derived above tell us only the *probability* that a given electron or hole state is occupied; to determine the concentrations (the volume density) of electrons or holes, we also need to know the number (or volume density) of available states for electrons or holes to fill. The product of the number of available states and the probability that a given state is occupied then yields a measure for the concentration of carriers at a given energy.

Density of States

The density of states (DOS) defines the number of available states for electrons and holes as a function of energy; the energy-dependence of the density of states derives directly from the band structure such that the allowable electron energies and the number of states available at that energy are closely linked. Assuming that the bands at the bandgap are parabolic (namely that energy varies as the square of momentum), the density of states for electrons, ρ_c [eV^{-1} cm^{-3}], and holes, ρ_v [eV^{-1} cm^{-3}], are given by

$$\rho_c = 4\pi\left(\frac{2m_e^*}{h^2}\right)^{\frac{3}{2}} \sqrt{E - E_c} \tag{3.9}$$

and

$$\rho_v = 4\pi\left(\frac{2m_h^*}{h^2}\right)^{\frac{3}{2}}\sqrt{E_v - E}, \tag{3.10}$$

where energy E is measured with respect to the conduction or valence band edges as defined in Fig. 3.4. The effective masses, m_e^* and m_h^*, differ for the two bands. We note that the units for the density of states, [energy^{-1} length^{-3}], imply that the density is an energy density as well as a volume density. The variation of the density of states with energy for the valence and conduction bands is shown in Fig. 3.5; it is clear that the density of available electron and hole states is minimum (namely zero) at the conduction and valence band edges (i.e., at E_c and E_v) and increases as we move further into the bands.

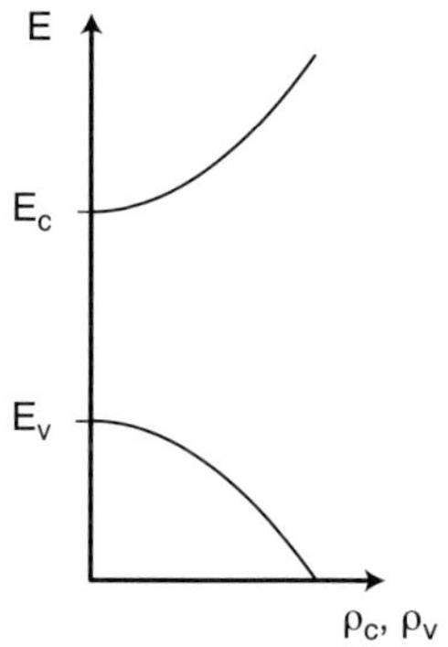

Fig. 3.5. Variation of the density of states in the valence and conduction bands, assuming that the energy varies parabolically with momentum at the band center

Based on these expressions, it is further useful to define an *effective* density of states for the conduction and valence bands, N_c [cm^{-3}] and N_v [cm^{-3}] as

$$N_c = 2\left(\frac{2\pi m_e^* k_B T}{h^2}\right)^{\frac{3}{2}} \tag{3.11}$$

and

$$N_v = 2\left(\frac{2\pi m_h^* k_B T}{h^2}\right)^{\frac{3}{2}} \tag{3.12}$$

where the latter definition assumes the existence of only a single valence band. We will employ these definitions in simplified expressions for the carrier concentration below.

Density of Carriers

From the density of states, Eq. 3.9 and Eq. 3.10, and the probability that these states are occupied by an electron or a hole, Eq. 3.7 and Eq. 3.8, the density of carriers at any given energy is then given as the product of the two, as shown schematically in Fig. 3.6. The carrier density thus increases from zero at the band edge to a maximum inside the conduction or valence bands.

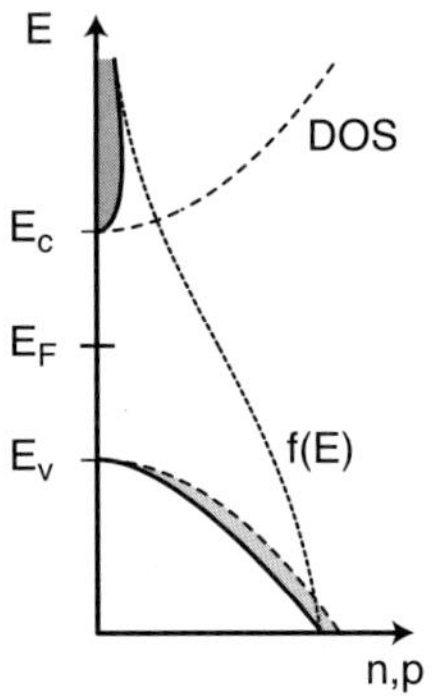

Fig. 3.6. Electron and hole concentration as the product of density of states (DOS) and occupation probability, $f(E)$; the upper and lower shaded regions represent the density of electrons and holes, respectively. In this example, the shape of the occupation probability is that which would be valid at relatively high temperatures

We may subsequently calculate the total electron and hole concentrations for all energies in the conduction and valence bands, n and p [cm^{-3}], as the integral over these energies, namely

$$n = \int_{E_c}^{\infty} \rho_c(E) f_c(E) dE \tag{3.13}$$

and

$$p = \int_{-\infty}^{E_v} \rho_v(E) f_v(E) dE \ . \tag{3.14}$$

Using the definitions of the effective density of states and the assumption that the Maxwell-Boltzmann distribution (Eq. 3.5) holds, these integrals may be evaluated to yield the carrier concentrations as a function of the Fermi energy, namely

$$n = N_c e^{\frac{E_F - E_c}{k_B T}} \tag{3.15}$$

and

$$p = N_v e^{\frac{E_v - E_F}{k_B T}} . \tag{3.16}$$

From the previous two expressions, it is clear that the position of the Fermi level defines the electron and hole concentrations, at least at thermal equilibrium.

Non-Equilibrium Carrier Distribution

For the non-equilibrium case, as we noted in the previous section, the conduction and valence bands have their own quasi-Fermi levels and the resultant carrier concentrations are then a function of E_{Fc} and E_{Fv}. The arguments presented above still hold, if we replace the Fermi level with E_{Fc} for electrons and E_{Fv} for holes. In that case, the electron and hole concentrations become

$$n = N_c e^{\frac{E_{Fc} - E_c}{k_B T}} \tag{3.17}$$

and

$$p = N_v e^{\frac{E_v - E_{Fv}}{k_B T}} \tag{3.18}$$

as is shown schematically in Fig. 3.7. We see that the carrier concentrations (the shaded areas in the figure) are larger than for the equilibrium case and that the energy distributions differ. As the previous two expressions relate the individual quasi-Fermi levels to the electron and hole concentrations, we may thus easily determine E_{Fc} and E_{Fv} when the carrier densities are known.

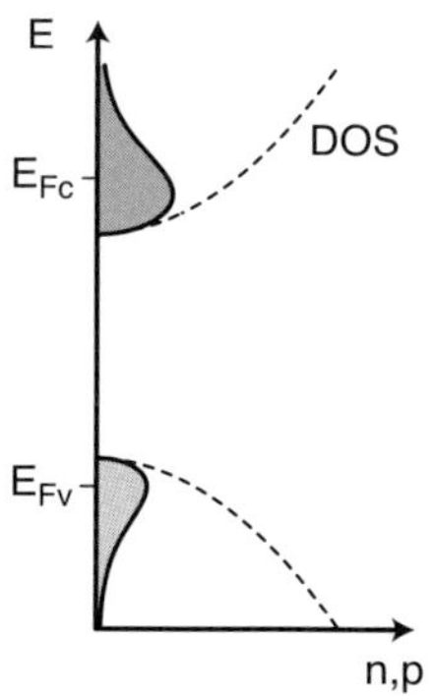

Fig. 3.7. Electron and hole concentration for the non-equilibrium case where individual quasi-Fermi levels are defined for electrons and holes; the upper and lower shaded regions represent the density of electrons and holes, respectively

Intrinsic Carrier Concentration

The product of electron and hole concentration, np, is a temperature-dependent constant for any given material, namely

$$np = n_i^2 \tag{3.19}$$

where n_i is the intrinsic carrier concentration; Eq. 3.19 is known as the law of mass-action and is valid for the Maxwell-Boltzmann limit. The intrinsic carrier concentration is a measure for the number of carriers present in a material at thermal equilibrium. Typical values for n_i for a few semiconductor materials are given in Table 3.3.

Table 3.3. Intrinsic carrier concentration at T = 300 K (Sze 1981)

Material	n_i [cm^{-3}]
GaAs	1.79×10^6
InP	8.7×10^6
Si	1.45×10^{10}
Ge	2.4×10^{13}

By expressing n_i explicitly from the definitions of p and n,

$$n_i = \sqrt{N_c N_v} e^{-\frac{E_g}{2k_B T}}, \tag{3.20}$$

we see that intrinsic carrier concentration is not only a function of the effective density of states but also exponentially depends on energy gap and temperature. Thus materials with a small bandgap have a high intrinsic carrier concentration, a phenomenon which becomes relevant when, for example, using narrow-bandgap semiconductors as photodetectors. The large numbers of carriers present due to n_i lead to a large background signal, a problem which, according to Eq. 3.20, may be addressed by decreasing the temperature.

Band Tails

In our consideration of the energy gap to this point, we have tacitly assumed that the conduction and valence bands are distinct features and that the density of states becomes non-zero at a well defined energies E_c and E_v. In reality, due to the effects of non-zero temperature and doping, the density of states in the bandgap exhibits so-called band tails as shown in Fig. 3.8 (Urbach 1953).

Whereas the density of states ideally varies as $E^{1/2}$, the band tails exhibit an exponentially decaying density into the bandgap. In addition, high doping levels can lead to local peaks in the density of states and the resulting band-tail states can have significant gain in a semiconductor laser. Doping levels in the range

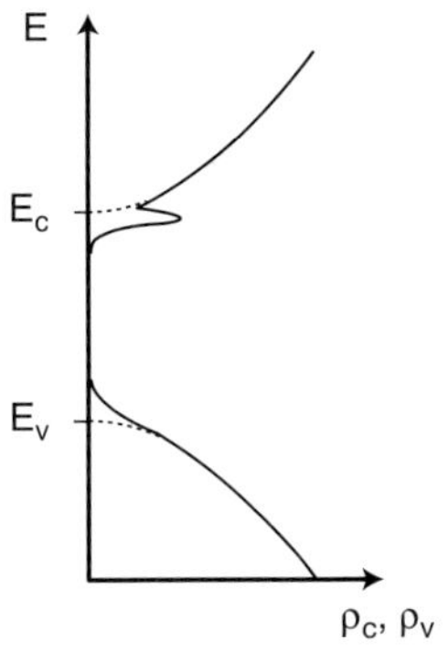

Fig. 3.8. Schematic representation of the band tails in the energy gap due to doping

5×10^{18} cm^{-3} can lead to a reduction in the effective bandgap by 0.05 to 0.15 eV. Although difficult to model save by approximation, consideration of the form and magnitude of band tails is essential for accurate calculation of semiconductor gain behavior.

3.1.4 Doping

The intrinsic carrier concentration represents the density of electrons and holes in an intrinsic semiconductor, meaning one which consists only of the semiconductor material itself. To increase the concentrations of electrons or holes, impurities may be incorporated into the semiconductor in a controlled manner. These impurity atoms are known as dopants; doped semiconductors may be of either n-type or p-type polarity, depending on the nature of the dopant atoms.

n-Type and p-Type

We may consider the example of silicon, which is found in column IV of the periodic table and thus has four valence electrons in its outer shell. Intrinsic silicon thus binds to four nearest neighbors to form the crystal lattice. Phosphorus is found in column V of the periodic table and has five valence electrons: incorporating P atoms into Si then results in a surplus of electrons, since only four per atom are required for bonding to the lattice; phosphorus is thus termed a *donor* ion for silicon. The electron concentration n is increased and phosphorus-doped silicon is thus an n-type semiconductor.

Conversely, boron, in column III of the periodic table, has three valence electrons, and thus one fewer than the four required for bonding to the silicon lattice; hence boron is termed an *acceptor* ion for silicon. Adding boron to silicon then results in a surfeit of holes, increasing p, so that boron-doped silicon is a p-type semiconductor.

The type of dopant ion employed depends on and varies with the semiconductor in question. For the laser material GaAs, for example, silicon acts as an n-type dopant and Zn or Mg are used for p-type doping.

Doping Levels

The incorporation of dopant atoms into a semiconductor crystal adds doping energy levels to the energy gap, as shown in Fig. 3.9. Acceptors generate an energy level E_a, close to the valence band and donors an energy level E_d close to the conduction band. The energy differences $E_a - E_v$ and $E_c - E_d$ are usually much smaller than k_BT, so that most of the dopant ions are ionized at room temperature.

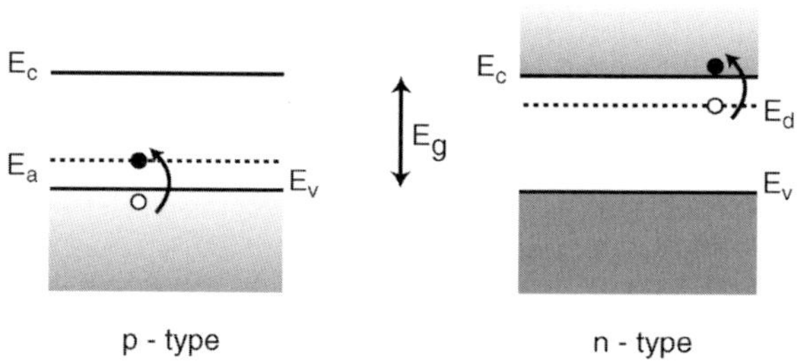

Fig. 3.9. Acceptor *left* and donor *right* energy levels due to doping of a semiconductor

For donor dopant concentration N_D or acceptor dopant concentration N_A, complete ionization of the dopants implies that

$$n \approx N_D \tag{3.21}$$

and

$$p \approx N_A \tag{3.22}$$

since the dopant concentrations are typically orders of magnitude higher than the intrinsic carrier concentration. The corresponding Fermi levels may then be determined from the dopant concentration for n-type dopants as

$$E_F = E_c - k_B T \ln\left(\frac{N_C}{N_D}\right) \tag{3.23}$$

and for p-type dopants as

$$E_F = E_v + k_B T \ln\left(\frac{N_v}{N_A}\right) . \tag{3.24}$$

We note that the previous two expressions are essentially identical to Eq. 3.15 and Eq. 3.16, where we have solved for the Fermi level and used the approximations of Eq. 3.21 and Eq. 3.22.

3.1.5 Carrier Transport

The electrons in the conduction band and the holes in the valence band are mobile. Either concentration gradients or applied electric fields will result in carrier transport and thus a current; the former mechanism results in diffusion current and the latter in drift current (Blatt 1968). In general, the total current in a semiconductor consists of the sum of both components.

Diffusion and Drift

Diffusion current density of electrons, $\boldsymbol{J}_n$ [A/cm^2], and holes, $\boldsymbol{J}_p$ [A/cm^2], may be expressed as

$$\boldsymbol{J}_n = qD_n\nabla n \tag{3.25}$$

and

$$\boldsymbol{J}_p = -qD_p\nabla p \tag{3.26}$$

respectively, where the minus sign accounts for the conventional definition of positive current flow. D_n [cm^2/s] and D_p [cm^2/s] are the diffusion coefficients, which we discuss further below. The concentration gradients, ∇n and ∇p , are thus the driving force behind diffusion current; such a gradient exists, for example, in the case where an n-type region directly abuts a p-type region of semiconductor.

When an electric field $\boldsymbol{E}$ is applied to a semiconductor, the energy bands and the Fermi level tilt in response to the field, as shown in Fig. 3.10. Due to the spatial variation of the Fermi level, electron and hole energy varies with position and a drift current results. The drift current densities for electrons and holes are then given by

$$\boldsymbol{J}_n = q\mu_n n\boldsymbol{E} \tag{3.27}$$

and

$$\boldsymbol{J}_p = q\mu_p p\boldsymbol{E} \tag{3.28}$$

for which the mobilities, μ_n [cm^2/Vs] for electrons and μ_p [cm^2/Vs] for holes, are the relevant material parameters. Thus high drift currents are generated by high electric fields, but also, for a given electric field strength, through the use of materials with a high mobility.

Diffusion constant and mobility are simply related through the Einstein relation as

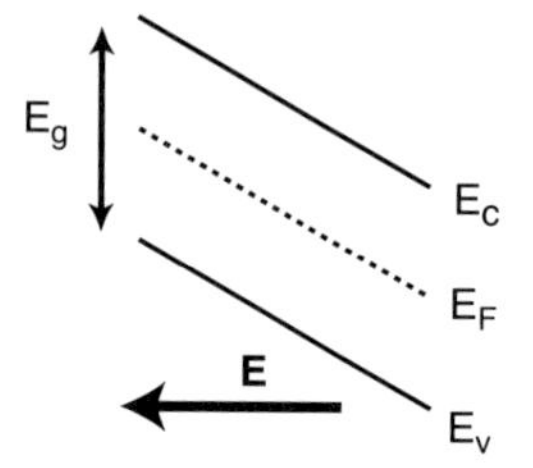

Fig. 3.10. The tilted energy bands and Fermi level in a semiconductor under an applied electric field, ***E***

$$\frac{D}{\mu} = \frac{k_B T}{q} \tag{3.29}$$

and typical values are summarized in Table 3.4. Mobility, and by extension diffusion constant, is a function not only of the material but also the impurity concentration, including dopants, and temperature. Moving electrons may lose energy through scattering, interaction with lattice vibrations (acoustic and optical phonons) and through the effects of charged impurities. All of these factors may reduce mobility and all are temperature-dependent.

Table 3.4. Typical values for electron and hole mobility for several semiconductor materials (Sze 1981; Nakamura 2000)

Material	μ_n [cm²/Vs]	μ_p [cm²/Vs]
InAs	33 000	460
GaAs	8 500	400
InP	4 600	150
Si	3 900	1 900
GaN	1 000	30
GaP	110	75

Current Flow and Resistance

As mentioned above, total current flow in a semiconductor is generally the sum of the drift and diffusion components. For homogeneous semiconductors, however, the drift component typically dominates over diffusion such that, far away from pn junction, current is often predominantly drift.

Including contributions from both electron and hole transport, the total drift current density is then given by

$$\boldsymbol{J} = q(\mu_n n + \mu_p p)\boldsymbol{E} = \sigma \boldsymbol{E} \tag{3.30}$$

where σ [$1/(\Omega\ \mathrm{m})$] is the conductivity, related simply to the resistivity ρ [Ω m] as $\sigma = 1/\rho$. Conductivity and resistivity are dimensionally-independent parameters, primarily a function of the material, doping levels and temperature. Once the cross-sectional area A [m^2] and length L [m] of a conducting path through a semiconductor are known, the resistance R [Ω] may then be found from resistivity or conductivity as

$$R = \frac{\rho L}{A} = \frac{L}{\sigma A}. \tag{3.31}$$

3.1.6 The pn Junction

When an n-type semiconductor directly abuts a p-type semiconductor, a pn junction results. The electrical properties of the pn junction are of central importance for pn diodes, transistors, photodiodes and, luckily for us, diode lasers.

Figure 3.11 shows a schematic view of an abrupt pn junction such as that which might be found in a laser heterostructure (see Sect. 3.3.3). In most electronic devices, pn junctions are formed by diffusion so that the transition between p- and n-regions is not usually abrupt, but the abrupt view is illustrative for a first analysis of the junction. At thermal equilibrium, no current flows through the junction, so the Fermi level must be flat. Since the Fermi level is at a different position in the bandgap for n-type and p-type materials, a flat Fermi level implies that the valence and conduction bands must bend through the transition region.

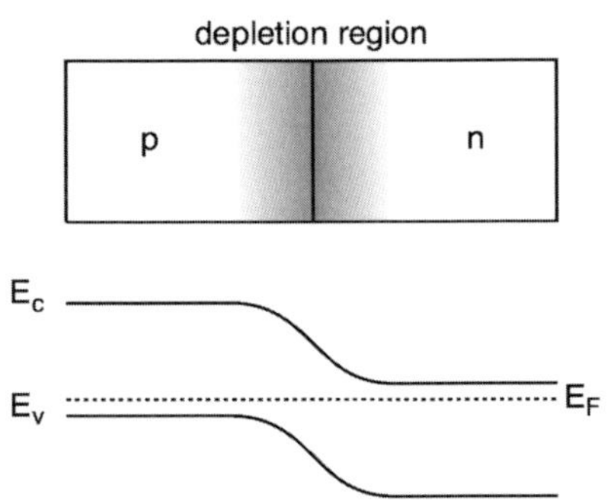

Fig. 3.11. Schematic depiction of a pn junction and the energy bands through it at thermal equilibrium; note that the Fermi level is flat

We may see how this thermal equilibrium situation is reached. Imagine that the two pieces of semiconductor, one p-type, the other n-type, are just put into contact. The p-type semiconductor on the left has a high density of holes but a low density of electrons; for the n-type semiconductor on the right, the reverse is true. This large concentration gradient gives rise to a large diffusion current flow (as we saw in Sect. 3.1.5), so that holes flow to the right and electrons to the left. This flow of mobile charge carriers away from the junction, however, uncovers the immobile ionized dopant ions (positively-charged donors and negatively-charged acceptors) which are normally neutralized by the mobile carriers. Thus a thin, so-called depletion region is formed at the junction, in which the mobile carriers have diffused away and the ionized dopant ions remain.

The uncompensated ionized donors give rise to an electric field which is established in the pn junction throughout the depletion region. This electric field (which points right-to-left in the example of Fig. 3.11) increases as more carriers diffuse away from the junction, until the drift current due to the electric field just balances the diffusion current due to the concentration gradient. At this point, thermal equilibrium is reached: no more electrons and holes flow across the junction, an electric field is established in a region depleted of carriers and the conduction and valence bands bend. Since net current flow is zero, the Fermi level is flat. The resulting bending of the conduction and valence bands thus leads to energy barriers for electrons and holes.

Depletion Region

The depletion region exists on both sides of the pn junction and is a volume depleted of electrons and holes. The width, W [m], of the depletion region may be found from the donor and acceptor concentrations, N_A and N_D respectively, on either side of the junction as (Sze 1981)

$$W = \sqrt{\frac{2\varepsilon_m}{q}\left(\frac{1}{N_A} + \frac{1}{N_D}\right)\phi_{bi}} \tag{3.32}$$

where ε_m [F/m] is the permittivity of the material and ϕ_{bi} [V] the built-in electrostatic potential across the junction, found from

$$\phi_{bi} = \frac{k_B T}{q}\ln\left(\frac{N_A N_D}{n_i}\right) . \tag{3.33}$$

The built-in potential is that across the junction at thermal equilibrium and is defined as

$$\phi_{bi} = \phi_n - \phi_p \tag{3.34}$$

where ϕ_n and ϕ_p are the potential on the n- and p-sides of the junction, respectively. Thus the potential on the n-side is larger than the potential on the p-side at thermal equilibrium.

The width of the depletion region depends on the doping level: a more lightly doped region has fewer dopant ions and thus requires a larger depletion region extent to generate a given amount of charge than a more heavily doped region.

Thus for an asymmetric pn junction, in which the p- and n-sides have different doping levels, the depletion region extends deeper into the more *lightly* doped side.

Biased pn Junctions

The pn junction is typically operated by the application of an external voltage, as shown schematically in Fig. 3.12. For the case where the applied voltage across the junction is greater on the p-side than on the n-side (i.e., $V_A > 0$ or $V_p > V_n$), the junction is *forward* biased; for the case where the applied voltage across the junction is smaller on the p-side than on the n-side (i.e., $V_A < 0$ or $V_p < V_n$), the junction is *reverse* biased.

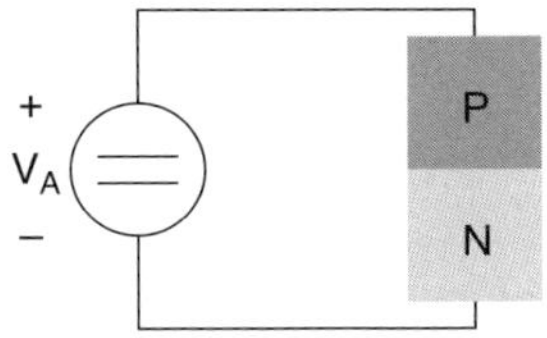

Fig. 3.12. Application of a voltage to a pn junction

For the forward bias case, $V_A > 0$, the bending of the energy bands becomes smaller and the energy barriers to electron and hole flow across the pn junction are reduced. As a result, carrier flow across the junction increases and a high density of carriers is found at the junction itself, a strongly non-equilibrium condition. As seen in Fig. 3.13, the Fermi level also splits into two quasi-Fermi levels inside the depletion region; the depletion region width decreases. This mode of operation is the typical one for laser diodes, where the pn diode is used to generate a local high density of electron-hole pairs.

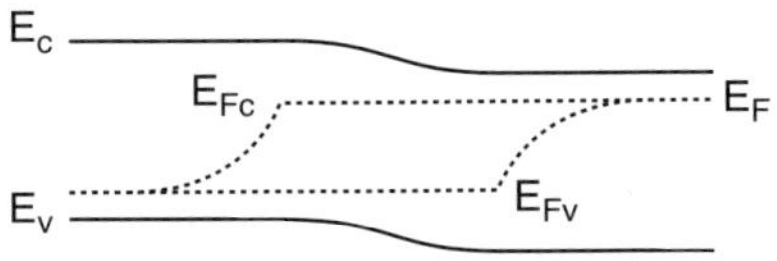

Fig. 3.13. Band structure in a pn diode under forward bias

For an applied reverse bias, the energy band bending increases and the energy barrier for electrons and holes grows; thus carrier flow remains small, but not zero, as we shall see below. The energy band diagram of Fig. 3.14 shows the bands and resulting quasi-Fermi levels clearly; in this case, the depletion region width increases with increasing reverse bias. Reverse bias is the typical mode of operation for pn-diode photodetectors, where generated photons are absorbed in the wide depletion region.

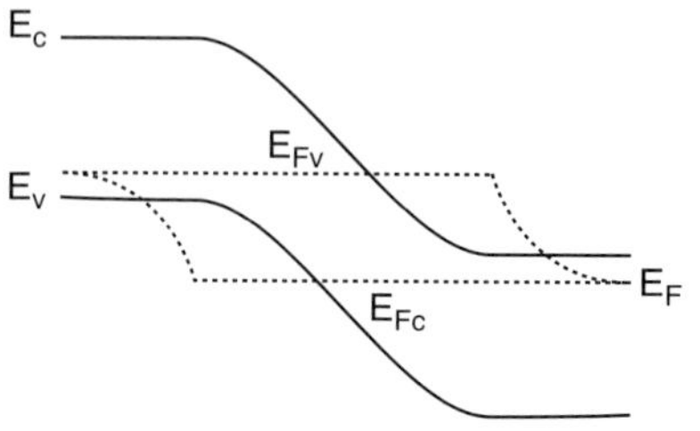

Fig. 3.14. Band structure in a pn diode under reverse bias

pn Junction IV Characteristics

The change from forward to reverse bias is not reflected in an abrupt change in current magnitude but rather through the exponential behavior of the Schockley equation (Schockley 1950). This relationship predicts that the current density through the pn junction, J, is given as a function of the applied bias, V_A, as

$$J = J_0\left(e^{\frac{qV_A}{k_B T}} - 1\right) \tag{3.35}$$

where the constant J_0 [A/cm^2]

$$J_0 = q\sqrt{\frac{D_p}{\tau_p}}\frac{n_i^2}{N_D} + q\sqrt{\frac{D_n}{\tau_n}}\frac{n_i^2}{N_A} \tag{3.36}$$

is known as the saturation current density. In the last expression, the parameters τ_n and τ_p are the electron and hole minority carrier lifetimes, respectively.

The resulting voltage/current (IV) characteristic is shown in Fig. 3.15. We may see that in the forward bias region the current increases exponentially with applied bias; as a consequence, the electron and hole concentrations in the vicinity of the pn junction increase rapidly, as required for achieving population inversion in a laser.

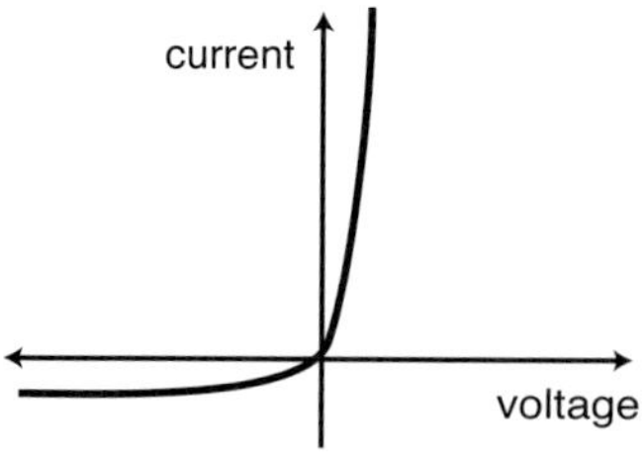

Fig. 3.15. Current/voltage characteristic of a pn diode

In reverse bias, the current is not zero but reaches its saturation value, as given by Eq. 3.36. Although there is a significant (and growing) barrier to electron and hole flow in reverse bias, some carriers achieve sufficient energy to surmount the barrier and contribute to a small, but non-zero current. As the electric fields across the junction increase with increasing applied reverse bias, avalanche effects will ultimately cause a rapid increase in current density and thus reverse-bias breakdown of the junction.

3.2 Optical Properties

Since semiconductor lasers are based on the absorption and emission of photons from semiconductors, the optical properties of these materials are also of primary importance for understanding device behavior (Klingshirn 1997). As for the electrical properties, the optical characteristics are defined to a large part by the bandgap of the material. Electronic transitions between the conduction and valence bands can lead to generation and absorption of photons, although non-radiative processes, in which the energy is lost to phonons, are also of importance in semiconductors.

3.2.1 Radiative Transitions

As we saw in Chapter 2, electronic transitions between different energy levels in a laser medium give rise to the generation or absorption of photons; in a semiconductor laser, the relevant energy levels are the valence and conduction bands. The energy difference between E_c and E_v is accounted for by the emission (for a $E_c \rightarrow E_v$ transition) or absorption (for a $E_v \rightarrow E_c$ transition) of a photon.

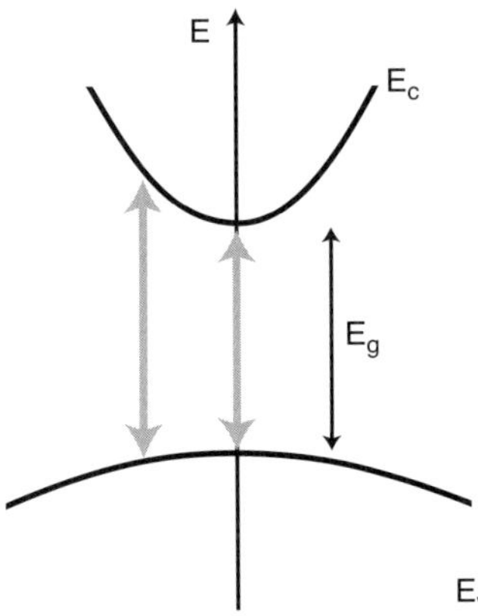

Fig. 3.16. The idealized bandgap of a direct-bandgap semiconductor showing some of the possible electronic transitions which give rise to absorption or emission of photons

Although, in general, electronic transitions between the energy bands and doping levels, band tails or impurity levels, as well as intra-band transitions within a single band, are important, we will restrict ourselves to transitions between idealized valence bands and conduction bands in the following discussion.

As seen in Fig. 3.16, the energy gap E_g defines the minimum electronic transition energy, if doping and bandtailing effects are ignored. If we move from the band center, for bands with a parabolic relation of energy to momentum, usually a good approximation near the Γ-point, the allowed electron energies then vary as

$$E_e = E_c + \frac{\hbar^2 k^2}{2m^{*2}_c} \tag{3.37}$$

and allowed hole energies as

$$E_h = E_v - \frac{\hbar^2 k^2}{2m^{*2}_v}. \tag{3.38}$$

We recall from Sect. 3.1.1 that the emission and absorption of photons is optimal for a vertical transition, namely one with $\Delta k = 0$, as indicated by the vertical arrows in Fig. 3.16. An electronic transition with $\Delta k \neq 0$ requires the addition of momentum to satisfy energy and momentum conservation. Photon momentum is typically several orders of magnitude smaller than the Δk required for momentum conservation, such that the transition can only take place through the addition of a phonon in the interaction. Since a transition involving an electron, a photon and a phonon is rather unlikely, indirect transitions typically conserve energy and momentum by the emission or absorption of phonons only and are thus non-radiative.

As a result, photon emission is very unlikely in indirect bandgap materials. Photon absorption does take place in materials where $\Delta k \neq 0$ and phonon interactions lead to momentum conservation; the absorbance is not as high and does not increase as rapidly from the bandgap energy as for a direct bandgap material. A notable example material is silicon, an indirect bandgap semiconductor from which excellent detectors are made but, in its native state, does not emit light efficiently. When considering light absorption and emission for semiconductor lasers, direct transitions are thus of primary importance such that we will restrict ourselves to these in the sections which follow.

3.2.2 Absorption

Assuming that electron and hole energies in the valence and conduction bands vary parabolically with energy, as given in Eq. 3.37 and Eq. 3.38, photons at energies $h\nu$, where

$$h\nu = E_e - E_h = (E_c - E_v) + \frac{\hbar^2 k^2}{2}\left(\frac{1}{m^{*2}_c} + \frac{1}{m^{*2}_v}\right), \tag{3.39}$$

may be emitted or absorbed in a direct transition. In the previous expression, ν [Hz] represents the photon frequency and m^*_c and m^*_v are the conduction and valence band effective masses, respectively. Thus photon transitions are allowed for photon energies $h\nu$ above the bandgap energy

$$(h\nu - E_g) = \frac{\hbar^2 k^2}{2}\left(\frac{1}{m^{*2}_c} + \frac{1}{m^{*2}_v}\right). \tag{3.40}$$

Equation 3.40 only tells us which energies are allowed for a transition and does not yet include any information concerning the availability of carriers for a transition at those energies. Since the density of electron and hole states varies with energy as $\sqrt{E}$ for non-degenerate carrier concentrations (Sect. 3.1.3), the absorption of photons, α [cm^{-1}], varies with incident photon energy as

$$\alpha = C_\alpha \sqrt{h\nu - E_g} \tag{3.41}$$

where the constant C_α [$\text{cm}^{-1}\ \text{eV}^{-1/2}$] is found from [Pankove 1976, Chap. 3]

$$C_\alpha \approx \frac{2\sqrt{2}q^2}{n_r c h^2}\left[\frac{1}{m^*_c}\left(\frac{1}{m^*_c} + \frac{1}{m^*_v}\right)^{-\frac{3}{2}}\right]. \tag{3.42}$$

In the previous expression, n_r represents the refractive index, where we use this notation here to avoid confusion with the carrier concentration n. The parameter α is known as the absorption coefficient which characterizes the decrease in optical intensity with propagation through a semiconductor material.

We recall from Eq. 2.71 (Sect. 2.2.8) that the electric field of an optical wave decreases exponentially with propagation through a material; the absorption coefficient characterizes this decrease and expresses the exponential decay of optical *intensity* S (recalling that $S \propto \boldsymbol{E}^2$) as

$$S = S_0 e^{-\alpha z} \tag{3.43}$$

for an initial intensity S_0 [W/cm^2] and distance z [cm]. Equation 3.43 is known as Beer's, Lambert's or Beer-Lambert's law and holds for all optical/material interactions. Although Equation 3.41 expresses α as a function of direct bandgap transitions, the absorption of light in a semiconductor (or in any material) is the sum of all photon absorption mechanisms, including transitions between impurity levels, band tails and so forth.

Absorption Transition Rate

As we discussed in Sect. 3.1.3, the concentration of carriers at any energy is given by the density of states (Eq. 3.9 and Eq. 3.10 which define ρ_c and ρ_v) and the Fermi function (Eq. 3.7 and Eq. 3.8 which define f_c and f_v) for occupation probability. Using these definitions, we may write an expression for the electron transition rate due to absorption of photons R_α [s^{-1} eV^{-1} cm^{-3}] at one particular energy as

$$R_\alpha = B_\alpha(1-f_c)\rho_c(1-f_v)\rho_v\Phi \tag{3.44}$$

where Φ [eV s cm^{-3}] is the photon energy density per volume and frequency interval and B_α [cm^6 s^{-2}] is a coefficient which includes the transition probability. The transition rate R_α defines the number of photons absorbed per unit volume per time per unit energy. In this relationship, the Fermi terms give the probability that there is no electron at that energy in the conduction band, and no hole in the valence band, so that photon absorption can may create an electron-hole pair at that energy.

3.2.3 Emission

When an electron falls from the conduction to the valence band, the resulting energy may be given off as a photon, yielding a radiative transition; the energy may alternatively be lost to phonons, due to the effect of non-radiative transitions discussed in Sect. 3.2.4. At thermal equilibrium, the rate of optical generation of electron-hole pairs is equal to the rate of radiative recombination, a condition known as the van Roosbroeck-Schockley relation (van Roosbroeck and Schockley 1954).

Spontaneous Emission

Given filled electron and hole states in parabolic conduction and valence bands, the generation of photons, G [cm^{-1}], then varies with incident photon energy in the same manner as the absorption, namely

$$G = C_e\sqrt{h\nu - E_g} \tag{3.45}$$

where the constant C_e [cm^{-1} $eV^{-1/2}$] is found from [Pankove 1976, Chap. 6]

$$C_e \approx \frac{2q^2}{n_r c h^2}\left[\frac{1}{m^*_c}\left(\frac{1}{m^*_c}+\frac{1}{m^*_v}\right)^{-\frac{3}{2}}\right] \tag{3.46}$$

and is thus roughly equal to the constant for absorption, C_α, defined in Eq. 3.42. This form of radiative transition leading to the generation of photons is known as spontaneous emission, since the transition is not stimulated by an external factor, such as another photon.

The allowed energies for photon emission, given by Eq. 3.45, have the same form as the absorption energies, given by Eq. 3.40. However, whereas the absorption spectrum is typically broad, since all combinations of energy states in the conduction and valence bands may play a role in a transition, photon emission typically takes place due to transitions of thermalized electrons near the bottom of the conduction band and thermalized holes near the top of the valence band. As a result, the emission spectrum tends to be narrow and emitted photon energies are close to but slightly larger than the bandgap energy.

Spontaneous Emission Rate

By including consideration of the density of states and the occupation probability of these states, as we did for the absorption rate, we may easily derive an expression for the transition rate of spontaneous emission, R_e [$\mathrm{s^{-1}\ eV^{-1}\ cm^{-3}}$]. Employing the definitions of Fermi level and density of states, R_e is given by

$$R_e = A_e f_c \rho_c f_v \rho_v \tag{3.47}$$

where the A_e [$\mathrm{eV\ cm^3\ s^{-1}}$] represents the transition probability and the product $f_c f_v$ gives the probability that an electron is present at a given energy in the conduction band and a hole in the valence band.

We may write a useful simplified form for spontaneous emission transition rate in a semiconductor subject to pumping. In this case, a large excess carrier concentration, n' and p', is present, such that the total carrier concentration, n and p, is given by

$$n = n_0 + n' \,,\ p = p_0 + p' \tag{3.48}$$

for background (equilibrium) concentration n_0 and p_0.

In this non-equilibrium case, the spontaneous emission rate per volume integrated over all energies, R_e' [$\mathrm{s^{-1} cm^{-3}}$] may be found from

$$R_e' = A_e' n' p' \tag{3.49}$$

where A_e' [$\mathrm{eV^2\ cm^3\ s^{-1}}$] is a slightly modified transition probability coefficient. Thus the spontaneous emission rate is, as we might expect, directly related to the concentration of *excess* carriers in the material.

Stimulated Emission Rate

We recall from Sect. 2.3.1 that the presence of a photon in an optical field moving through a semiconductor may induce an electronic transition and give rise to stimulated emission; a second photon is generated with the same energy and phase as the first. The rate of stimulated emission, R_s [$\mathrm{s^{-1}\ eV^{-1}\ cm^{-3}}$], may be found from the carrier concentrations in the conduction and valence bands and the intensity of the optical field, namely

$$R_s = B_s f_c \rho_c f_v \rho_v \Phi(E) \tag{3.50}$$

in which Φ [eV s cm^{-3}] represents the photon energy density and B_s [cm^6 s^{-2}] is a transition probability.

3.2.4 Non-Radiative Transitions

Transitions of an electron from the conduction band to the valence band need not necessarily result in the generation of a photon. Indeed, in many materials the predominant transfer of energy for such a transition is not a photon but to phonons and thus ultimately lattice vibrations or heat. In a laser, such non-radiative transitions represent the loss of electrons from the conduction band, thereby reducing laser efficiency since harder pumping is required to reach inversion.

Non-radiative transitions often result from the existence of additional energy levels in the bandgap, possibly due to impurities, defects or flaws in the crystal lattice, surfaces and other imperfections. Impurities may either be intentionally incorporated (as in the case of dopants) or inherently present in the material; crystalline defects may be generated during crystal growth, particularly in epitaxially grown heterostructure layers. These non-idealities in the crystal typically lead to the formation of deep level traps, energy levels near the center of the bandgap, which efficiently swallow electrons or holes. The resulting Schockley-Read-Hall recombination is non-radiative and phonons alone account for energy and momentum conservation.

Recombination Lifetime

Non-radiative recombination thus results in a loss of carriers from the conduction band without the concomitant generation of photons, in contrast to what we would like in a laser. We may characterize recombination through a recombination lifetime, for both radiative transitions, τ_r [s] and non-radiative transitions, τ_{nr} [s]; these lifetimes give a mean expected time for an electron to undergo either a radiative or non-radiative transition. The radiative lifetime is frequently referred to as the spontaneous lifetime, τ_{spon}, the expected time to the spontaneous emission of a photon.

Since the total recombination of carriers is generally the sum of radiative and non-radiative events, we may express the overall minority carrier lifetime, τ, as

$$\tau = \left[\frac{1}{\tau_r} + \frac{1}{\tau_{nr}}\right]^{-1} \tag{3.51}$$

which corresponds to the electron or hole lifetime used, for example, in the definition of pn diode reverse current in Eq. 3.36.

Which of the two mechanisms, radiative or non-radiative recombination, dominates the overall carrier lifetime depends on which of the two is *smaller*. For a dominance of radiative transitions, we require that $\tau_r < \tau_{nr}$, such that an electron is more likely to lose its energy to a useful photon than to waste it on a phonon.

Auger Recombination

Narrow bandgap materials, such as those used for long-wavelength lasers, are prone to a loss of electrons through the Auger mechanism (Agrawal & Dutta 1986, Chap. 3.3). In this process, an electron and hole recombine, where the energy is transferred to another electron in the conduction band which is subsequently raised to a higher level in that band. This second electron eventually thermalizes to the bottom of the band, giving off its excess energy to phonons.

The Auger process exists in a number of variations depending on the band structure, the portions of the bands involved in the transition and whether or not the assistance of phonons is required for the transition. For the non-equilibrium case in a material, in which n' excess minority carriers are present for example, the Auger recombination rate is found to be proportional to

$$R_{Auger} \propto n'^3 . \tag{3.52}$$

One or more of these Auger mechanisms may be relevant in the materials used for near- and mid-infra-red lasers such that this represents the dominant path for non-radiative recombination.

3.2.5 Gain

A laser requires optical gain implying, as we saw in Sect. 2.3.3, that the photons in an optical field moving through the semiconductor should preferably give rise to stimulated emission as opposed to absorption. We know that the rates of stimulated emission and absorption are given by Eq. 3.50 and Eq. 3.44, respectively, so that optical amplification may be expected when

$$R_s > R_\alpha \tag{3.53}$$

or

$$B_s f_c \rho_c f_v \rho_v \Phi(E) > B_\alpha (1 - f_c) \rho_c (1 - f_v) \rho_v \Phi \quad . \tag{3.54}$$

Since $B_s = B_\alpha$, the previous expression implies that

$$f_c f_v > (1 - f_c)(1 - f_v) \tag{3.55}$$

or

$$f_c + f_v > 1 \tag{3.56}$$

for stimulated emission to dominate absorption.

Recalling the definitions of the Fermi functions (Eq. 3.7 and Eq. 3.8), it can easily be shown that the condition of Eq. 3.56 requires the difference in electron and hole energies, which is equal to the photon energy, must be less than the Fermi level difference, namely

$$h\nu = E_e - E_h < E_{Fc} - E_{Fv} . \tag{3.57}$$

Since a photon must have at least the bandgap energy E_g to result in a transition of any kind, we thus see that the amplification of light in a semiconductor only takes place for photons in the energy range

$$E_g < h\nu < E_{Fc} - E_{Fv} . \tag{3.58}$$

Gain Spectrum

Once the condition of Eq. 3.58 is fulfilled, we may calculate the gain spectrum, the optical gain in a pumped semiconductor as a function of photon wavelength and pump level. Accurate gain calculations are in general quite involved (Chow and Koch 1999) but we may provide a simplified example here to give a flavor of the type of relationships expected.

The gain of a direct bandgap material, γ [cm^{-1}], may be expressed as a function of the Fermi functions and the photon frequency ν [Hz] as

$$\gamma(\nu) = \frac{\lambda^2}{8\pi^2\tau}\left[\frac{2}{\hbar}\frac{m^*_c m^*_v}{m^*_c + m^*_v}\right]^{\frac{3}{2}}\sqrt{2\pi\nu - \frac{E_g}{\hbar}}(f_c + f_v - 1) \tag{3.59}$$

where m^*_c and m^*_v are the conduction and valence band effective masses and τ [s] is the spontaneous emission lifetime of the carriers as given in Eq. 3.51 [Yariv 1991]. Based on material parameters, this last equation may be used to calculate the spectral dependence of gain as a function of carrier concentration, where the latter is a function of the pump level in the material.

Example 3.2

Assignment: *Calculate the gain spectrum for GaAs as a function of injected carrier concentration.*

Solution: *To apply Eq. 3.59, we need first to determine a few material constants. From Table 3.1, we find* $m^*_c = 0.068m_0$ *and* $m^*_v = 0.48m_0$ *for electron mass* $m_0 = 9.1\times10^{-31}$ *kg. The spontaneous lifetime may be taken to be* $\tau_{sp} = 3$ *ns (Adachi 1992) and the energy gap* $E_g = 1.424$ *eV. The wavelength* λ *is that in the material,* $\lambda = \lambda_0/n$ *, for refractive index* $n = 3.5$ *, and is related to the photon frequency as* $\nu = c/\lambda$ *.*

Next we need to determine the quasi-Fermi levels for the carrier concentration (n and p) assumed. Using Eq. 3.13 and Eq. 3.14, recalling the definitions of the Fermi functions in Eq. 3.7 and Eq. 3.8 as well as the density of states in Eq. 3.9 and Eq. 3.10, we may evaluate the integrals by integrating over the entire conduction band (Eq. 3.13) and the entire valence band (Eq. 3.14). We thereby select a value for quasi-Fermi level such that the result of the integral yields the desired value for n or p; since we cannot evaluate the integral analytically, this calculation is thus an iterative procedure, where we tweak the value for E_{Fc} *or* E_{Fv} *to obtain the desired n or p. Given* E_{Fc} *and* E_{Fv}*, we may plug these values back into the Fermi functions (Eq. 3.7 and Eq. 3.8) and then finally evaluate Eq. 3.59 as a function of wavelength. A typical result is shown in Fig. 3.17 where the gain* γ *has been calculated for four carrier concentrations. We see that the characteristic begins abruptly at the right side of the curve, at the bandgap wavelength, slightly below 870 nm. For carrier concentration* 1.2×10^{18} cm^{-3}, γ *is barely above zero before becoming negative: we thus denote this carrier concentration as the "transparency carrier concentration", since that carrier level leads to the case where the material neither absorbs nor amplifies light (i.e.,* $\gamma = \alpha = 0$*) and is thus transparent.*

For higher carrier concentrations, $\gamma > 0$ *for a wider wavelength range, to shorter and shorter wavelengths. In addition, the peak gain increases with increasing carrier concentration. We may thus see the effect of increasing pumping on the gain clearly from this example.*

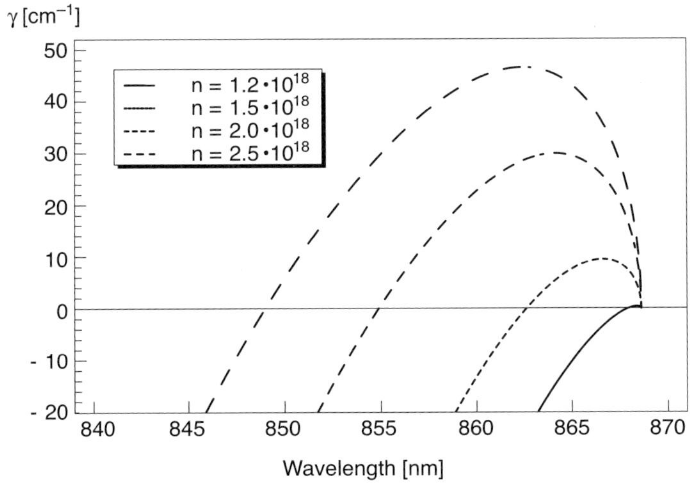

Fig. 3.17. Calculated gain for GaAs as a function of photon wavelength for four carrier injection levels. Wavelengths with $\gamma > 0$ are amplified in the medium

3.3 Semiconductor Laser Materials

Semiconductor lasers are fabricated using direct-bandgap semiconductor materials, those which efficiently absorb and emit photons; in practice, these are typically compound semiconductors. Whereas the fabrication of these materials is more complicated than that of everyone's favorite semiconductor, namely silicon, the variability of refractive index and energy gap as a function of composition renders the compound semiconductors of unparalleled utility in the fabrication of complex laser structures.

We provide a brief overview of the materials used for fabrication of semiconductor lasers here, and discuss some of the basic material structures, namely heterostructures and quantum wells, which form the basis of a laser. An extensive literature (For example: Casey and Panish 1978; Adachi 1992; Nakamura and Chichibu 2000) covers many material aspects in considerably greater detail.

3.3.1 III–V Semiconductors

Compound semiconductors are crystalline solid-state alloys formed from the combination of two or more elements. The majority of these combine elements from the columns III and V or columns II and VI of the periodic table. The former are imaginatively called the III–V semiconductors and represent the most popular family of compounds for use in lasers. Variations in composition of these materials allow access to a wide range of energy gaps implying a broad spectrum of emission and absorption wavelengths. For this reason, a considerable number of material systems has been subject to development and availability of semiconductor lasers at new wavelengths is often contingent on the existence of high-quality semiconductor material with the corresponding bandgap. The advances in III–V nitride technology leading to the availability of blue laser diodes is a recent example (Nakamura 2000).

Binary Materials

As seen in Fig. 3.18, a number of elements may be combined to yield a semiconductor alloy. Taking one element each from column III and column V yields a binary compound semiconductor, of which GaAs and InP are amongst the best known. The crystal structure of the resultant material is almost always zincblende, GaN being a notable exception in that its predominant form is wurtzite.

Table 3.5 presents a few popular binary III–V semiconductors along with their energy gap and refractive index. The materials listed in the table are all direct bandgap; some binary compounds may be indirect and thus of limited utility for lasers. The II-VI materials, fabricated from elements of these two columns of the periodic table, are also of interest for lasers, particularly for very short (blue) or very long (mid-IR) wavelengths. At long wavelengths, the II-VI materials are also very popu-

IIB	IIIA	IVA	VA	VIA
	B	C	N	
	Al	Si	P	
Zn	Ga	Ge	As	Se
Cd	In	Sn	Sb	Te
Hg				

Fig. 3.18. A portion of the periodic table showing the elements from the columns II, III, V and VI most popular in the composition of compound semiconductors

lar for diode-based photodetectors. We will, in correspondence to their popularity and market relevance, focus on the III–V materials here.

Table 3.5. A few popular binary III–V semiconductors with the energy gap and refractive index measured at the bandgap wavelength

Material	E_g [eV]	n
GaN	3.39	2.22
GaAs	1.424	3.66
InP	1.351	3.45
GaSb	0.726	3.82
InAs	0.36	3.52

Ternary Materials

The material properties of the III–Vs become really interesting if we assemble an alloy using two elements from one column and one from the other, leading to a ternary material. In this case the composition may be altered by varying the relative mole fractions of the elements from the same column; the result is a material in which numerous material parameters may be tuned to varying degrees (Blakemore 1982).

There are 18 possible ternary systems for III–V semiconductors, of which more than half have been studied and developed in detail. A popular example is AlGaAs which combines aluminum and gallium from column III with arsenic from column V. Explicitly writing the mole fraction of aluminum, x, we then have $Al_xGa_{1-x}As$, where $0 \leq x \leq 1$; the mole fraction represents the percentage of aluminum in the alloy. Thus $x = 0$ results in the binary GaAs, $x = 1$ in AlAs and $0 \leq x \leq 1$ some composition of $Al_xGa_{1-x}As$.

By varying the mole fraction, x, in a ternary material, one may, in general, change the energy gap, refractive index and lattice constant (Adachi 1985). For $Al_xGa_{1-x}As$, for example, the energy gap at 300 K varies with x as

$$E_g = 1.424 + 1.247x \text{ [eV] for } 0 \leq x \leq 0.45 \tag{3.60}$$

$$E_g = 1.424 + 1.247x + 1.147(x - 0.45)^2 \text{ [eV] for } 0.45 \le x \le 1. \quad (3.61)$$

In the case of AlGaAs, the semiconductor becomes indirect for $x > 0.45$. It should be noted that the energy gap for semiconductors also varies strongly with temperature (Varshni 1967; Thurmond 1975).

Furthermore, the refractive index of $Al_xGa_{1-x}As$ also varies with x; at a wavelength of $\lambda = 900$ nm ($E = 1.38$ eV), n varies as (Casey and Panish 1978, p. 45)

$$n \approx 3.59 - 0.71x + 0.091x^2 . \quad (3.62)$$

Equations of the form of Eq. 3.62 are known as Sellmeier equations, of which a considerable number have been derived for a wide spectrum of materials including InGaAsP (Fiedler and Schlachetzki 1987) and InP (Pettit and Turner 1965).

The lattice constant, a_0 [m], is a measure of the atomic spacing in the crystal lattice of the semiconductor. AlGaAs has the attractive feature that a_0 varies minimally for $0 \le x \le 1$ at room temperature and is given by (INSPEC 1990)

$$a_0\big|_{Al_xGa_{1-x}As} = a_0\big|_{GaAs} + 7.8\times10^{-13}x \text{ [m]}. \quad (3.63)$$

As a result, all compositions of AlGaAs (including GaAs and AlAs) are virtually lattice-matched to one another, since the lattice mismatch between GaAs (0.5654 nm) and AlAs (0.5662 nm) is less than 0.15%; as a result, these materials may be grown epitaxially on one another with good resultant crystal quality. This last aspect will be seen to be of great relevance when we discuss heterostructures in Sect. 3.3.3 and is one of the reasons that the GaAs/AlGaAs system has been so extensively studied and applied. It should be noted that, in general, the lattice parameter does vary with composition for many ternary and quaternary compounds.

Amongst the ternary materials, other laser-relevant compounds include InGaAs, GaInP, InGaN, GaInSb and AlGaSb. We will see the relevance of some of these when we outline the wavelength ranges accessible by these materials below.

Quaternary Materials

The atomic mix-and-match game can be taken further when two elements from column III and two from column V are combined to yield a quaternary compound semiconductor. In this case two mole fractions are defined, x and y, and the composition may be correspondingly more complex. In particular, some materials require a fixed ratio of x to y to guarantee a certain lattice parameter and not all combinations of x and y are necessarily possible. As for the ternary materials, changing mole fraction again changes energy gap, refractive index and lattice constant.

A popular quaternary III–V material is InGaAsP, the semiconductor on which most near-infrared telecommunications lasers are based. Writing the mole fractions explicitly, we have $In_{1-x}Ga_xAs_yP_{1-y}$ where x is the gallium mole fraction and y the arsenic mole fraction. For this material, setting $x = 0.45y$ yields a value for a_0

which permits lattice matching to InP (Pearsall 1982), the substrate material of choice. In this case, the energy gap at 300 K for InGaAsP varies with y as

$$E_g = 1.35 - 0.72y + 0.12y^2 \text{ [eV]} \tag{3.64}$$

and the refractive index at the bandgap wavelength with y as

$$n \approx 3.4 + 0.256y - 0.095y^2 . \tag{3.65}$$

Similar expressions and material parameters in tabular form are available in the literature (Adachi 1992).

3.3.2 Wavelength Ranges

The relationships between the composition of ternary materials, the resultant bandgap and associated lattice constant are perhaps best illustrated by a graphical representation as that of Fig. 3.19. The bandgap energy and wavelength are plotted as a function of lattice constant for some of the more popular material systems. The points correspond to a binary material and the lines between them the associated ternary compound varying in mole fraction. The area of the shaded region corresponds to the compositional variation of InGaAsP (Moon et al. 1974).

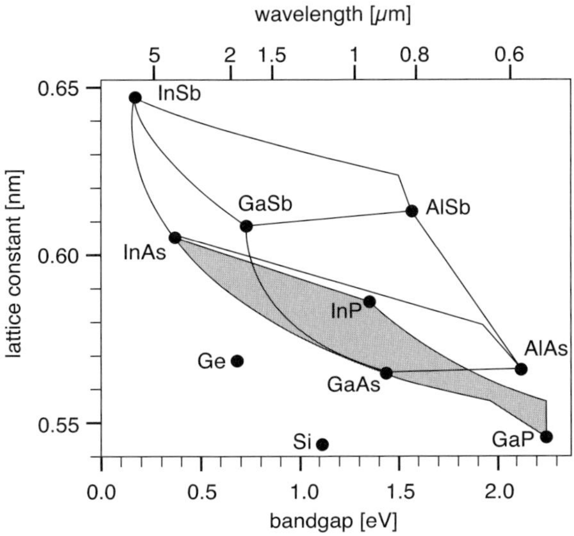

Fig. 3.19. Bandgap energy and wavelength as a function of lattice constant for a selection of popular III–V ternary materials, with silicon and germanium shown for comparison; the shaded region is InGaAsP

Considering, for example, the AlGaAs system, represented by the almost horizontal line between GaAs and AlAs, we see that the bandgap wavelength changes by almost 500 nm whereas the lattice constant is almost unchanged. InGaAs, on the other hand, extends the wavelength range to the mid-infrared, but with a large concomitant change in lattice constant. The composition of InGaAsP lattice matched to InP is a line (not shown) extending through the shaded area to the left of InP.

As mentioned above, the primary reason that such a broad spectrum of materials has been subject to development is wavelength: a semiconductor laser emitting at a certain wavelength requires a material with a bandgap energy close to that wavelength[2]. Figure 3.20 summarizes the accessible wavelength ranges of a few III–V and II-VI materials, limited to the regimes where the materials given have a direct bandgap. Important material systems, from the applications point of view, include AlGaAs for lasers in the near-infrared ($\lambda \approx 800$ nm), InGaAsP for telecommunications lasers ($\lambda \approx 1.55$ μm) (Agrawal and Dutta 1986), AlGaInP for red lasers ($\lambda \approx 670$ nm) (Roentgen et al. 1991; Yoshikawa et al. 1995) and GaInN for green and blue lasers ($\lambda \approx 400$ nm) (Ponce and Bour 1997; Nakamura 2000). The II-VI compounds, such as ZnSe and CdZnS, also represent a system used for blue laser diodes (Mensz 1994; Zeng et al. 1998), but have been eclipsed by the nitrides due to the latters' considerably longer laser lifetime. The IV-VI materials, PbSnTe, PbSSe or PbSnSe, also known as the lead salts, have been employed for lasers with very long wavelengths (up to $\lambda = 50$ μm), albeit only when operated at low temperatures (Shani 1988).

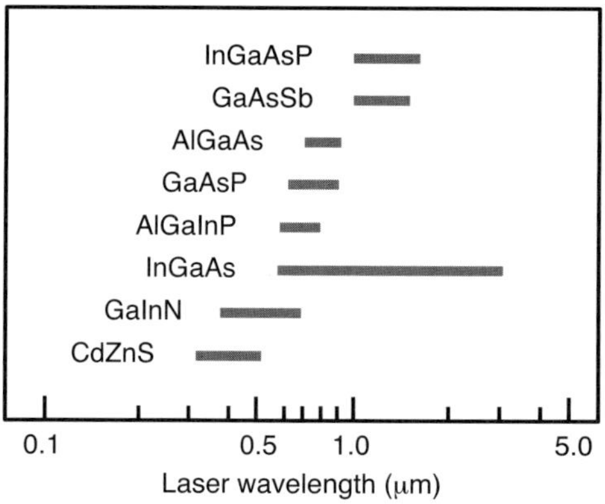

Fig. 3.20. The accessible wavelength ranges of a few important III–V and II-VI material systems

---

[2] An exception are the quantum cascade lasers, considered in Sect. 4.6.

3.3.3 Heterostructures

We have seen that the compound semiconductors may vary in composition and that composition may result in a variation of energy gap, refractive index and lattice constant. Using ternary and quaternary materials, these material parameters may in some cases be tuned continuously so that limited range of energy gap and refractive index values is accessible using a single material.

For certain variations in composition, the lattice constant may change little or not at all. As a result, materials of differing composition and thus differing energy gap and/or refractive index may be lattice-matched to one another. Multi-layer structures of various materials and with high crystalline quality may therefore be fabricated: these are known as heterostructures and the interface between two materials, a heterojunction. Heterostructures form the material basis for virtually all semiconductor lasers (Panish 1975).

The Laser Core

The great advantage of using heterostructures as the basis for a laser may be illustrated by the structure of Fig. 3.21. Using the GaAs/AlGaAs system as an example, a nominally undoped GaAs layer, the laser active region or core, is situated between two AlGaAs layers, the upper typically p-doped and the lower typically n-doped. The resulting p-i-n diode configuration is known as a double heterostructure as it consists of two GaAs/AlGaAs heterojunctions.

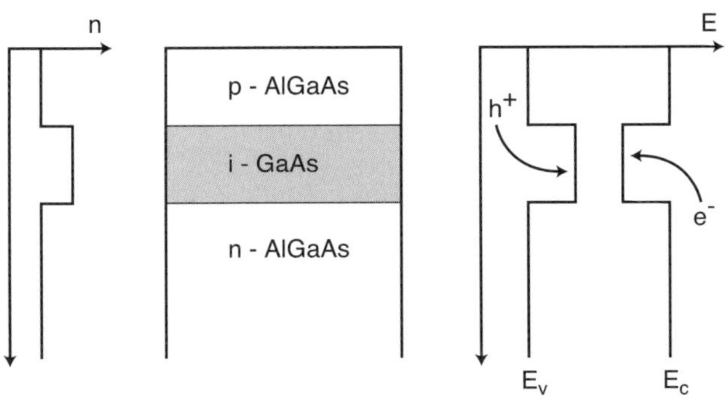

Fig. 3.21. Example heterostructure using the GaAs/AlGaAs system; the laser core region, with smaller energy gap and larger refractive index, is shaded

Nature is kind to the heterostructure engineer using this material system: the GaAs laser core has both a lower energy gap and at the same time a higher refractive index than the AlGaAs cladding layers. As a result, electron hole pairs generated by current injection into the heterostructure diode find it energetically favorable to fall into the low–E_g core region and recombine radiatively there; the high density of carriers which results from the existence of this energy well implies that a high photon density is generated in the active region.

The higher relative refractive index of the active region then further implies that the generated photons are also to a large part confined to the active region, since total internal reflection at the heterojunction boundaries results in waveguiding in the GaAs region. We will study the nature and utility of the waveguide laser structure in Sect. 3.4. Optical waveguiding results in a further concentration of the optical field such that population inversion and therefore threshold gain is reached for relatively low pumping levels. Indeed, the incorporation of the double heterostructure into a semiconductor laser was the key step that allowed room-temperature lasing in semiconductors, achieved in 1970 (Alferov et al. 1970).

Heterostructure Systems

The GaAs/AlGaAs system used in the example of Fig. 3.21 is a simplified version of a well studied, well established laser heterostructure. By choosing compositions which allow lattice matching, a variety of semiconductor heterostructures are commonly used for lasers; the substrate wafer is typically GaAs or InP, as these are available as bulk material, fabricated using similar bulk crystal growth techniques to those used for silicon (Campbell 2001).

In addition to AlGaAs, the AlGaInP system is usually grown on GaAs. InP is the substrate of choice for InGaAsP, InGaAsSb and InGaAs, where the latter is lattice matched for the composition $In_{0.53}Ga_{0.47}As$. The GaN system is not lattice matched to any commonly used binary semiconductor; the best approximations are SiC (3.5% lattice mismatch), spinel ($MgAl_2O_4$ with 9.5% lattice mismatch) and sapphire (Al_2O_3 with 13% lattice mismatch). The use of clever growth techniques, involving multi-layer lattice matching structures, has allowed laser-quality GaN to be grown on these substrates despite the lattice mismatch (Nakamura and Chichibu 2000).

Typical laser heterostructures then vary according to the material system employed to reach the desired wavelength range. GaAs lasers for the near infra-red region often use an $Al_{0.3}Ga_{0.7}As$ core surrounded by $Al_{0.8}Ga_{0.2}As$ cladding layers, all grown on GaAs. Telecommunications lasers in the 1.55 μm wavelength range might employ $In_{0.72}Ga_{0.28}As_{0.6}P_{0.4}$ for the active region, and an InP cladding, all grown on InP. Red lasers frequently use a GaInP active region, lattice matched to $(Al_{0.7}Ga_{0.3})_{0.52}In_{0.48}P$ and the GaAs substrate. Finally, blue lasers might be fabricated using GaN or InGaN active regions, an AlGaN cladding and the entire structure with a lattice matching network grown in Al_2O_3. A real laser, or for that matter, even a real LED, will generally employ considerably more complex layer sequences, but often based on the material systems just described.

Epitaxial Growth

Fabrication of the complex heterostructures needed for lasers requires a means for growing single crystal layers of precisely known composition and exactly controlled thickness on an appropriate substrate. This growth is accomplished by epitaxy, where this term refers to thin-film crystal growth by means of any number of specialized techniques. As mentioned above, successful epitaxy requires lattice matching between the materials of the substrate and the growing layer, although specialized techniques for overcoming small mismatches in lattice constant have been developed.

Three epitaxial growth techniques have established themselves for the growth of heterostructures for optoelectronic applications. The oldest is liquid phase epitaxy, LPE, which uses a melt of the material to be grown and the substrate is pulled back and forth over the melt by mechanical means (Astles 1990). Each traversal of the melt results in the deposition of a thin film of the molten material, resulting ultimately in a layer of the desired thickness. Multi-layer heterostructures require repeated applications of the technique using a different material melt. LPE is conceptually and technically relatively simple, although layer thickness and stoichiometry control is limited. Due to its low cost, high growth rates and low overhead, it is the technique of choice for growth of simple LED structures.

At the other extreme, perhaps, is molecular beam epitaxy, MBE (Herman and Sitter 1989). This technique requires a high-vacuum chamber, in which the substrate is mounted and heated to temperatures between 500 and 700°C. Material for the epitaxial layer to be grown is emitted from a number of thermal effusion cells, one for each element required. Thus growth of AlGaAs, for example, requires three effusion cells, one each for aluminum, gallium and arsenic, where the elements are present in extremely high purity. By heating the cells, atoms are thermally emitted from the source and thus form an "atomic beam" onto the substrate, and thereby result in epitaxial crystal growth. Material composition and quality, as well as layer thickness, are extremely well controlled, such that monolayer accuracy in film thickness is routinely achieved. Unfortunately, growth rates are slow (about 1 monolayer per second) and the equipment investment is significant, such that MBE is generally used for electronics and research applications, and only occasionally for volume production of optoelectronic devices.

The third approach for epitaxy, metal-organic vapor phase epitaxy (MOVPE), also referred to as metal-organic chemical vapor deposition (MOCVD), represents a mean between the previous two techniques and is the approach most likely to be found in the production of laser substrates (Razeghi 1989). For MOVPE, the substrate is placed in a growth chamber, heated (500–750°C) and typically spun using a planetary mechanism; the chamber is frequently near ambient pressure, although low-pressure forms of MOVPE have also been employed. The atomic species for epitaxial growth are delivered in the gas phase, where a very controlled gas flow of the appropriate precursors is directed over the substrate surface. The rate and composition of the gas flow is accurately controlled by mass flow controllers, one for each precursor, thereby allowing accurate determination of film composition and thickness.

Precursors for the components of most III–V materials are available. The column III materials derive from, for example, tri-methyl gallium or tri-ethyl gallium (TMGa or TEGa), TEAl or TMAl for aluminum, TMSb for antimony and TEIn for indium. The hydrides (AsH_3 and PH_3) are used for column V materials such as arsenic and phosphorus; NH_3 may be the nitrogen source for GaN. Epitaxial growth takes place on the substrate surface, where the high temperatures result in cracking of the precursor and subsequent deposition of the desired material. Growth rates of 10–30 nm/min are reached, with thickness accuracy only slightly worse than for MBE. Film stoichiometry is also good, although the non-negligible concentration of carbon following cracking of the precursors may result in incorporation into the epitaxial film. Nevertheless, MOVPE is a very well controlled, well characterized and reproducible epitaxial growth technique and dominates the production of laser substrates.

3.3.4 Quantum Wells

The availability of high-accuracy epitaxial growth techniques, allowing the reproducible deposition of films with precisely controlled thickness in the sub-nanometer range, opens a wide range of new opportunities for fabrication of heterostructures with useful electrical and optical properties. When the dimensions of the layer decrease to the length scales of the electron wavelength, in the nanometer to tens of nanometers range, quantum effects begin to dominate the behavior. Decreasing the size of semiconductor structures to these levels in one, two or three dimensions results in quantum wells, quantum wires and quantum dots, respectively. Whereas research on the latter two has resulted in novel and interesting semiconductor laser structures, we will take a closer look at quantum wells here, as these have become standard features in laser heterostructures since their first demonstration in 1975 (van der Ziel 1975; Yariv 1991, Chap. 16; Zory 1993).

The heterostructures we have considered are planar structures, thus quasi-infinite in two dimensions with finite thickness in the third. Let us take a heterostructure sandwich consisting of a layer with thickness t and small bandgap surrounded on both sides by a second material with larger bandgap. By decreasing t into the nanometer regime, the center layer essentially becomes a two dimensional potential well, as shown in Fig. 3.22, and is thus termed a quantum well.

The two-dimensional nature of the quantum well significantly changes the optical and electrical properties of the material with respect to the bulk three-dimensional material case; carrier lifetime, transition rates, polarization, exciton effects and ultimately optical gain are thus affected (Bastard 1988). We consider two characteristics of particular relevance for lasers, namely the generation of discrete electron energy levels and the two-dimensional density of states.

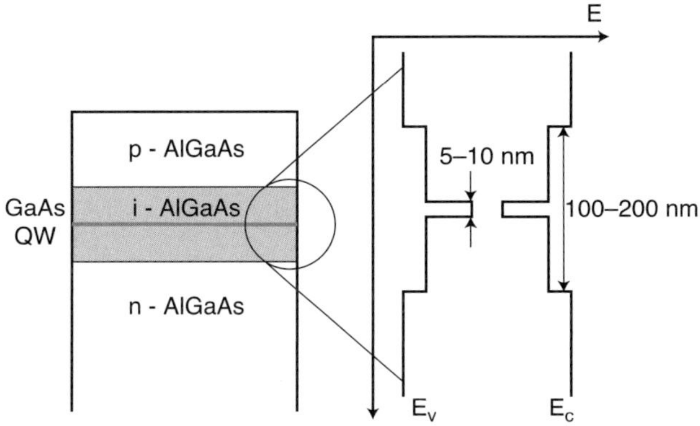

Fig. 3.22. A quantum well in the active region of a typical semiconductor laser substrate

Energy Levels

As a result of the reduced dimensionality of the quantum well, the allowed electron energies become quantized; this result is exactly that of the particle-in-a-box calculation done in introductory quantum mechanics. The number of discrete energy levels, or sub-bands, in a quantum well of thickness t and energy depth E_w (the difference in energy gap between the well and the outer layers) may be found from (Mitin 1999)

$$n_{max} = 1 + \mathrm{int}\left[\sqrt{\frac{2m^* E_w t^2}{q\pi^2\hbar^2}}\right], \tag{3.66}$$

where m^* is the effective mass of electrons in the band; the function int gives the integer part of its argument.

In the limit that the energy well is infinitely deep, we may then approximate the energy of the sub-bands as

$$E_n = \frac{\hbar^2\pi^2}{2m^* t^2} n^2 \text{ for } n = 1,2,3,\ldots, \tag{3.67}$$

where n represents the number of the energy level. We see that the spacing between energy levels increases as n increases; the resultant large energy difference, with respect to the thermal energy, implies that most electrons will thermalize to the lowest energy level in the well.

For a so called Type-I heterostructure, where the valence and conduction bands both give rise to an energy well in the quantum well, the electron and hole energies are both subject to the discrete energy levels in the well. Transitions between the upper and lower bands then take place between the sub-bands, not E_c and E_v as in a bulk semiconductor; we will see the utility of this below.

Density of States

In addition to the appearance of discrete energy bands in the well, as we have just seen, the two dimensional structure also gives rise to a different form for the density of states. Due to the discretized energy levels, the two-dimensional density of states, ρ_{QW} (E) [eV^{-1} cm^{-2}] takes on a step-like form, namely

$$\rho_{QW}(E) = \frac{m^*}{\pi^2 \hbar^2} \sum_n \upsilon(E - E_n) , \tag{3.68}$$

where $\upsilon(E - E_n)$ represents the step function and the sum is taken over all n energy levels in the quantum well. Note that the dimensions for two-dimensional density of states are per energy per area, in contrast to the three-dimensional case, per energy per volume (Eq. 3.9).

The two- and three-dimensional density of states are compared in Fig. 3.23. The bulk case has the square-root dependence we saw previously, whereas the two-dimensional case for a quantum well shows discrete steps at the allowed energy levels in the quantum well. A true numerical comparison is difficult, as the two- and three-dimensional systems require that we measure density in different ways. We see, however, that the density of states in a quantum well exhibits pronounced differences in form when compared to the bulk case, particularly near the bottom of the band. The two-dimensional case shows a considerably stronger variation with energy, particularly of course at the steps. For higher energies, the two forms begin to coincide.

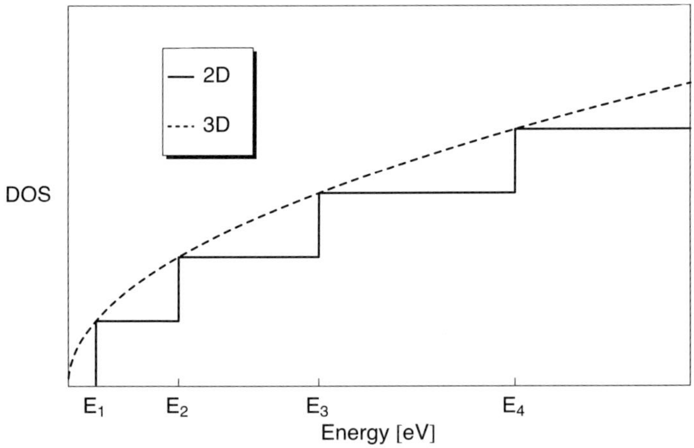

Fig. 3.23. The density of states for a bulk material (3D) and a quantum well (2D)

Quantum Wells in Lasers

We mentioned above that the optical and electrical properties of a quantum well are quite different from those in bulk material; gain and minority carrier lifetime, for example, may be considerably enhanced over non-quantum-well lasers. In addition, optical transitions for absorption and emission of photons now take place between the discrete energy levels rather than the conduction and valence bands such that structural factors (i.e., quantum well thickness) may be used to tune the emission wavelength of a laser.

Eq. 3.67 gives the allowed energies for electrons in the quantum well measured with respect to the energy at the bottom of the well. Given that a quantum well has sub-bands in both conduction and valence bands, we may write an expression for the total energy difference between the lowest energy levels in the two sub-bands as

$$E_{h\nu} = E_g + \frac{\pi^2 \hbar^2}{2t^2}\left(\frac{1}{m^*_v} + \frac{1}{m^*_c}\right) . \tag{3.69}$$

$E_{h\nu}$ thus represents the photon energy for emission or absorption in a quantum well; the quantum well leads to an increase in the transition energy above the energy gap.

Since the transition energy is a strong function of the quantum well width, t, we see that the well width may be used as a tuning parameter for determining the emission wavelength of a laser. Thus the choice of material system gives the wavelength range and subsequently the width of a quantum well made from these materials defines the ultimate transition wavelength.

In real semiconductor lasers, the quantum well region often has a more complex structure than the simple single well considered here. Using continuous spatial variation of material composition, regions with a graded refractive index may be used to optimize electron capture and confinement factor (see Sect. 3.4.3) of the optical field in the region of the quantum well (Tsang 1987). In addition, a typical laser active region employs not one but several quantum wells, regularly distributed at the center of the active region (McIlroy 1985). The three to five wells commonly used increase the gain available in the laser and are of particular value for high operating currents.

Example 3.3

Assignment: *Design the active layer structure for an 806 nm quantum well laser.*

Solution: *We will select a material system and then design a quantum well thickness to give the desired wavelength.*

By consulting Fig. 3.20, we see that 806 nm could be reached with a number of materials; we choose the GaAs/AlGaAs system since it is one of the best established and if an epitaxialist can do one material system, it's typically that one.

Let's take $Al_{0.3}Ga_{0.7}As$ for the core material in which we will place a GaAs quantum well. From Eq. 3.60 (and the useful relationship of Eq. 2.73), we see that the energy gap of GaAs is 1.424 eV (implying a transition wavelength of 871 nm) and the energy gap of $Al_{0.3}Ga_{0.7}As$ leads to a transition wavelength of 690 nm. We note that the energy gap difference between GaAs and $Al_{0.3}Ga_{0.7}As$ is 0.374 eV.
Since we want a transition wavelength of 806 nm, we intend to employ a quantum well where the lowest energy level spacing will give us the desired wavelength. From Eq. 3.69, including the energy gap of GaAs, we find that a quantum well thickness of 7.5 nm will lead to a transition wavelength of 806 nm. By comparing the sum of the quantum well transition energies (Eq. 3.69 without the energy gap of GaAs), namely 0.11 eV, with the energy gap difference between GaAs and $Al_{0.3}Ga_{0.7}As$, 0.374 eV, we see that the quantum well energies are smaller. Thus, although the quantum well is far from infinitely deep, the lowest discrete energy levels are indeed confined in the well.
We thus propose the use of a 7.5 nm thick GaAs quantum well in an $Al_{0.3}Ga_{0.7}As$ core region to make an 806 nm laser. For the design of the rest of the laser core, we need to consider its waveguiding aspects and we will continue in Example 3.4.

3.4 Waveguides

The active region of a semiconductor laser, as we saw in Sect. 3.3.3, consists of a double heterostructure: the thin middle layer has a lower gap energy and, by happy coincidence, a higher refractive index than the surrounding layers. This latter fact implies that the active region also acts as an optical waveguide, confining the optical field in the thin light-emitting layer of the laser. The concentration of emitted photons implies a reduction in the threshold gain and an improved pumping efficiency, as well as a reduction in the overall active region volume, resulting in a lower laser threshold as compared to lasers in which such confinement is absent.

A waveguide is a structure for guiding light: through reflections at the boundary, an optical field is confined to the waveguide layer and propagates within this layer (Marcuse 1974; Tamir 1988; Hunsperger 2002). A two-dimensional slab waveguide confines light in the vertical direction, leading to a planar region of guided optical field. The one dimensional stripe waveguide confines light in the vertical and one lateral direction, resulting in an "optical wire". Propagation in a waveguide is distinct from that in free space, as we shall see below: light is transmitted in discrete modes, with strong wavelength and structural dependencies of transmission properties.

Most semiconductor lasers are ultimately stripe waveguides. Confinement in the vertical direction is through the layer sequence of the double heterostructure, where the III–V semiconductors have long been known to be useful for this function

(Garmire 1973). Lateral confinement is often, but not always, achieved by means of structuring the slab waveguide through a process technology such as etching. The result is a long, thin laser waveguide active region, with typical dimensions $0.2 \times 5 \times 200\ \mu m^3$, which forms the core of most semiconductor lasers. We will provide an overview of waveguides in semiconductors in the following sections, referring the reader to the literature (Zappe 1995) for a more extensive treatment.

3.4.1 Waveguiding: The Ray Optic Model

The waveguide may be analyzed by one of two useful mathematical models: the ray optic model and the electromagnetic model. The former treats propagation through a waveguide as that of discrete light "rays" which reflect at the boundaries; this model provides a relatively intuitive understanding of the waveguiding mechanism and we will consider it in this section. The electromagnetic model yields expressions for the fields in the waveguide and thus allows an analysis of the optical intensity distribution. While perhaps giving less of a feel for why guiding takes place at all, the utility of the field calculations will become evident in Sect. 3.4.2.

The Slab Waveguide

Let's consider the structure of a slab waveguide as shown in Fig. 3.24: the guide or core layer with refractive index n_g is sandwiched between an upper cap layer, with n_c, and a lower substrate layer with n_s; the latter two are referred to as the cladding layers and may be of identical materials such that often $n_c = n_s$. The thickness of the core layer is given by t. This structure should be reminiscent of the double heterostructure considered in Sect. 3.3.3 where the core might be GaAs and the cladding (cap and substrate) AlGaAs.

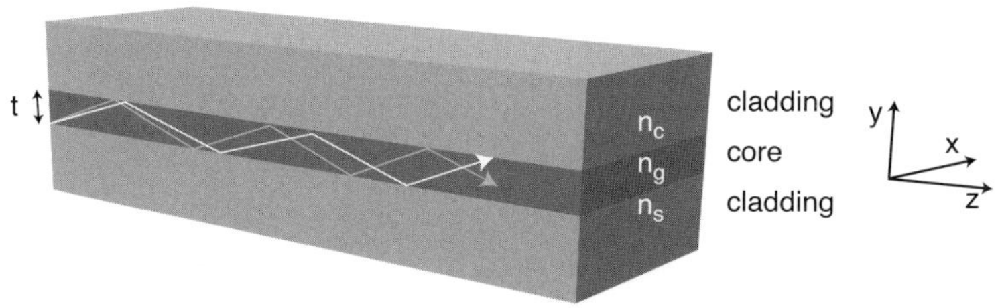

Fig. 3.24. Typical slab waveguide structure showing the propagation of two distinct modes in the ray-optic representation

In the ray optic model, light is represented as a distinct ray which reflects at the boundaries between the core and the cladding layers. The reader will recall from basic optics (Hecht 1987) that when the incident angle at the boundary is greater than the critical angle, θ_c, given by

$$\theta_c = \sin^{-1}\left(\frac{n_{clad}}{n_{core}}\right) \tag{3.70}$$

the light is totally internally reflected and none is transmitted into the cladding region. Confinement by total internal reflection is the mechanism by which light is transmitted through macroscopic glass light guides, for example, in which light propagation is as through bulk material but may be bent around corners.

When the dimensions of the guide layer or core decrease and approach those of the wavelength of the transmitted light, the transmission behavior changes: most importantly, propagation becomes discrete. Only certain discrete values of reflection angle, θ, are allowed and these are determined by a phase matching condition of light propagating through the guide. Taking the phase shift due to reflections at the boundary and due to various path lengths through the material (Zappe 2003), a characteristic equation relating the reflection angle and the structure of the waveguide may be derived; for the TE polarization[3], this characteristic equation takes the form

$$\frac{4\pi t}{\lambda}\cos\theta - 2\tan^{-1}\sqrt{\frac{\sin^2\theta - \frac{n_s^2}{n_g^2}}{\cos^2\theta}} - 2\tan^{-1}\sqrt{\frac{\sin^2\theta - \frac{n_c^2}{n_g^2}}{\cos^2\theta}} = 2m\pi \tag{3.71}$$

for the waveguide dimensions and refractive indices given in Fig. 3.24; λ is the wavelength in the material and m is an index, $m = 0,1,2,\ldots$. Similar expressions may be derived for TM polarized optical fields.

Modes

The expression in Eq. 3.71 may be solved, typically numerically, to yield allowed values of θ for given waveguide dimensions, material constants, wavelength and the mode index m. Each mode thus has a distinct value for θ for a given wavelength. As waveguide thickness t becomes smaller, the maximum value of m which still leads to a solution decreases; when only $m = 0$ gives a solution, the waveguide is said to be single-mode.

A typical waveguide or laser structure has a number of different "modes". The modes which result from a solution of Eq. 3.71 for a slab waveguide are *transverse* modes, in the y direction. For a stripe waveguide, considered in Sect. 3.4.4, a solution of a similar characteristic equation in the x direction leads to *lateral* modes. Finally, *longitudinal* modes in the z direction result from the resonant condition of the finite-length waveguide which forms the laser cavity (see Eq. 2.120 in Sect.

---

[3] For TE, the electric field is always parallel to the material interfaces; a definition is given in Fig. 3.25 of Sect. 3.4.2.

2.5.2); the modes referred to as "lasing modes", shown in Fig. 2.21, are longitudinal modes.

Of interest in the design of a waveguide, whether as part of a semiconductor laser or an integrated optical circuit, is the *z* component of the propagation constant, in the direction of light propagation. Once the value for reflection angle θ has been determined by solution of Eq. 3.71, the *z* component of the propagation constant may be simply found from

$$k_z \equiv \beta = n_g k_0 \sin\theta \ . \tag{3.72}$$

Each mode thus has a distinct value for β and thus propagation velocity down the waveguide. The value $n_g \sin\theta$ is frequently referred to as the waveguide effective index, *N*, so that

$$k_z \equiv \beta = N k_0 \ . \tag{3.73}$$

The effective index has a value which lies between the physical refractive index of the core and that of the cladding and may be interpreted as the refractive index a material would have to have to yield the same propagation conditions as the waveguide structure.

Example 3.4

Assignment: *Let's continue with the analysis of the laser active region we began in Example 3.3. Recall that we calculated that an* $Al_{0.3}Ga_{0.7}As$ *active region requires 7.5 nm quantum wells to make a* $\lambda = 806$ *nm laser; this active region represents the waveguide core.*

We will now analyze the $Al_xGa_{1-x}As$*-based waveguide: let's assume a 500 nm thick core with* $x = 0.30$*, the upper cladding with* $x = 0.85$ *and the lower cladding with* $x = 0.75$*. The waveguide is operated at* $\lambda = 806$ *nm. Determine the number of TE modes and then the reflection angle and z component of the propagation constant for all these modes. Finally, determine the necessary waveguide thickness so that only one mode is supported.*

Solution: *Solving this problem amounts to solving Eq. 3.71. Since we are given* λ*, t and the composition of the waveguide layers, we may begin by assuming that Eq. 3.62 represents a good approximation to the refractive index for our slightly shorter wavelength. Applying this, we obtain* $n_g = 3.39$*,* $n_c = 3.05$ *and* $n_s = 3.11$*.*

We now need to haul out our favorite numerical equation solver (Mathematica®, MatLab®, MathCad®, Maple® etc.) and solve Eq. 3.71 numerically, first with $m = 0$*. This first solution will provide a value for* θ *for the lowest order mode. We then repeat the exercise for* $m = 1$*. For* $m = 2$*, the solver will not return a solution such that we realize that the waveguide supports two modes.*

For $m = 0$*, the solution yields* $\theta = 80.1°$*: from Eq. 3.72, this value gives* $\beta_1 = 26.0\ \mu m^{-1}$ *and the waveguide effective index may be found from*

Eq. 3.73 to be $N = 3.334$. Repeating the procedure for $m = 1$ yields $\theta = 70.5°$ and thus $\beta_2 = 24.9\ \mu m^{-1}$ and $N = 3.191$. It is thus clear that the two modes supported in this waveguide will propagate with slightly different propagation constants.
To determine the necessary thickness of the waveguide so that only a single mode is supported, we take Eq. 3.71, set $m = 1$, and repeatedly solve the equation while decreasing t until no solution is found. Doing this, we find that only the $m = 0$ mode is supported for $t \leq 341$ nm. Thus a single transverse mode laser would probably use a core thickness of approximately 340 nm or less using the material system assumed.

3.4.2 Electromagnetic Model

Propagation of light in a waveguide may also be analyzed by solution of Maxwell's equations in the three-layer structure. While this approach may provide somewhat less intuitive feel for waveguide propagation, the rigorous mathematical analysis allows determination of the electromagnetic fields and thus intensities in the guide, which is essential for a complete analysis of the modes emitted from a laser.

Using the waveguide dimensions shown in Fig. 3.25, where the waveguide thickness $t = 2h$ and the origin in the $\boldsymbol{y}$ direction is defined at the waveguide center, we may write the field solution for the special case of an $\boldsymbol{x}$-polarized electric field with temporal frequency ω propagating in the z direction as

$$\boldsymbol{E} = \boldsymbol{E}_x(y) e^{-jk_z z} e^{j\omega t} . \tag{3.74}$$

This special case represents the TE polarization, where the z-propagating electric field vector points in the $\boldsymbol{x}$ direction, and $\boldsymbol{E}_x$ varies only in the $\boldsymbol{y}$ direction.

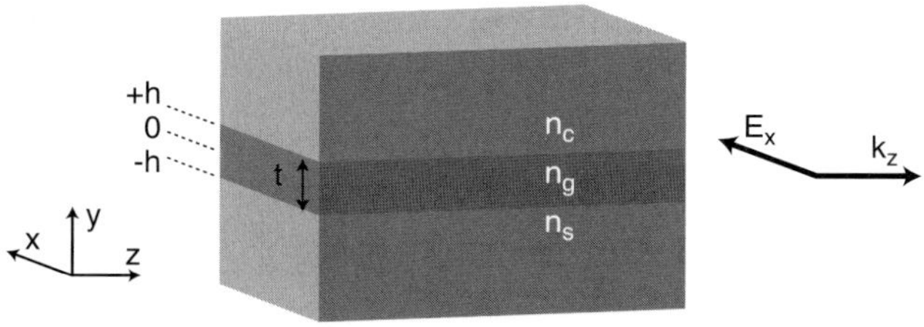

Fig. 3.25. A slab waveguide showing the definition of thickness as $t = 2h$; the electric field is shown oriented for TE polarization (in the $\boldsymbol{x}$ direction) with propagation in the z direction

Field Solutions

The form of $\boldsymbol{E}_x$ is a function of the portion of the waveguide in which the field is considered. For the case of an ideally guided mode, the fields vary as

$$\boldsymbol{E}_{\boldsymbol{x}}(y) = \boldsymbol{E}_{c0} e^{-k_{yc}(y-h)} \quad \text{for } y \geq +h \tag{3.75}$$

$$\boldsymbol{E}_{\boldsymbol{x}}(y) = \boldsymbol{E}_{g0} \cos(k_{yg} y + \phi) \quad \text{for } -h \leq y \leq +h \tag{3.76}$$

$$\boldsymbol{E}_{\boldsymbol{x}}(y) = \boldsymbol{E}_{s0} e^{k_{ys}(y+h)} \quad \text{for } y \leq -h \tag{3.77}$$

such that an oscillating solution is found in the waveguide core (Eq. 3.76) and exponentially decaying solutions in the upper and lower cladding layers (Eq. 3.75 and Eq. 3.77). These latter characteristics are in contrast to the ray-optic model, in which no optical field was assumed to exist in the cladding.

In the previous expressions, the propagation constants in the **y** direction may be determined from the free space propagation constant k_0 and the modal propagation constant β as defined in Eq. 3.72. For the case where $k_0 n_g > \beta > k_0 n_s$ (assuming that $n_s \geq n_c$), which is the situation when the optical field propagates in the core region but not in the cladding, the **y**-directed propagation constants are given by

$$k_{yc} = \sqrt{\beta^2 - k_0^2 n_c^2} \tag{3.78}$$

$$k_{yg} = \sqrt{k_0^2 n_g^2 - \beta^2} \tag{3.79}$$

$$k_{ys} = \sqrt{\beta^2 - k_0^2 n_s^2}\,. \tag{3.80}$$

By applying the appropriate boundary conditions for continuity of electric and magnetic fields at the core/cladding boundaries, the field solutions given above can be shown to yield the same characteristic equation for propagation in a waveguide as in the example of Eq. 3.71. This characteristic equation may then be solved for β (or equivalently θ, or k_{yc}, k_{yg}, k_{ys}) to yield the values for the propagation constants for any given mode.

Electric Field and Intensity Distribution

We may thus summarize the procedure for determining the field distribution in a waveguide: the characteristic equation, such as the one in Eq. 3.71, is solved to yield an allowed value for θ for a given mode with index m. The z-directed propagation constant β is determined from Eq. 3.72, and from this the **y**-directed propagation constants from Eq. 3.78 – Eq. 3.80. These are in turn plugged into the expressions for the $\boldsymbol{x}$-directed electric field (Eq. 3.75 – Eq. 3.77) in the three regions (upper cladding, core and lower cladding), which then yield the complete field solution, for propagation in the z direction, given by Eq. 3.74.

Doing this for two modes ($m = 0$ and $m = 2$) leads to the expected electric field variation as shown in Fig. 3.26; the shaded region represents the extent of the waveguide core. Recalling that the optical intensity is proportional to $\boldsymbol{E}^2$, Fig. 3.27

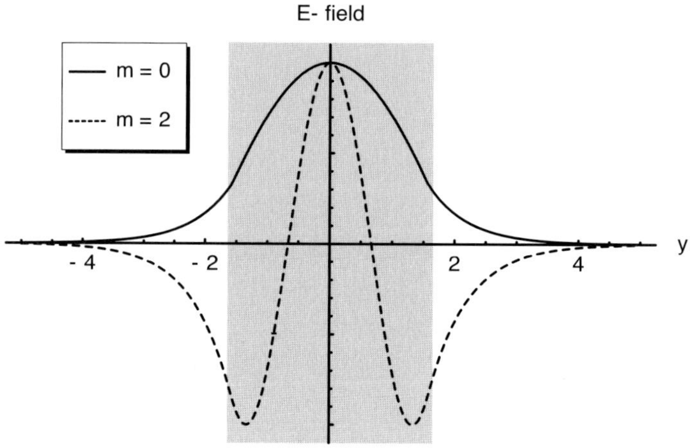

Fig. 3.26. The electric field distributions for the $m = 0$ and $m = 2$ modes in a waveguide; the dimension of the waveguide is given by the shaded region

shows the variation in intensity with transverse position in the waveguide. We see that the lowest order mode ($m = 0$) has a single peak in the center of the waveguide, whereas the higher order mode shows three peaks.

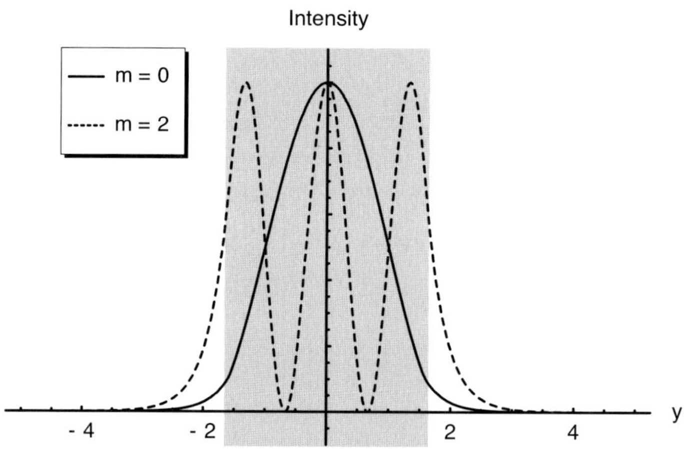

Fig. 3.27. The optical intensity distributions for the $m = 0$ and $m = 2$ modes in a waveguide; the dimension of the waveguide is given by the shaded region

From the previous characteristics, we see clearly that the field decays exponentially into the cladding layers. The extent of electric field penetration into the cladding is given by the exponential coefficients, namely the y-directed propagation constants defined in Eq. 3.78 and Eq. 3.80. A larger value for k_{yc} or k_{ys} implies a more rapid decay of field with distance from the interface: the penetration depth is then typically defined as $1/k_{yc}$ or $1/k_{ys}$, and this is thus the point where the field has decayed to $1/e$ of its value at the interface.

Example 3.5

Assignment: *Calculate the penetration depth of the electric field into the upper and lower cladding regions for the laser waveguide analyzed in Example 3.4.*

Solution: *We base our calculation on the values for β derived in Example 3.4. Using Eq. 3.78 and Eq. 3.80, we can calculate the y-directed propagation constants for the two supported modes:*

$m = 0$: $k_{yc} = 10.5\ \mu m^{-1}$ *and* $k_{ys} = 9.4\ \mu m^{-1}$

$m = 1$: $k_{yc} = 7.3\ \mu m^{-1}$ *and* $k_{ys} = 5.6\ \mu m^{-1}$

Simply taking the reciprocal of these values, leaves the penetration depths:

$m = 0$: $1/k_{yc} = 96$ *nm and* $1/k_{ys} = 138$ *nm*

$m = 1$: $1/k_{yc} = 106$ *nm and* $1/k_{ys} = 178$ *nm*

From these results, we see two important characteristics of the electric field at and beyond the edge of the core: 1) higher order modes penetrate more deeply into the cladding region than lower order modes; and 2) a smaller index step between core and cladding also results in larger mode penetration into the cladding (recall that in our model structure, n_s (3.11) is closer to n_g (3.39) than n_c (3.05) is to n_g).

3.4.3 Confinement

From the previous discussion, and consideration of the field solutions in a waveguide, it is clear that not all the optical power is found in the waveguide core, as was implicitly assumed in the ray-optic model. A measure for the portion of the intensity which is confined to the waveguide core is given by the confinement factor, Γ, defined by the integral

$$\Gamma = \frac{\int_{-h}^{h} \boldsymbol{E}_x^2(y)dy}{\int_{-\infty}^{\infty} \boldsymbol{E}_x^2(y)dy}, \qquad (3.81)$$

in this case for a transverse mode, $\boldsymbol{x}$-polarized and bounded in the $\boldsymbol{y}$ direction by a waveguide with thickness $t = 2h$.

The confinement factor has a maximum value of unity, which would represent optimal confinement of the mode. As Γ decreases, the portion of energy found in the cladding layers increases and the mode is said to be less well-confined or less well-guided. In general, higher order modes are less well confined than lower order modes (which we saw explicitly in Example 3.5, where the penetration depth was larger for the higher-order mode) and as a mode moves toward cutoff, Γ also decreases. For a single-mode structure, decreasing the waveguide thickness also reduces confinement for the fundamental mode.

For most semiconductor laser structures, Γ is maximized to assure that the laser mode is maximally confined to the waveguide, which typically corresponds to the active region of the laser as well. Poorly confined modes not only suffer from reduced gain, and thus higher threshold, but may also be subject to higher optical losses due to imperfections, impurities and roughness of structures outside the waveguide core.

3.4.4 Stripe Waveguides

The slab waveguide confines light in one dimension: propagation of light is in a plane defined by the two-dimensional layers. Whereas this structure forms the basis for one family of semiconductor lasers, the broad-area laser (Sect. 4.2), most semiconductor lasers are so-called stripe lasers: these employ a stripe waveguide, in which light is confined in two dimensions, as shown in Fig. 3.28. Such a structure represents an "optical wire", with an optical modal structure in both the $\boldsymbol{x}$ and $\boldsymbol{y}$ directions.

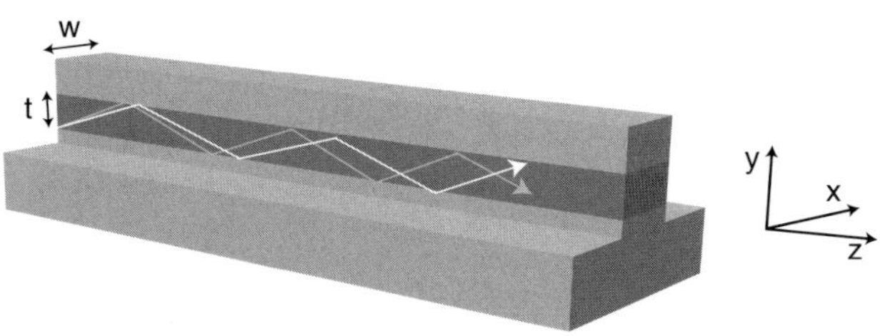

Fig. 3.28. Typical stripe waveguide structure showing the propagation of two modes in the ray-optic representation; in this ridge structure, the slab waveguiding layers have been removed in the regions outside the waveguide stripe

There are a number of approaches for fabricating stripe waveguides (Zappe 2003); we will consider some of the most relevant structures in Chapter 4. Semiconductor laser waveguides typically begin with a slab heterostructure as described in the previous two sections. The lateral structure of the waveguide stripe is then defined using etch or implantation techniques. The perhaps simplest approach is to remove the slab waveguide layers outside the stripe waveguide region; this case, yielding a *ridge* waveguide (Takeuchi and Oe 1989), is the one shown schematically in Fig. 3.28. Variations on this structure include burying the stripe under a further epitaxial layer, a technology which reduces scattering losses at the waveguide sides (Dutta et al. 1991). The waveguide core region may also be partially etched, yielding a *rib* stripe waveguide (Reinhart, Logan and Lee 1974) and by structuring only the upper cladding layers, a *strip-loaded* stripe waveguide (Kapon and Bhat 1987) may be defined.

By oxidizing (Caracci et al. 1992) or intermixing (Kumar et al. 1993) the regions outside the waveguide stripe, the slab waveguide may be locally destroyed, leaving only the stripe region intact. Such a process may be used to form stripe waveguides without recourse to etching technology and thus resulting in a planar surface. Finally, by fabricating a metal contact stripe on a laser heterostructure, the current injection region of the laser may be defined along the stripe. High current injection generating locally enhanced gain then leads to gain guiding under the contact stripe, thereby inducing a stripe laser in a nominally slab waveguide structure. We will discuss this and related structures in greater detail in Chapter 4.

Lateral Waveguide Modes

The definition of a waveguide stripe implies that the propagation constants in the lateral (x) direction also become discrete. As a result, a stripe waveguide may have one or more lateral modes in addition to the transverse modes we considered in Sect. 3.4.2. The modal and propagation analysis of a two-dimensional stripe waveguide is somewhat more complex than that of the one-dimensional slab waveguide, but employs the same basic concepts (Kogelnik 1988). Using a characteristic equation, the allowed modes may be determined, where in the lateral direction the waveguide is often defined by etched or implanted features rather than by alternating materials of a heterostructure. Although solving the stripe waveguide "by hand" as we did above for the slab waveguide is in general quite difficult, we may approximate the two-dimensional modal solution as the product of the two one-dimensional solutions, as shown in Fig. 3.29. More extensive and accurate waveguide modeling is best done by numerical methods, using a variety of software packages available commercially.

Fig. 3.29. Calculated two-dimensional mode solutions for a stripe waveguide showing the $m = 0$ mode *left* and $m = 1$ mode *right*

The measured intensity distribution in a series of waveguides of varying width is shown in Fig. 3.30. As the waveguide increases in width, higher order modes in the lateral direction are supported: this transition from single-mode to dual-mode

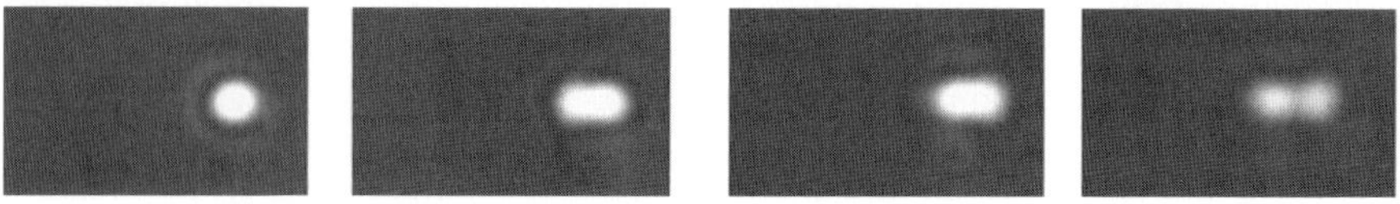

Fig. 3.30. Measured intensity distribution from a III–V-based stripe waveguide used for fabrication of a semiconductor laser; from left to right, the waveguide width is 1.5 μm, 2.0 μm, 2.5 μm and 3.0 μm

behavior is clearly seen in the measured intensity. The higher order mode manifests itself as a distinct pair of intensity peaks; we see that this higher order mode is already supported for the 2 μm wide waveguide, and dominates the behavior for the 3 μm waveguide.

4 Laser Diode Structures

Whereas the previous chapter provided an overview of some of the quantum physics and material science of laser diodes, we now move from the mists of theory to practical and functional laser diode chips. In this chapter, we will engage in a survey of the semiconductor laser structures which translate the abstract phenomenon of lasing into useful devices.

Beginning with the light-emitting diode, essentially a frustrated laser, we will examine the most important edge-emitting laser diodes, including the Fabry–Perot and distributed feedback structures. We continue with the vertical cavity laser, which emits from the wafer surface and represents a relatively new development with great potential for microsystem applications. Finally, the quantum cascade laser will be presented, a complex device presently moving from research into industrial application. For each of these laser types, we will examine the most popular and useful device structures, present unique aspects of their operating principle, consider important aspects concerning their fabrication and finally discuss their relevance to optical microsystems. A more detailed look at the operating characteristics of these structures is then the subject of Chapter 5.

A variety of further laser types, including whispering-gallery micro-disk, quantum dot, quantum wire (Kapon 1999) and photonic crystal-based (Painter et al. 1999) semiconductor lasers are the subject of considerable research from the laser physics as well as potential applications points of view (Gourley 1994). Due to their relative immaturity coupled with market unavailability, we will omit these types from the present discussion.

4.1 Light-Emitting Diodes

Light-emitting diodes (LEDs) are not lasers, but are made of the same materials as and share numerous features in their device structure with semiconductor lasers. Indeed, some highly developed LED structures must be carefully designed to ensure operation as an LED and to keep them from lasing. The market for LEDs has grown rapidly, due primarily to great improvements in output power, available wavelengths and efficiency. Volume applications now include large-format video displays, traffic signals as well as automotive and domestic illumination. Microsystems benefit from the availability of LEDs insofar as compact, low-power and very economical light sources with a variety of spectra are now obtainable for use in systems with constricted spatial requirements.

4.1.1 Structure

In its simplest form, the LED consists of a pn junction fabricated in a direct-bandgap semiconductor. Light emission is due to spontaneous emission, which is a spatially isotropic process; as a result, many of the developments in LED performance have derived from increasing the generation efficiency of photons as well as improving probability that a generated photon may escape from the block of semiconductor and contribute to useful illumination.

Homojunction LED

A basic LED is shown in Fig. 4.1 in which a top-emitting device with an absorbing substrate is considered. The structure consists of a planar pn junction, a small surface contact and large-area back-side contact. The pn junction is close to the surface, with a typical depth between < 1 μm and several μm; chip thickness is typically on the order of 350 μm. Emission of photons is from the surface and edges of the chip.

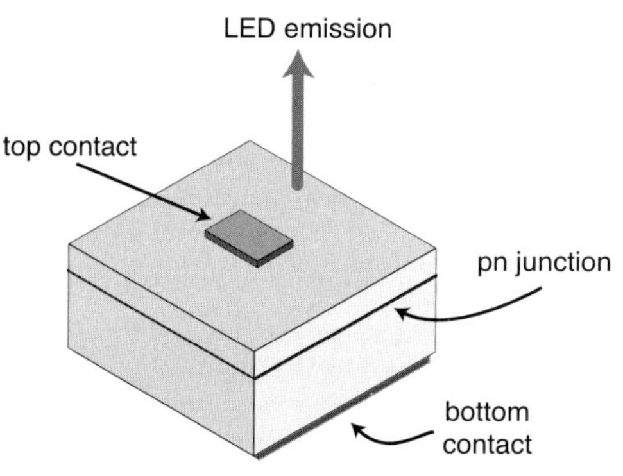

Fig. 4.1. A top-emitting LED with a large planar pn homojunction and an absorbing substrate; emission is from the top surface

The LED in this example is based on a homojunction, meaning that only a single material is used for the pn junction. Typical substrates might be GaAs for emission in the near infrared (λ ~ 870 nm) or GaP (which is indirect bandgap) for emission in the green (λ ~ 540 nm); Table 4.1 lists a few further relevant materials. There is no optical or electrical confinement implying that light generation takes place in a diffuse region near the junction. The top-emitting LED may also employ a heterojunction, as demonstrated in the following structural example, to improve light generation efficiency.

Whereas this device structure is appealing due to its fabrication simplicity, considerable light is lost due to absorption in the substrate. Photon emission is roughly equal in all directions, such that only those photons emitted upwards and toward the edges of the chip are likely to escape the substrate; the rest are absorbed.

Heterojunction LED

Improvement in LED efficiency may be attained through the use of a heterojunction and a transparent substrate, as shown in the example of Fig. 4.2. In this case the pn junction is formed as a heterojunction, implying that the active region is of a different material than the rest of the substrate, typically with lower energy gap and higher refractive index (see Sect. 3.3.3). The resulting optical and electrical confinement may increase the generation of efficiency of photons within the active region. Typical dimensions of the active region are ≤ 1 μm in thickness with a position within several microns of the surface.

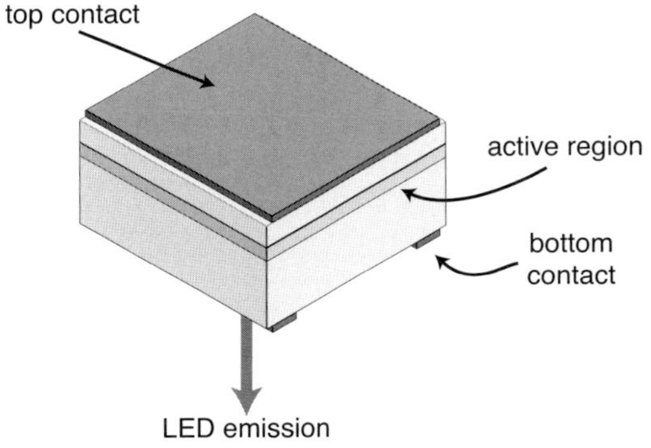

Fig. 4.2. A bottom-emitting, heterostructure LED; the active region represents a pn heterojunction and the substrate is transparent at the emission wavelength. The broad-area top contact also acts as a mirror for generated photons

The bottom-emitting LED is fabricated on a substrate which is transparent at the emission wavelength; a typical example might be an InGaAsP active region fabricated between InP cladding layers on an InP substrate. This type of LED for the mid-infrared range ($\lambda \sim 1500$ nm) emits through the substrate since InP is transparent at the wavelength emitted through direct transitions in the InGaAsP layer. A further limitation on LED efficiency is due to Fresnel reflection at the semiconductor-air boundary. The use of anti-reflection coatings as well as shaping of the semiconductor chip itself (to reduce total-internal-reflection effects) is often required to yield high-efficiency light emission.

The top contact, which covers the entire surface, is often made using a metal multi-layer structure and thus acts as a mirror to reflect photons downward for emission; the bottom contact is structured to allow light transmission. For transparent but electrically-insulating substrates, such as Al_2O_3 for InGaN green LEDs (Nakamura et al. 1995), the lower contact to the pn diode is made from the surface using an appropriately etched recess. Red-orange AlGaInP LEDs, in contrast, are grown on an absorbing GaAs substrate, removed by epitaxial liftoff and transferred to a transparent GaP substrate.

Burrus LED

The limitation of the absorbing substrate on LED emission efficiency may be mitigated by using the so-called Burrus LED structure (Burrus 1971) shown in Fig. 4.3. This device is fabricated on an absorbing substrate which is subsequently etched away below the active region allowing emission from the back-side. With this type of configuration, a GaAs or AlGaAs active region, with emission wavelengths in the near infrared region, may be epitaxially grown on a GaAs substrate; generated photons, including those reflected from the top contacts, may then efficiently be emitted from the back of the LED.

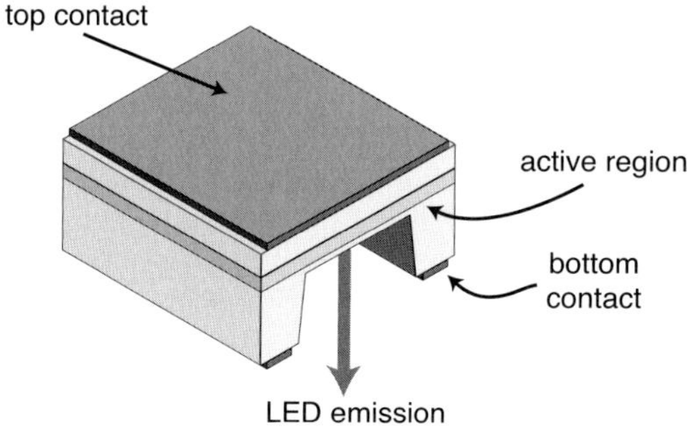

Fig. 4.3. The Burrus LED using a heterostructure active region and an absorbing substrate; the substrate is etched away allowing efficient bottom emission

Resonant Cavity LED

The emission intensity and directionality of the LED may be enhanced by modification of the active region and its surroundings to give rise to stimulated emission and thus a higher photon intensity. The resonant cavity LED, shown in Fig. 4.4, employs an epitaxial Bragg mirror to reflect a portion of the generated optical field back into the heterostructure active region (Schubert et al. 1992; Wirth et al. 2001); this mirror typically consists of 10 periods of a Bragg reflector, resulting in a reflectance of about 0.5. The large-area top contact is highly reflecting ($R \geq 0.95$), such that these two mirror structures form a resonant cavity with an enhanced photon density. The structure is reminiscent of a vertical cavity laser (which we will discuss in Sect. 4.5), although the quality of the resonator here is much reduced in comparison.

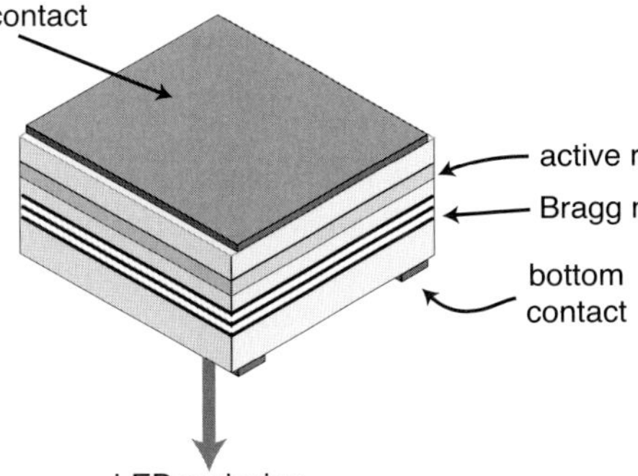

Fig. 4.4. A resonant cavity LED using an AlGaAs quantum well active region and an epitaxial Bragg mirror to reflect a portion of the generated photons upward; the top contact is reflecting and emission is downward through the substrate

The cavity resonance yields a high photon density and thus stimulated emission will begin to contribute to photon generation. The result is a higher intensity, improved beam quality and reduced spectral width when compared to conventional LEDs. The emission spectrum of the resonant cavity LED may be on the order of several nanometers and the intensity increase achievable with this structure may be as high as a factor three.

The resonant cavity structure may also be used to improve the performance of detectors (Ünlü and Strite 1995), where captured photons are more likely to result in a generated electron. In addition, the technique has been demonstrated using organic materials (Nakayama, Itoh and Kakuta 1993), currently of great interest for the fabrication of extremely low-cost organic light emitters.

Superluminescent LEDs

High-intensity LED emission may also be generated using structures which are essentially inhibited lasers. If we reduce the quality of the cavity of a stripe laser structure, such as those we will examine in the following sections, lasing will be inhibited but stimulated emission may still be a dominant source of photons. The laser structure thus acts as an amplifier for spontaneous emission, but laser resonance is not achieved. The resulting superluminescent diode then emits with a high intensity and with improved beam properties over conventional broad-area LEDs; on the other hand, the low level of coherence implies that the device may exhibit lower noise and is generally less sensitive to feedback reflections than a real laser diode.

Considerable care needs to be taken to ensure that the cavity of a superluminescent diode does indeed not result in lasing. Typical approaches to accomplish this include anti-reflection coatings on the laser facets, bent waveguides (Lin and Juang 1996), window structures which couple light out of the cavity with minimal back-reflection (Kashima, Matoba and Takano 1992) or integrated absorption regions in the cavity (Okamoto et al. 1998). These waveguide-based structures are of particular advantage for coupling the LED emission into fibers or waveguide structures; spot-size converters may be integrated into the superluminescent LED to optimize coupling into a single-mode fiber core, an operation which is considerably more difficult for conventional LED structures.

4.1.2 Operating Principle

LEDs are conceptually simple. Light emission relies on spontaneous emission due to direct-bandgap transitions in materials appropriate for the wavelength desired, with amplified spontaneous emission playing a role in resonant cavity and superluminescent devices. With the exception of the latter two structures, a pn homo- or hetero-junction is all that is required for light emission. As with laser diodes, LEDs are current-driven devices and the output power is directly related to the level of injected current.

As a result, the emission of a LED is generally spectrally broad (50 – 100 nm) and, with the exception of the superluminescent or resonant cavity LED, of lower intensity (optical power per area) than a laser diode. We will consider the output characteristics in greater detail in Chapter 5.

4.1.3 Fabrication Issues

Since an LED, in its simplest form, requires only a pn junction with contacts to it, mass-production of these devices is possible with a relatively simple, low-cost technology. A primary issue is the active-region material: a direct-bandgap semiconductor with a bandgap energy yielding the wavelength of interest must be available. A sampler of popular materials used for LEDs of various wavelengths is given in Table 4.1; available emission wavelengths range from the mid-infrared to the ultraviolet (Basrur et al. 1997).

Through various techniques, LED emission with a very broad spectrum or with two or more spectral peaks may be generated. While retaining the advantages of a semiconductor light emitter, the spectrum may be tailored to a variety of illumination applications. Emission of a "white" spectrum is often desirable, where "white" implies a mixture of colors which results in a shade of white. For example, through the stacking of two InGaN/GaN-based LEDs emitting in the blue and green portions of the spectrum, a broad emission spectrum which can be electrically tuned has been demonstrated (Chen et al. 2002). An alternative, presently well-established industrially, is through the use of an inorganic fluorescence conversion layer

Table 4.1. Common materials used for fabrication of LEDs with emission wavelength; all materials listed are direct-bandgap

Material	Peak wavelength [nm]	Color
InGaN	390 – 530	UV – green
AlGaInP	590 – 620	yellow
GaAsP	650	red
GaAs	860	infrared
$In_{0.76}Ga_{0.24}As_{0.55}P_{0.45}$	1300	infrared
$In_{0.65}Ga_{0.35}As_{0.79}P_{0.21}$	1550	infrared

(typically Ce-doped YAG, $Y_5Al_5O_{12}$:Ce^{3+}) inside the LED package (Baur, Schlotter and Schneider 1998) or deposited on the LED chip directly. Using a blue GaN-based LED as the pump wavelength, the combination of blue and fluorescent yellow permits the generation of "white" LED emission with high efficiency. While primarily developed for automotive and domestic lighting applications, white LEDs also have a range of microsystem applications.

The LED epitaxial layers are typically grown by LPE or MOVPE, the former being the technique of choice for high-volume production of simpler LED structures due to the high growth rates and low system costs. Standard metallization processes are employed and the completed chips are separated by sawing since, with the exception of specialized structures, no demands are placed on the crystalline facets.

A primary issue in overall LED device performance is packaging. The end-user is interested in the photon intensity ultimately delivered and that quantity is strongly affected by the LED mounting and package technology. We have seen that the LED chip may emit light from the top, side and/or bottom surfaces and the challenge of packaging technology is to collect all of these photons and direct them to the user's application. Lens and reflector structures, fabricated at low cost, are often an integral part of an LED package and it is estimated that a further 30 – 50% improvement in overall output efficiency of current devices may be achieved by optimization of these features.

4.1.4 Microsystem Aspects

For illumination applications in an optical microsystem where spectrum, total power and power density are not subject to demanding constraints, LEDs represent an attractive alternative to thermal emitters or laser diodes. High efficiency, small size and low power consumption when compared with light bulbs allow LEDs to replace these components in microsystems where the very broad spectrum of a thermal emitter is not required. On the other hand, reduced cost, lower sensitivity to feedback and improved noise performance are criteria which work to the advantage of the LED in applications for which laser diodes are considered but the particular features of the latter are not essential.

The emission properties of the LED are often an advantage for coupling into the optical components of a microsystem. Using symmetric emission regions, the near-field beam quality may be better than that of an edge-emitting semiconductor laser implying that coupling into micro-lenses, for example, is facilitated. The packaging, if present, again plays a major role in determining beam quality.

Microsystem assembly techniques for LED-based optical microsystems depend on the structure employed. LEDs are available as chips, with typical dimensions on the order of $300 \times 300 \times 100\ \mu m^3$, allowing mounting on micro-machined platforms. The next larger size uses a surface mount device (SMD) concept, with millimeter dimensions, a small package which may include reflector structures and may be integrated with micro-lenses or lens arrays. The macroscopic 3 mm and 5 mm plastic packages, with long leads and lensed apertures, are suitable for larger microsystems (or milisystems) where the possible disadvantage of larger size may be offset by considerably improved beam quality due to the lensed package.

Example 4.1

Assignment: *Given a GaAs LED structure with a pn junction 1 µm below the surface, a 350 µm thick substrate and a background loss at the emission wavelength of* $10\ cm^{-1}$*, estimate the limitation of emission efficiency due to absorption in the material and the effect of the surfaces. For simplicity, consider emission from only one point on the pn junction. Would a Burrus-type structure improve the performance?*

Solution: *Absorption losses are exponentially related to absorption coefficient and path length. We are given that* $\alpha = 10\ cm^{-1} = 10 \times 10^{-4}\ \mu m^{-1}$*, so that the absorption losses for a path length L are proportional to* $e^{-\alpha L}$*. The path length to the top surface is 1 µm and to the bottom (350 – 1) µm resulting in transmission efficiencies of 0.999 and 0.705, respectively.*

Optical loss at the surfaces is due to Fresnel reflection at the GaAs/air boundary and the fact that light incident on the surface at an angle above the critical angle returns into the substrate due to total internal reflection. Let us consider these effects separately.

The reflectance (reflection coefficient for power) of a normally-incident light beam onto an interface is related to the refractive indices of the materials on either side of the boundary as

$$R = \left(\frac{n_{GaAs} - n_{air}}{n_{GaAs} + n_{air}}\right)^2$$

and thus gives the loss due to Fresnel reflection. Assuming vertically-incident light (not strictly true) and given $n_{GaAs} = 3.59$ *and* $n_{air} = 1.0$*, we obtain* $R = 0.318$ *and thus a transmittance of* $1 - 0.318 = 0.682$*.*

Finally, the critical angle for light impinging from GaAs onto an air interface is given by

$$\theta_c = \sin^{-1}\left(\frac{n_{air}}{n_{GaAs}}\right)$$

which evaluates to $\theta_c = 16.2°$. Only light incident within an "escape cone" whose vertex is defined by this angle may escape from the substrate. Rays incident onto the surface with $\theta > \theta_c$ are totally internally reflected, such that the emission efficiency due to this factor is given by (assuming spherically uniform emission from a point on the pn junction)

$$\eta_{crit} = \frac{4 \times \theta_c}{360°} \approx 0.18 .$$

We may thus take the product of absorption loss, transmittance and critical angle effect to yield the emission efficiency from the top surface

$$\eta_{top} \approx 0.122$$

and from the bottom surface

$$\eta_{bottom} \approx 0.086 .$$

If emission from the pn junction is equal in both directions, upwards and downwards (usually a good assumption), then the total emission efficiency is given by

$$\eta_{total} = \frac{1}{2}\eta_{top} + \frac{1}{2}\eta_{bottom} \approx 0.104 .$$

Thus even for unity internal quantum efficiency, only about 10% of the light makes it out of the LED chip from the top and bottom surfaces.
Would a Burrus-type LED structure help? Let's assume we etch all but 10 μm of the GaAs substrate away below the junction. The path length is then reduced from 349 to 10 μm yielding an emission efficiency from the lower surface

$$\eta_{bottom} \approx 0.121$$

and the new total efficiency

$$\eta_{total} \approx 0.122 .$$

We thus have an improvement of less than 2%. The Burrus structure does help overall efficiency a bit, but the primary loss mechanism in our example is due to reflection at the boundary, not absorption. Thus the minimal improvement achievable probably does not warrant the extra technological complexity.

4.2 Broad-Area Lasers

With slight improvement to the optical cavity, the LED structure may be made to lase. By adding high-quality mirrors to the block of semiconductor which forms an LED, the resulting resonant cavity yields a broad-area laser; this laser emits in a broad stripe with very little or no lateral confinement. Broad-area lasers generally have a complex modal structure (Lang, Larsson and Cody 1991) and relatively poor beam quality (Shah et al. 1990), but are typically used in applications which require high absolute power and high optical intensity.

4.2.1 Structure

As seen in Fig. 4.5, the structure of the broad-area laser is reminiscent of the LED, in that large-area contacts are used for current injection and no lateral structuring is required. Vertically, a double heterostructure is typically used, where the active region may contain one or more quantum wells to improve laser gain and accurately define the emission wavelength. The cavity is defined by two parallel mirrors, typically cleaved semiconductor surfaces, which in turn define the emission axis.

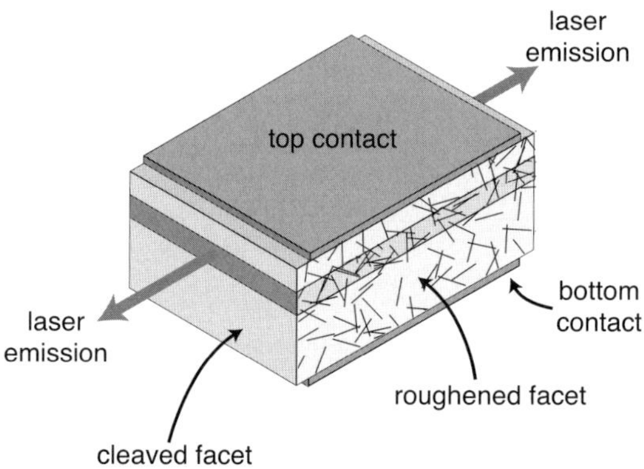

Fig. 4.5. A broad-area laser with a cavity defined by two cleaved mirrors; emission is from the heterostructure active region and takes the form of a wide stripe with no lateral confinement. The orthogonal facets are roughened to inhibit lasing in the direction normal to the intended beam

Broad-area lasers may be fabricated as single chips, as shown in the figure, or as long bars in which the laser region is defined by a patterned upper metal layer. The pumped region is restricted to that under the contacts but is typically quite wide so that emission is still from a thin wide stripe. Numerous such metal stripes may be placed along a laser bar, yielding a one-dimensional broad-area laser array. Total optical output power from such arrays may reach the several tens of Watts level in commercially available devices.

Typical device dimensions vary widely between manufacturers and depend in part on the emission properties desired. Cavity length is frequently between 200 and 1 000 μm, contact (and thus laser) width 100 – 200 μm and chip thickness for an unthinned substrate 350 μm, for a thinned substrate 80 – 100 μm. The thickness of the heterostructure active region is the same as for most LED and laser structures, and varies between 200 nm and several micrometers.

4.2.2 Fabrication Issues

Fabrication of the broad-area laser is relatively simple and only marginally more involved than the production of an LED. The epitaxial layers which define the active region, with or without quantum wells, are typically grown by MOVPE or MBE. The entire wafer is then metallized on both sides, where the top-side metallization may be patterned to define the individual lasers. Separation into long bars or individual chips completes the process.

Laterally, the laser may be defined by the dimensions of the chip or by the size of the upper metal contact. For the former case, chips are separated by cleaving, a process which generates laser-quality mirror facets, albeit in both directions. To keep the laser from lasing in the direction normal to the desired axis, the cleaved facets frequently need to be roughened to depress the quality of the cavity in one direction. If chip separation is through sawing, facet quality is inherently poor and the problem does not arise. Due to the simplicity of their fabrication process, broad-area lasers are frequently fabricated from one wafer of an epitaxial growth run to verify the quality of the crystalline layers and the laser emission wavelength.

4.2.3 Operating Principle

The broad-area laser is a Fabry–Perot laser with no or little lateral guiding. In contrast to the LED, the resonant cavity is designed to allow laser resonance such that superluminescence gives rise to lasing at sufficiently high pumping. Current injection is through a large-area contact on the surface which leads to uniform pumping over the entire device area.

Broad-area lasers generally lase in several longitudinal modes and with many lateral modes; emission can be spatially non-uniform and exhibit complex near-field and far-field profiles. In addition, hole burning effects and filamentation due to self-focussing in the gain region (Chow and Amano 2001) may lead to modal instabilities, which makes coupling into fibers or other micro-optical components difficult.

Temperature effects are also an issue: the large planar pumped volume results in significant heat generation, leading to significant temperature variations within the device. As a result, broad-area lasers are often operated using short current pulses (typically ≤ 1 μs) rather than continuous current injection. Pulse length is chosen so that heating effects during device operation are reduced, improving emission stability and modal properties during laser emission.

4.2.4 Microsystem Aspects

The broad-area laser is generally a relatively high-power device, both in terms of optical emission power and electrical input power. Macroscopic applications are numerous: pumping of solid-state (YAG) lasers, commercial laser printing, machining of plastics and powering optical amplifiers in telecommunications systems are among the examples in which broad-area lasers currently play a significant industrial role. Due to the availability of high optical power from a compact semiconductor laser, with its advantages in terms of reliability and efficiency in comparison to most types of macroscopic lasers, this laser type will continue to enjoy development for new applications.

In microsystems, the broad-area laser is of utility for applications in which optical power is required with relatively few concomitant demands on spectral quality. The device structure is only large when compared to other semiconductor lasers: it is still possible to derive hundreds of milliwatts of optical power from a chip on the order of 1 mm^3 in volume.

Two issues need careful consideration when incorporating a broad-area laser into an optical microsystem: beam quality and packaging. The former results from the fact that laser emission is from a relatively large area with numerous lateral and longitudinal modes. These give rise to a far-field intensity distribution with numerous irregular lobes, making coupling into an optical system a challenge. Whereas numerous approaches have been suggested to improve beam quality (Lang, Larsson and Cody 1991), the microsystem designer should consider using a high-power stripe laser diode rather than a broad-area laser diode in systems for which high-power as well as optical coupling are both required.

The packaging aspect of broad-area lasers revolves about the need to get considerable electrical power into the system and the generated heat back out. For truly high power applications, device wiring and heat sinking become non-trivial issues. Laser sub-mounts made from materials with good thermal conductivity (Cu, Si), often with diamond as a heat-spreader, are thus generally to be preferred to plastics.

4.3 Fabry–Perot Stripe Lasers

The stripe laser confines the lasing region laterally under a waveguide stripe. In contrast to the broad-area laser, the stripe laser emits from a small facet at the end of the waveguide, so that single lateral and transverse mode operation is possible. The Fabry–Perot stripe laser uses cleaved or etched mirrors to define the laser cavity and thus represents the simplest form; distributed feedback lasers use the same or similar waveguide structures for lateral confinement but a different technique for achieving optical feedback as will be seen in the next section.

Fabry–Perot stripe lasers represent the most popular semiconductor lasers and it is this type of structure which is sold in quantities of several hundred million per year, primarily for optical data storage applications. There is a host of techniques for achieving lateral waveguiding representing a spectrum of process complexities; we will examine a few of the more popular structures.

4.3.1 Structure

The stripe laser uses the same type of laser active region layers as the broad-area lasers; indeed, these laser types generally differ only in their lateral structure, not in their epitaxy. The active region is typically a double heterostructure with one or more quantum wells, grown by MOVPE or by MBE. The material used for the active region layers depend on the desired emission wavelength, as we saw in Sect. 3.3.2, and the substrate is chosen to be lattice matched to the active region materials. A typical structure for a near-infrared GaAs-based laser diode might be a 200 nm thick $Al_{0.3}Ga_{0.7}As$ laser core with three 8 nm GaAs quantum wells surrounded by 1 μm $Al_{0.8}Ga_{0.2}As$ cladding layers, all grown on GaAs. Detailed structures may be more complicated.

The Fabry–Perot cavity itself is defined by cleaved or etched mirrors, essentially yielding a planar semiconductor surface normal to the laser stripe. These mirrors reflect, to first order, independently of wavelength, such that wavelength selection of the laser is through a combination of the position of the gain peak and the resonant condition of the longitudinal cavity. Surface coatings may be used to enhance the reflectivity of one or both mirrors to adapt the laser to a particular application; increasing reflectivity of one facet mirror may improve overall efficiency since typically emission from only one facet is used.

Ridge Waveguide

Given the epitaxial layers, any one of a number of approaches may be used for achieving lateral guiding in a stripe laser; the techniques outlined in Sect. 3.4.4 are generally applicable to laser diode structures. The ridge structure uses an etched waveguide to define the lateral region as shown in Fig. 4.6. The upper cladding region above the laser core is etched away outside the laser stripe, the etch generally stopping before it reaches the layers of the laser active region. The optical field is confined laterally by the slight increase in effective refractive index of the waveguide under the ridge with respect to that under the etched regions. This structure leads to so-called weak guiding; a strongly guided structure, which would result when the laser active layers are etched through, is generally not used for lasers, as the deep etch can lead to optically and electrically active damage at the edges of the laser core.

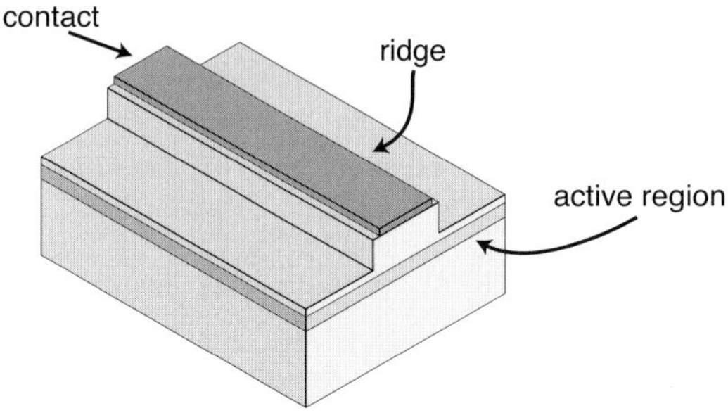

Fig. 4.6. A ridge Fabry–Perot laser with a waveguide etched into the upper cladding region

The ridge width varies according to the demands placed on laser performance, particularly with respect to its modal behavior. Narrow ridges (typically ≤ 2 μm) are required for single-lateral-mode lasers, for which the output powers may be limited due to the reduced active volume of the laser. Wider ridges (typically up to 5 μm) may be used in lasers where single-lateral-mode operation is not required but a higher power level is an advantage. The ridge width also affects the mode size, such that applications in which coupling from the laser diode into a single-mode fiber or external waveguide also require an optimized laser ridge width to allow adequate mode matching.

The active layer thicknesses are similar to those in a broad-area laser, namely a core layer of 100 – 300 nm, cladding layers of 1 μm and the entire active region within a few microns of the substrate surface. The ridge depth is typically ≤ 1 μm as it is only etched into the upper cladding layer. Laser length may also vary according to the intended laser application and is typically in the range 100 – 500 μm. Typical substrate thicknesses are 350 μm but may be thinned to 100 μm for small chips. The size of the entire laser chip is determined by the need to handle it during packaging or in a hybrid assembly step for the fabrication of an optical microsystem; a typical size might be $300 \times 300 \times 300\ \mu m^3$.

Ion-Implanted

Optical waveguiding in a stripe laser may also be achieved through ion implantation rather than etching of the cladding layer. A shown in Fig. 4.7, the laser stripe is defined by implantation of impurities into the regions outside the stripe. Implantation of oxygen, for example, may lead to the formation of oxide complexes, which effectively destroy the laser active layer and its waveguiding properties. Alternatively, high-energy implantation of H^+ ions or heavier atoms may generate sufficient crystalline damage outside the laser stripe, restricting lasing and light confinement to the unimplanted region.

The ion-implanted laser features a planar laser surface; save for the top-side contact metallization, there is no surface topography. A planar surface is advantageous for subsequent processing, since the metal contacts do not need to be aligned to a

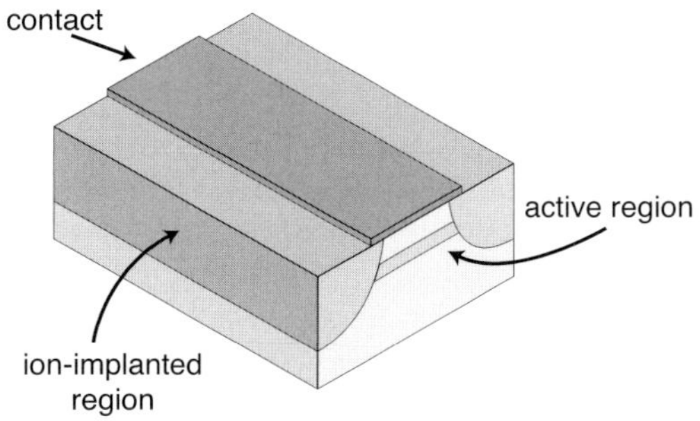

Fig. 4.7. Stripe laser defined by ion implantation into the cladding and active regions

narrow etched ridge. For the end user, a planar surface also implies a somewhat greater ease in chip packaging and mounting: when hybridly integrated into an optical microsystem, wire bonding is eased due to the larger surface contact and flip-chip mounting, for example, is facilitated since the laser does not need to be balanced on one or more etched ridges.

The ion-implanted laser structure generally has a larger mode size than the etched ridge waveguide laser, since lateral optical confinement is smaller. By extension, single-mode operation is easier to achieve, even with wider laser stripes. As a result, the near-field modal behavior of this type of laser is often favorable when used in applications which require laser-fiber coupling.

Gain-Guided

The gain-guided laser represents perhaps the simplest type of semiconductor stripe laser. The gain-guided structure, shown in Fig. 4.8, does not use a fabricated physical structure to achieve lateral guiding but instead the change in optical properties of the laser layers when subject to strong optical pumping. The laser structure consists of a patterned metallization stripe on the surface, whose dimensions are typically defined by an insulator layer deposited on the chip surface; the back-side contact is a large-area metallization, as with all the laser structures under consideration.

When subject to high current injection, the region under the top-side contact is pumped, reaches inversion and thus exhibits optical gain. Despite some current spreading, the regions of the laser active layer not directly under the contact stripe are not pumped and are thus absorbing. The real refractive index in the pumped region is increased with respect to the unpumped region, resulting in guiding of the generated light. Thus the gain region is concomitantly the waveguide region and small, but sufficient, lateral confinement is achieved. The "waveguide" thus only exists when the laser lases.

Due to their fabrication simplicity, gain guided lasers are attractive for high-volume applications; indeed, most of 10^8 semiconductor stripe lasers sold annually for

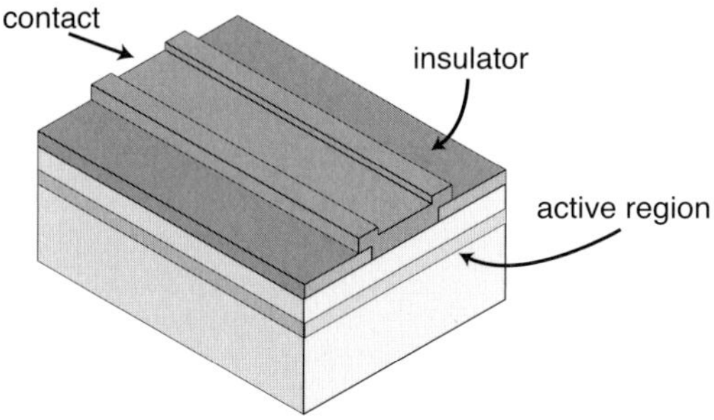

Fig. 4.8. A gain-guided stripe laser for which the lasing region is defined by the top-side metallization

optical data storage applications are of this type. Due to the weak lateral guiding, gain-guided lasers are almost always single-lateral-mode, even for relatively wide laser stripes. The lateral "waveguide" width is a function of the lateral distribution of injected current, which typically spreads to an area considerably wider than the contact width. As a result, the width of the guiding region and thus the mode width are somewhat poorly defined and may change with injection conditions. In general, we may state that the mode size is larger than that of ridge waveguide or ion-implanted lasers. In addition, modal instabilities may occur due to changes in the induced waveguide structure with pump current or temperature; hole-burning effects may also limit the mode output quality, as these tend to be more serious in gain-guided structures.

Buried Heterostructure

Telecommunications lasers are typically subject to heightened demands on mode size and stability and thus require a laterally well defined laser active region. At the expense of considerable process complexity, the buried heterostructure laser has established itself as a useful device configuration for InP-based long-wavelength lasers for this application (Matsumoto et al. 1994). Shown schematically in Fig. 4.9, the buried heterostructure laser employs an etched waveguide stripe which is subsequently buried by epitaxial regrowth of material to ultimately yield a planar surface. In contrast to the etched ridge waveguide laser, the etched stripe is in this case etched through the laser core layers; the etched surface is then covered by the regrown crystalline materials.

Mode confinement in the buried-heterostructure laser is based on strong confinement due to the index step between the laser heterostructure active layers and the regrown layers. The latter are typically doped oppositely to the original heterostructure lasers, so that a pn diode forms between the two. This pn diode is reversed biased during laser operation, so that injected current is tightly confined to the ridge region, thereby improving overall laser efficiency. Since the regrown materials are similar to those of the heterostructure, the refractive index step between the ridge

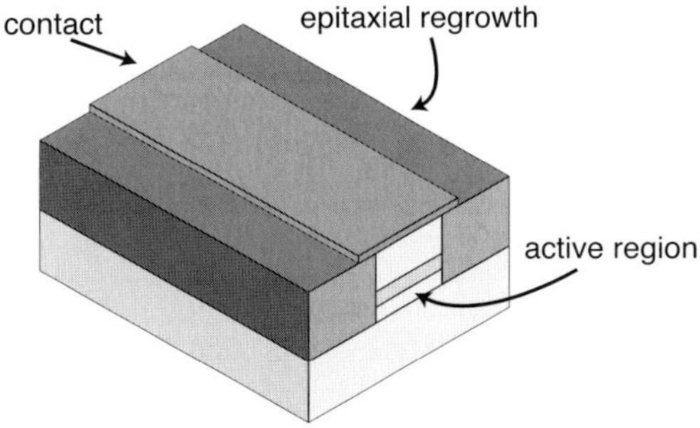

Fig. 4.9. The buried heterostructure stripe laser used for InP-based telecommunications lasers

and the lateral cladding layers is relatively small, resulting in good confinement but easily achieved single-lateral-mode behavior. Stable current injection conditions then combine with this latter characteristic to yield stable lateral modes under varying device operating conditions and a mode size well matched to optical fibers. The latter characteristics are essential for most telecommunications applications.

4.3.2 Operating Principle

As for the broad-area lasers, the Fabry–Perot stripe lasers operate according to the basic laser principles outlined in Sect. 2.6: current pumping leads to stimulated and amplified light emission at a wavelength defined by the bandgap of the material in the laser active layers, which leads to laser oscillation as the optical field reflects from two parallel mirrors which form the Fabry–Perot cavity. The optical field is confined vertically by the index steps of the active region heterostructure and laterally by the structural (or in the case of the gain-guided laser, induced) definition of the laser stripe. The cleaved mirrors reflect to first order independently of wavelength and a combination of the peak wavelength of the gain spectrum and resonant wavelength of the cavity determines the longitudinal laser mode. Lateral mode width and structure is defined by the width and nature of lateral guiding where hole burning and temperature effects are also of relevance.

4.3.3 Fabrication Issues

More than the LEDs or broad-area lasers discussed above, the stripe lasers rely strongly on standard (and sometimes non-standard) microfabrication technology. Many of the processing techniques developed for microelectronics fabrication and some of the specialized techniques resulting from advances in MEMS technology may be applied to the fabrication of stripe lasers. Photolithography, dry and wet etching, thin-film deposition and patterning are amongst the processes on which mass-fabrication of these lasers rely.

Fabrication complexity for stripe lasers varies strongly with structure, as we saw in the selected examples in Sect. 4.3.1. The simplest structures, the gain-guided and ion implanted lasers, require only patterned metallization and, in the case of the latter device, selective implantation. As with all these laser structures, the back-side metallization is an unpatterned large-area contact; the top-side metallization is patterned in correlation with the desired stripe structure, frequently using a lift-off process. The metallization systems are typically multi-layer alloys and the composition employed depends on the surface material and doping polarity: for GaAs, n-type contacts may use NiGeAu and p-type contacts TiPtAu.

The ridge waveguide laser uses an etched stripe to define the laser laterally. The stripe is dry or wet etched, where the former process may be more controlled and yield better sidewalls and the latter lower surface damage. The etch step is generally followed by deposition of surface passivation, an electrically insulating layer which also prevents surface degradation due to reaction of the semiconductor with environmental factors such as humidity and oxygen; Si_3N_4 or SiON are typically used. Contact areas are subsequently opened on the ridge surface and the top-side metal is deposited and patterned. The narrow typical ridge widths, $2 - 5$ μm, coupled with the surface topography resulting from the ridge etch, may limit the accuracy with which the last two steps are accomplished.

The Fabry–Perot cavity is formed by crystalline facets, normal to the plane of the wafer and normal to the laser stripe. In the majority of cases, a cleaving step is used to generate the facet, while at the same time separating the lasers into bars. Cleaving involves making an incision into the wafer surface using a diamond scribe and subsequently breaking the wafer at that point; the material cleaves along a crystalline plane, yielding a surface of excellent planarity and with roughness on the atomic level. Surface coatings for protection of the laser facet or, as suggested above, for augmenting the reflectance are often subsequently applied immediately after cleaving to prevent facet degradation.

Some applications, such as lasers integrated with other optical functions on a single chip or lasers fabricated from certain materials, require an etched facet. It is difficult to reach the quality of cleaved facets using an etch process: etch facet verticality, roughness and surface damage are all factors which adversely affect laser operation. Dry etch processes based on chemically-assisted ion beam etching (CAIBE) have been shown to yield adequate results, yet these processes are not as yet standard in most fabrication laboratories.

The generation of laser chips from the cleaved bars may also be accomplished through cleaving, whereas sawing is an industrially more viable alternative. The individual chips, with typical size $300 \times 300 \times 300$ μm^3, may then be used as components in a hybrid optical microsystem. More macroscopic applications may employ the lasers mounted in standard packages, typically either TO headers or on open submounts for specialized applications. In a TO package, the laser is mounted on a support normal to the emission direction and wire-bonded. Emission from the rear facet is typically directed onto a monitor photodiode and the entire submount is capped and filled with an inert gas, typically Ar.

4.3.4 Microsystem Aspects

Since Fabry–Perot stripe lasers are manufactured in large quantities with a wide variety of physical structures, emission wavelengths and power levels and thereby exhibit the attractive features of all laser diodes, such as small size and high efficiency, they are usually the light source of choice for incorporation into high-performance optical microsystems. Available as chips commercially, laser diodes may be incorporated into hybrid micro-optical systems using mounting techniques which we will discuss in Chapter 7; flip-chip is perhaps the most popular and advanced of these, and the process is available industrially. Microsystem designers with access to laser diode fabrication facilities may further design the physical structure of the diode to facilitate integration into a hybrid system, where components such as physical alignment features, specialized bond-pad configurations and customized chip sizes may be of utility in the optimization of assembly techniques and subsequent system performance.

For use in a microsystem, the stripe laser has the advantage of a small mode size, which often permits efficient coupling into waveguide or optical fiber structures. By using a butt-coupling approach, in which the optical component is essentially placed in the near-field of the laser diode emission, mode matching is frequently sufficiently good to allow adequate coupling efficiencies, whereas specialized laser diodes with mode expanders have been demonstrated which further optimize this function.

The astigmatic nature of the beam renders coupling into free-space micro-optical systems more difficult. Optimized beam collimation can usually only be achieved through the use of multiple, asymmetric lenses but adequate results may be generated using single micro-lens systems. As most free-space optical microsystems are of limited physical size, the effects of the asymmetric beam profile may often be less dramatic than in macroscopic systems.

Stripe lasers, however, have their price. Compared to LEDs or, in some cases, even miniaturized incandescent light bulbs, the cost of using even an inexpensive stripe laser in an optical microsystem is higher. For high-performance applications, this may not be an issue, but particularly low-cost optical sensor microsystems have stringent cost restrictions. As a result, laser diodes are typically used only if the spectral or optical intensity demands warrant their use. The only reason that the optical microsystem used for reading the CD or DVD surface is so cheap, while incorporating a laser diode, is that it is sold in quantities of 100 million, a figure that few other microsystem applications are likely to match.

Example 4.2

Assignment: *Given a 1 μm wide laser waveguide, emitting at* $\lambda_0 = 780$ *nm, compare the number of supported lateral modes in an etched ridge structure with the number supported in a buried heterostructure. Assume the ridge to be GaAs and the material for epitaxial regrowth* $Al_{0.1}Ga_{0.9}As$*.*

Solution: *We may approximate the number of supported modes by analyzing the waveguide in one dimension (laterally) using the characteristic waveguide equation (Eq. 3.71) in the same manner as we did in Example 3.4 of Sect. 3.4.1. We solve the characteristic equation repeatedly using our favorite numerical software, incrementing the index m to a maximum value* m_{max} *until no further solution is found; the total number of modes supported is then* $m_{max} + 1$*.*

Using the refractive index values for GaAs (3.59) for n_g *and air (1.0) for* $n_c = n_s$*, and recalling that the wavelength is that in the material* (λ_0 / n_g)*, we find that Eq. 3.71 has solutions up to* $m_{max} = 8$*, so that 9 modes are expected to be possible in the 1 μm wide etched ridge.*

The buried heterostructure has the GaAs ridge surrounded by $Al_{0.1}Ga_{0.9}As$ *with refractive index* $n_c = n_s = 3.52$ *found from Eq. 3.62. The index step is thus* $\Delta n = 0.07$*, and Eq. 3.71 may be solved up to* $m_{max} = 1$*, implying 2 supported modes. It is thus clear that the weaker lateral guiding which results from the reduced refractive index step in the buried heterostructure when compared to the etched ridge effectively reduces the number of possible lateral modes.*

4.4 Distributed Feedback Lasers

The lasers considered in the previous sections had optical cavities defined by a Fabry–Perot resonator. If the planar parallel mirrors are replaced by a periodic grating structure, the resonance of this grating at a particular wavelength may also be used to generate feedback into the laser cavity. The distributed feedback laser which results has numerous advantages over the Fabry–Perot laser, particularly since the wavelength-selective grating mirror stabilizes the emission wavelength and more easily leads to single-mode operation. These spectral features are important performance aspects for optical telecommunications systems and distributed feedback lasers are indeed the optical workhorse for this application.

4.4.1 Structure

Distributed feedback lasers use the same basic waveguide stripe structure as the Fabry–Perot lasers of Sect. 4.3. The lateral guiding mechanisms are the same, where the buried-heterostructure concept is the one most frequently used for telecommunications lasers due primarily to the materials employed to reach the desired wavelengths. The cleaved or etched facet mirrors are replaced by a grating structure etched into or close to the waveguide; Bragg reflection of the optical field from this grating generates optical feedback into the laser cavity, as we will see in more detail below (Kogelnik and Shank 1972; Nakamura et al. 1973). Since reflection from the grating occurs only at a single wavelength, we may regard this structure as a wavelength-selective mirror.

Two basic types of lasers may be fabricated using this principle: the distributed Bragg reflector, or DBR, laser with grating mirrors outside the laser cavity and the distributed feedback, or DFB, laser, for which the cavity and grating structure are concomitant. The nomenclature for the latter is somewhat confusing, since "distributed feedback lasers" refers to the entire class of grating-based lasers as well as this particular structure.

The DBR Laser

The schematic drawing of Fig. 4.10 shows the essential structure of the distributed Bragg reflector, or DBR, laser. This laser structure shares a number of features with its Fabry–Perot cousins: the epitaxial layers are typically a double heterostructure, with a thin active layer and vertical guiding region. Lateral guiding in this example is achieved by an etched ridge waveguide and laser emission is again from a crystalline facet which is typically anti-reflection coated so as to suppress any Fresnel reflection and thus mirror action.

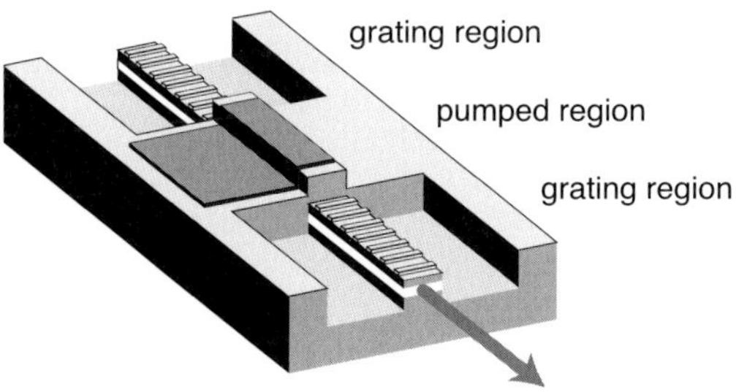

Fig. 4.10. Schematic representation of a distributed Bragg reflector (DBR) laser. The grating and pumped regions are separated

The DBR laser in this case has three sections: a center pumped region, surrounded on either side by reflecting grating regions. The pumped region has a contact to the upper layers of the laser pn diode and it is here that current is injected, light is generated and amplified. In the reflector regions, a grating is etched onto the surface of the waveguide, typically close to but not into the laser active layer. The optical field in the waveguide thus sees a periodic variation in refractive index which gives rise to the reflection of a portion of the optical field, based on the mechanism we will discuss in Sect. 4.4.2. The grating is typically placed close to the laser active region to allow good coupling, such that a large etched recess as shown in the figure is often of utility. The grating region may be buried by a dielectric or through epitaxial re-growth.

Typical DBR laser dimensions vary according to the desired output characteristics, primarily output power. The pumped gain section is typically 200 – 500 μm long and the grating reflector sections 50 – 200 μm. The length of at least one of the grating sections must be short enough to provide sufficient reflectance to allow laser resonance but enough transmission to permit emission of a portion of the generated optical field. The length of grating required depends strongly on the coupling coefficient between the optical mode and the grating (see Sect. 4.4.2) and varies significantly between laser and grating designs. Waveguide width and active layer thickness are comparable to those for standard Fabry–Perot lasers.

The DFB Laser

The DFB laser combines the pumped and grating regions. In contrast to the DBR laser, the grating is placed directly into the pumped area and feedback is generated throughout the laser. Thus the DFB laser does not have a discrete "mirror" but a laser cavity which generates optical feedback throughout its length.

The structure is shown in the schematic view of Fig. 4.11. The grating is again placed on the waveguide, close to the active region, but exists throughout the device length. The corrugation structure is fabricated after the active layer heterostructure is epitaxially grown; in subsequent epitaxy steps, the remaining upper cladding layers are grown, such that the grating is buried in the device. The same considerations concerning grating shape, position and coupling efficiency as for the DBR laser apply here to the DFB laser.

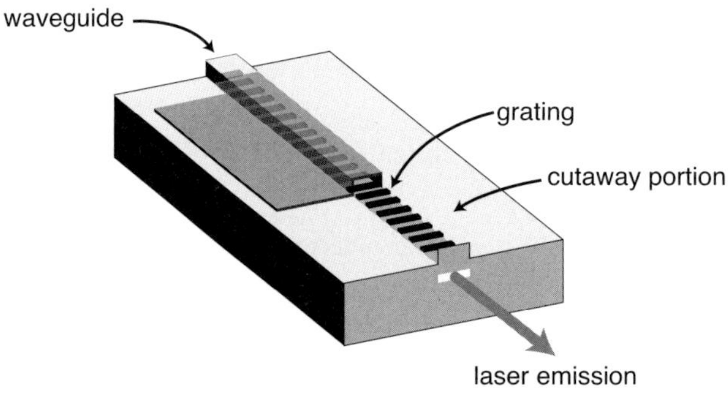

Fig. 4.11. Schematic representation of a distributed feedback (DFB) laser. The pumped and grating regions are concomitant such that the grating is buried inside the semiconductor. A cutaway view shows the grating in more detail

The overall laser dimensions are comparable to those of the DBR laser, certainly with regard to waveguide width and epitaxial layer thicknesses. Laser length is typically on the order of 300 to 600 μm. Anti-reflection coatings are frequently required at the laser facets, as parasitic reflection may unduly affect the modal structure and stability of the laser.

Tunable Lasers

There is considerable interest in the fabrication of tunable distributed feedback lasers, allowing emission at a user-specified wavelength within a broad band; telecommunications and sensor applications would benefit greatly from availability of such components (Kobayashi and Mito 1988). Whereas tunable lasers have been demonstrated to have attractive optical properties, the added complexity involved in their manufacture (Hillmer et al. 1995; Delorme et al. 1997) leads to relatively expensive fabrication processes.

Most of the tunable distributed feedback lasers are three- or four-section DBR lasers: the gain section is followed by a phase-matching section and subsequently the grating sections. Both the phase-matching and grating sections are contacted, such that the electric field generated by an applied voltage gives rise to a change in the material refractive index due to the electro-optic effect (Okuda and Onaka 1977). The grating section often has a variable grating period or a continuous chirp (linear variation in Λ) which permits access to a range of electrically-tunable wavelengths (Hillmer et al. 1995). Continuous tuning ranges of over 40 nm have been demonstrated (Öberg et al. 1995) but the fabrication complexity has limited the industrial applicability of these lasers. Much current work thus focusses on simplifying tunable laser processing to yield affordable devices (Schreiner et al. 2001) and the use of sampled gratings (Majewski et al. 2002).

4.4.2 Operating Principle

The family of distributed feedback lasers uses Bragg reflection from periodic grating structures to generate feedback into the optical cavity. The generation and amplification of photons is based on the same mechanisms as in Fabry–Perot lasers, thus only the establishment of resonance in the cavity uses a different physical phenomenon. To understand the nature of resonance in a DFB structure, it is necessary to examine both reflection from a periodic structure as well as the coupling between forward and backward traveling modes which results from the interaction of these modes with the grating.

Bragg Reflection

Distributed feedback lasers are based on grating structures, which have numerous uses in a variety of optical applications (Hutley 1982). Gratings are periodically repeating structures with spatial periods on the order of the wavelength; reflections from each of the grating periods may interfere with one another giving rise to useful spatial intensity distributions based on interference, in general in three dimensions. These phenomena are highly wavelength-dependent since the phase matching required for interference is dependent on the relationship between grating period and excitation wavelength.

For the case of distributed feedback lasers, the grating analysis is generally one-dimensional. For a stripe waveguide with effective refractive index N (see Eq. 3.73) and a periodic refractive index variation due to a grating with period Λ, an optical field propagating through the waveguide will be reflected back into the direction in which it came (reflection angle π) given that the wavelength is related to the structural parameters by

$$\lambda = \frac{2N\Lambda}{m} . \tag{4.1}$$

In this expression, m is an integer number representing the grating order; for most DFB laser gratings, $m \le 3$. The above relation expresses the so-called Bragg condition, named for William Bragg (1890–1971), in which an interference resonance gives rise to a counter-propagating optical field.

If we recall from Eq. 3.72 that the propagation constant for an optical field in a waveguide, $k_z \equiv \beta$, is given by

$$k_z \equiv \beta = n_g k_0 \sin\theta = N k_0 = N\frac{2\pi}{\lambda_0} , \tag{4.2}$$

where N is the effective index of the waveguide mode, we may re-write the Bragg condition as

$$\beta_0 = \frac{m\pi}{\Lambda} . \tag{4.3}$$

For this value of z-directed propagation constant, the resonance condition is fulfilled and interference results in the generation of both forward and backward traveling modes.

The strength of the reflection, and thus ultimately the reflectance of the grating mirror, is a function of numerous parameters. The essential considerations are the overlap of the optical field with the grating and the grating shape; these are absorbed into a coupling coefficient κ [cm^{-1}] which defines how strongly the optical field propagating in the waveguide is affected by the grating. Grating corrugation depth, spacing above the waveguide core, tooth shape, index step, modal effective index and grating order all come into a calculation of κ. Typical values, for a square-wave corrugation of 100 nm depth, are $\kappa \sim 200$ cm^{-1} for a first order grating, $m = 1$, and $\kappa \sim 100$ cm^{-1} for a third order grating, $m = 3$. Detailed calculations are complicated (Agrawal and Dutta, 1986) but necessary for accurate determination of coupling efficiency.

The grating coupling coefficient directly allows the definition of the grating reflectivity. For a DBR laser, we may relate the reflectivity r to κ as (Coldren and Corzine 1995, Chap. 3.5)

$$r \approx \tanh(\kappa L_g) \tag{4.4}$$

for grating length L_g, which for small values of κL_g leads to $r \approx \kappa L_g$.

Coupled Modes[1]

For propagation of a mode close to the resonance condition, such that $\beta \sim \beta_0$, an initially forward-propagating mode will couple with the grating and give rise to a mode propagating backwards (Yariv 1973). As a result, two coupled counter-propagating modes exist in the laser cavity giving rise to optical feedback. If the modal propagation constant β is de-tuned from the resonance condition by an amount $\Delta\beta$, given as

$$\Delta\beta = \beta - \beta_0 \tag{4.5}$$

we may then write the total propagating electric field in the waveguide as the sum of forward and backward components. Explicitly, we have for the total field $\boldsymbol{E}_z$

$$\boldsymbol{E}_z = \boldsymbol{E}_F(z)e^{j\beta_0 z} + \boldsymbol{E}_B(z)e^{-j\beta_0 z} \tag{4.6}$$

where

$$\boldsymbol{E}_F(z) = \boldsymbol{E}_{F0}e^{j\Delta\beta z} \tag{4.7}$$

and

$$\boldsymbol{E}_B(z) = \boldsymbol{E}_{B0}e^{-j\Delta\beta z} . \tag{4.8}$$

The latter two fields, the forward and backward propagating components of the total field, may be related by the following coupled mode equations:

$$\frac{d}{dz}\boldsymbol{E}_F(z) = j\Delta\beta\boldsymbol{E}_F(z) + j\kappa\boldsymbol{E}_B(z) \tag{4.9}$$

$$\frac{d}{dz}\boldsymbol{E}_B(z) = -j\Delta\beta\boldsymbol{E}_B(z) - j\kappa\boldsymbol{E}_F(z) . \tag{4.10}$$

From the previous expressions, we see that the coupling between the two components is through the grating coupling coefficient κ. We may write the general solutions to Eq. 4.9 and Eq. 4.10 as

$$\boldsymbol{E}_F(z) = \boldsymbol{E}_{F1}e^{j\beta' z} + \boldsymbol{E}_{F2}e^{-j\beta' z} \tag{4.11}$$

and

---

[1] This treatment follows that of (Agrawal and Dutta, 1986, Chap. 7.3)

$$\boldsymbol{E}_B(z) = \boldsymbol{E}_{B1}e^{j\beta' z} + \boldsymbol{E}_{B2}e^{-j\beta' z}\,, \tag{4.12}$$

where the parameter β' is defined as the difference between the de-tuning and the grating coupling coefficient, namely

$$\beta' = \pm\sqrt{\Delta\beta^2 - \kappa^2}\,. \tag{4.13}$$

From this last expression, we may derive the reflection coefficient for the DFB, $r(\beta')$, as

$$r(\beta') = -\frac{\kappa}{\beta' + \Delta\beta}\,. \tag{4.14}$$

We see from this definition that, as we would expect, the reflection coefficient is directly related to the coupling coefficient, and that $r(\beta') = 0$ if $\kappa = 0$. Equation 4.14 also allows us to determine the sign to choose in Eq. 4.13: we take the sign such that the reflection coefficient $|r(\beta')| \leq 1$ (Agrawal and Dutta, 1986, Chap. 7.3.1).

Using the reflection coefficient, may then write the forward and backward propagating fields in the DFB structure in the following useful form:

$$\boldsymbol{E}_F(z) = \boldsymbol{E}_{F1}e^{j\beta' z} + r(\beta')\boldsymbol{E}_{B2}e^{-j\beta' z} \tag{4.15}$$

and

$$\boldsymbol{E}_B(z) = \boldsymbol{E}_{B2}e^{-j\beta' z} + r(\beta')\boldsymbol{E}_{F1}e^{j\beta' z}\,. \tag{4.16}$$

The remaining unknowns, $\boldsymbol{E}_{F1}$, $\boldsymbol{E}_{B2}$ and $r(\beta')$ are determined from the boundary conditions at the end of the laser, such that the facets play a role in the calculation.

In summary, we see from solution of the coupled mode equations that the presence of a grating in the DFB laser cavity leads to counter-propagating modes; Equation 4.6, combined with Eq. 4.15 and Eq. 4.16, gives the total electric field in the cavity, in which both forward and backward propagating modes exist and are coupled to one another through the grating. The cavity itself thus acts as a "continuous reflector", feeding a portion of the generated optical field back into itself. Whereas grating structures in real DFB lasers tend to have more complex features, such as a $\lambda/4$ shift to remove a degeneracy in the lasing modes and assure strictly single-mode operation, the previous analysis allows a rudimentary understanding of how feedback in a DFB laser is achieved.

4.4.3 Fabrication Issues

Distributed feedback lasers, both the DBR and the DFB structure, are considerably more complex to fabricate than run-of-the-mill Fabry–Perot stripe lasers. Most of this added complexity is due to the grating mirrors: the grating shape, position, and period must be accurately determined to yield a laser with the desired output characteristics. In addition, the DFB laser structure requires epitaxial regrowth above the etched grating to complete the upper laser layers and the DBR frequently needs to employ some mechanism for making the grating waveguide regions transparent at the laser wavelength. These process aspects combine to make the fabrication of distributed feedback lasers an expensive prospect.

DFB Fabrication

Fabrication of a DFB laser, which has established itself as the key component for optical telecommunications, begins with epitaxial growth of lower heterostructure layers, up to and including the laser active region and perhaps a portion of the upper cladding. The heterostructure dimensions are similar to those used in Fabry–Perot lasers, with an active layer thickness of several 100 nm. Following etch of a stripe waveguide, the grating is etched into the layers just above the laser core: the spacing between the grating and the core as well as the depth of the grating features themselves are essential factors in determination of the coupling coefficient, κ. Whereas it is also possible to fabricate the grating below the core (such that the epitaxial growth of the core layers is done after grating etch rather than before), completing the core first allows accurate determination of the waveguide effective index, N; from Eq. 4.2, we saw that this factor determines the Bragg condition and thus the laser wavelength.

Grating definition needs to be accomplished with high accuracy and is thus usually a critical process step. Definition in photoresist is either through direct writing using electron-beam or through flood-exposure in a holographic exposure system; the former allows greater flexibility in determining the grating period, which may be varied between devices, at the expense of low throughput. Holographic exposure, while a parallel process, yields the same grating period over the entire wafer and changing periods between wafers requires re-calibration of the system.

Exposure in photoresist only defines the period of the grating; typical periods for first-order gratings ($m = 1$) range from about 100 nm, for the 800 nm wavelength range, to about 230 nm, for the telecommunications wavelengths around 1.55 μm. Higher order gratings have correspondingly longer periods: third order for an 800 nm laser would require a period of only about 300 nm, much easier to define holographically than the 100 nm required if a first-order grating were used.

The grating shape, depth and position above the active region are defined by the subsequent development and etch process, which may be quite refined and difficult to control. A dry-etch process is frequently used, with intermediate steps to allow transfer of a desired grating tooth shape into the semiconductor. The result is then typically a grating structure with a depth on the order of 100 nm, at a spacing of 50 – 100 nm from the laser core.

Processing of the DFB laser then continuous with one or more further epitaxy steps which bury the grating and complete the upper cladding and contact layers of the laser. Essential for successful completion of this step is cleaning of the surface, including the etched grating, since impurities on the surface will result in poor quality of the epitaxially grown films. Multi-step epitaxy is thus anything but trivial, yet has become standard fabrication technology for DFB lasers. The resulting structure is then typically the buried heterostructure of Fig. 4.9, including the grating as shown in Fig. 4.11; the surface is typically planar, facilitating top-surface contact metallization.

DBR Fabrication

Manufacture of a DBR laser places considerably fewer demands on epitaxy but gives rise to a few other technological difficulties which will keep the process engineer on her toes. Epitaxy is typically completed in a single step, yielding the same laser heterostructure as we have employed previously; a possible variation might be the inclusion of etch-stop layers to accurately define the grating depth and position. A grating region is then defined on either side of the laser pumped region; this step may occur before or after the laser waveguide is defined, depending on the process. The example in Fig. 4.10 shows the grating in a large recessed region outside the active layer; this option is one approach which allows uniform grating exposure and accurate vertical placement (Hofstetter et al. 1994) but a variety of other configurations has been demonstrated. Grating definition uses the same techniques as described for the DFB laser, and the same considerations with respect to period, shape, depth and position, apply here.

The difficulty in DBR grating fabrication arises when considering the effects of optical absorption in the un-pumped grating section. Unlike for the DFB laser, the DBR laser grating mirrors are attached to the pumped region as purely passive waveguide sections. Since this waveguide is fabricated from the same heterostructure layers as the pumped region, the passive sections exhibit strong optical absorption at the same wavelength at which the pumped region emits. As a result, the grating mirror sections reflect but also absorb the laser emission, significantly limiting performance.

A number of ways for getting around this problem have been proposed and demonstrated. Conceptually simple but technologically challenging is the replacement of the absorbing waveguide by one transparent at the laser wavelength. The heterostructure layers are etched away and a set of transparent waveguide layers, frequently a dielectric sandwich, is deposited in their place; the grating is then defined in this new waveguide. Serious difficulties arise, however, in the coupling of the original waveguide (still present in the active region) to the new grating waveguide, as an accurate vertical layer placement and a low-loss transition region are difficult to assure in fabrication.

An alternative is to selectively alter the composition the waveguide to make it transparent at the laser emission wavelength. Key to the success of such a process is the quantum well; as we saw in Sect. 3.3.4, the width of the quantum well defines the transition energy and thus the emission and absorption wavelength of photons. Whereas the quantum well width is defined during epitaxial growth of the laser active region, anneal processes may be used to alter the shape of the quantum well after growth such that the quantum well ground states are raised in energy. The extent of this diffusion process may be controlled by conditions on the wafer surface: certain surface dielectrics, through generation and diffusion of vacancies at high temperatures, result in enhanced quantum well diffusion in the heterostructure under the surface whereas others inhibit the process. Thus a spatially-selective diffusion process, known as vacancy-enhanced disordering or quantum-well intermixing (Ooi et al. 1997; Hofstetter, Maisenhölder and Zappe 1998), may be used to blue-shift the absorption wavelength in certain defined regions of the wafer while keeping other regions unshifted. The result is a waveguide section which is transparent at the emission wavelength and it is in this section that the grating is fabricated. This technology is mature and is applied industrially to DBR laser and photonic integrated circuit fabrication, as we will see in Sect. 7.3.1.

4.4.4 Microsystem Aspects

We have seen that distributed feedback lasers have desirable spectral properties but are complex and expensive to manufacture. For use in microsystems, the same assembly, mounting and mode-matching considerations we outlined for Fabry–Perot lasers apply; the laser chip size, mode size and shape and power requirements are similar. Optical communications systems, insofar as they require the highly-tuned spectral properties, will employ distributed feedback lasers in compact, hybridly-integrated modules for data transmission. The assembly is typically based on a metal sub-mount, on which the laser is fixed using conductive epoxies for electrical and thermal contact. The remainder of the module often consists of an optical isolator, collimation lens and fiber pigtail; the latter component is often positioned using active alignment techniques. Such assemblies are relatively large insofar as microsystems are concerned, but high-accuracy positioning and good long-term stability are required for industrial applications. The assembly technology must be capable of being automated so as to allow reasonable fabrication costs.

Distributed feedback lasers do play a unique role in monolithically-integrated optical microsystems, which we will discuss in Sect. 7.3. Monolithic systems combine all optical functions, including the laser as light source, on a single chip; the grating-based laser mirrors of the DBR or DFB laser allow the monolithic attachment of a further waveguide circuit including other passive or active optical components. Since the Fabry–Perot laser requires a cleaving step to generate the mirrors, the resultant separate laser chip precludes the manufacture of a monolithic system. Such monolithic integrated optical systems represent a special type of optical microsystem, with advantages in system size and robustness at the expense of complex technology, price and, occasionally, performance.

Example 4.3

Assignment: *Determine the period of a first-order DBR grating for a double heterostructure-based GaAs/AlGaAs laser consisting of a 340 nm thick core with* $x = 0.30$*, an upper cladding with* $x = 0.85$ *and a lower cladding with* $x = 0.75$*. Operating wavelength is* $\lambda = 806$ *nm. Repeat for a third-order version of the grating. Discuss how you would fabricate these.*

Solution: *The attentive reader may recall that this laser structure is the same one we considered from a waveguide-analysis point of view in Example 3.4 of Sect. 3.4.1. We found there that a core thickness of < 341 nm would yield a single-mode waveguide, such that our given structure is expected to be single transverse mode, essential for making a DBR laser out of it.*

To determine the grating period (using Eq. 4.1), we first need to find the waveguide effective index, N; thus a numerical solution of the characteristic equation (Eq. 3.71) will again be necessary. Determining the refractive indices of the material using the approximate relation of Eq. 3.62, yields $n_g = 3.39$*,* $n_c = 3.05$ *and* $n_s = 3.11$*. Including the core thickness and the wavelength in the material, the characteristic equation is solved to yield a waveguide effective index of* $N = 3.30$*.*

Rewriting Eq. 4.1 then yields the simple expression for the grating period

$$\Lambda = \frac{m\lambda_0}{2N}$$

for grating order m. Plugging in our values for wavelength and effective index, we than have, for a first order grating $\Lambda = 122$ *nm and for third order,* $\Lambda = 366$ *nm.*

Fabrication of a first order grating with a period on the order of 100 nm can practically only be done using electron-beam lithography; holographic techniques could be used for the third order grating.

4.5 Vertical Cavity Lasers

The broad-area, Fabry–Perot and distributed feedback lasers are all so-called edge-emitting lasers: laser emission is ultimately from a facet normal to the wafer surface and the laser cavity is thus defined in the plane of the substrate. The vertical cavity laser (VCL) or vertical cavity surface emitting laser (VCSEL) as it is commonly and redundantly known, turns the cavity 90° and emits from the wafer surface (Wilmsen, Temkin and Coldren 1999). The laser resonator is defined by Bragg mirrors, albeit formed by epitaxially-grown multi-layer structures rather than etched features. The laser waveguide is defined by etching or implantation and may, in contrast to edge-emitting lasers, be circularly symmetric and thus emit a symmetric Gaussian mode.

The VCL is a relative newcomer to the semiconductor laser scene: its structure was first proposed in 1979 (Soda et al. 1979); first pulsed demonstration was in 1984 (Iga et al. 1984); and the first CW version was not demonstrated until 1988 (Koyama, Kinoshita and Iga 1988). However, due to its small size, very low threshold currents, excellent near-field properties and the ability to fabricate closely packed one- or two-dimensional arrays, the VCL is becoming increasingly attractive for use in hybrid optical microsystems. Coupling to fibers, waveguides or free-space optical components is considerably easier than for edge-emitting lasers and the potentially small chip size has advantages for integration in highly-compact optical systems. Cost may also become an issue: the fabrication and testing process for VCLs is generally simpler than for other laser types and it is expected that mass-market prices may drop below those for comparable edge-emitters (Choquette and Hu 1997; Whitaker 1999).

4.5.1 Structure

The structure of the vertical cavity laser is quite different from edge emitting lasers (Jewell et al. 1991). The cavity is oriented normal to the wafer surface; feedback is generated by two Bragg mirrors formed by alternating epitaxial layers; the length of the laser active region is typically only one wavelength long; and lateral guiding is achieved by a waveguide structure formed normal to the wafer surface. The VCL laser structures are typically quite demanding from the epitaxy point of view but relatively simple when subsequent processing is considered.

Epitaxial Bragg Mirrors

The Bragg mirrors are based on epitaxially-grown multi-layers using two materials with differing refractive indices. Most VCLs begin with a lower Bragg mirror, epitaxially grown on an appropriate semiconductor substrate, consisting of 30 – 50 pairs of alternating material layers. The materials are chosen to have the maximum possible difference in refractive index at the wavelength of interest, thereby giving rise to the distributed Bragg reflection. The thicknesses of the individual layers are designed to be $\lambda/4$ for laser emission wavelength in the material $\lambda = \lambda_0/n$, given material refractive index n. The laser active region, often containing one or more quantum wells, is grown above this mirror at a thickness typically chosen to be λ or $3\lambda/2$. The structure continues with the top Bragg mirror, again alternating layers of material with thickness $\lambda/4$, typically 20 – 30 pairs. A thin semiconductor contact layer completes the structure. If an n-doped substrate is used, the lower mirror layers are n-doped, the laser cavity nominally undoped and the upper mirror layers p-doped.

Epitaxy of a VCL structure thus yields a pair of Bragg mirrors and a laser cavity over the entire wafer. Laser emission is vertical, upwards out of the surface but also into the substrate. If the substrate is transparent at the operating wavelength, the

VCL may be operated as bottom-emitting. Most of the variations in VCL structure result from different techniques to confine the optical field laterally, where a number have established themselves as technically and commercially viable.

Mesa

The mesa VCL, alternatively known as an air-post or etched-post VCL, uses an etch process to define the laser laterally (Choquette et al. 1991). This type of structure represents the oldest VCL configuration, although, in comparison with the other types discussed below, it has high optical loss, poor thermal dissipation and degradation concerns have been voiced (Takeshita et al. 2002). As is shown in the three-dimensional schematic sketch of Fig. 4.12, the laser waveguide is etched from the surface; the etched post is typically circular with a diameter which varies from about 5 μm to several tens of microns. Contact to the upper p-doped layers is frequently through a contact ring on the mesa surface: the opening in the ring then defines the laser emission aperture. The metal layers are directed down the sides of the etched mesa to allow the definition of a large area contact beside the laser post. Contact to the n-doped bottom layers is through a large-area contact on the substrate back-side. Current flow to the active region is through the upper and lower mirrors, such that the doping levels must be sufficiently high to allow reasonably low series resistance.

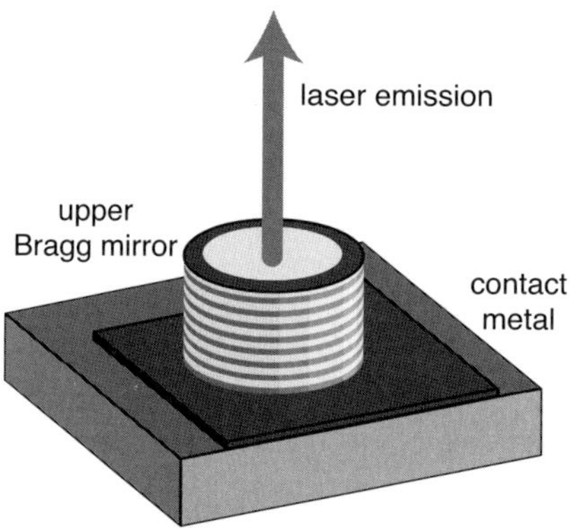

Fig. 4.12. Schematic three-dimensional sketch of a mesa VCL where the top Bragg mirror has been etched to just above the laser active region

The cross-sectional view of a mesa VCL in Fig. 4.13 shows clearly that the upper mirror has been completely removed outside the vertical waveguide region. The mesa etch is generally stopped above the active region, to avoid problems with etch-induced damage which adversely affects laser performance. The current flow and the optical field are thus both strongly confined in the VCL post. The current confinement is an advantage, as current spreading directly above the active region is minimal and the region of high pump intensity is laterally well-defined. Strong confinement of the optical field leads to good overlap between injected carrier and generated photons but also implies that multiple lateral modes are supported, even for small mesa radii.

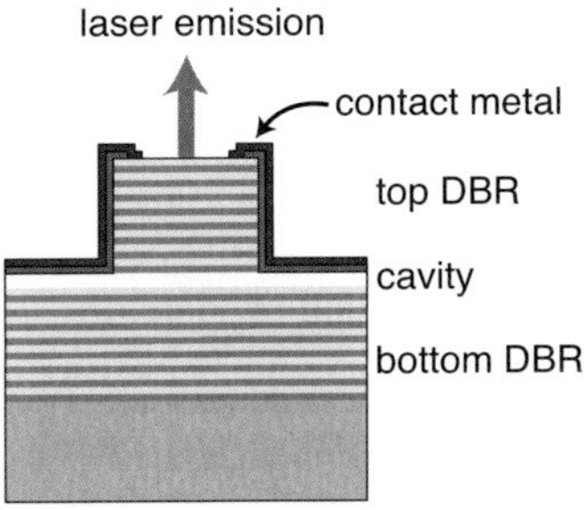

Fig. 4.13. Cross-sectional view of a mesa VCL

Implanted

Lateral confinement in a VCL may also be achieved through implantation outside of the laser waveguide (Lear et al. 1994; Morgan et al. 1995). Implantation of impurities or atomic species which leads to destruction of the Bragg mirrors as well as electrical conductivity is effective in defining the laser region. Proton implantation, for example, may lead to the structure shown in Fig. 4.14: the implanted region is electrically non-conducting, so that current injection is limited to the unimplanted region. As a result, current injection is laterally defined as is thus the pumped region. The lateral confinement of current flow leads to increased temperature at the center of the laser and thus thermal lensing (Tell et al. 1990); this latter effect leads to optical guiding of the generated light.

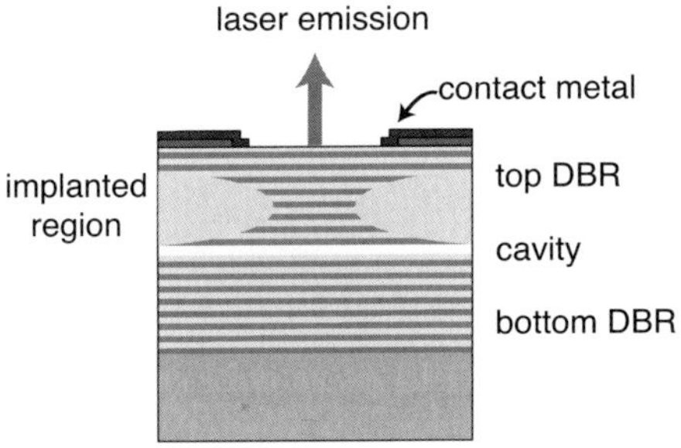

Fig. 4.14. Cross-sectional view of a proton- or ion-implanted VCL; the implanted region is electrically insulating

The implanted VCL is the one most popular for commercial manufacture. The weak optical confinement relaxes the dimensional restrictions for operation in a single lateral mode such that single-mode lasers with relatively large aperture may be fabricated; there is however some indication that new degradation mechanisms may limit lifetime (Herrick and Petroff 1998). In addition, the planar surface results in more flexibility in wiring, subsequent processing, integration of micro-optical structures and flip-chip mounting.

Buried Heterostructure

The strong optical confinement of the etched mesa VCL, making single-lateral-mode operation difficult, may be alleviated by refilling the etched region with a semiconductor, leading to the buried heterostructure VCL shown in Fig. 4.15. The VCL waveguide is etched in the same manner as for the mesa VCL; a second epitaxy step then results in refill of the etched region using a material with a bandgap higher than the materials of the laser post (Choquette et al. 1993). As a result, current confinement inside the laser is optimized, but the refractive index step between post and outer cladding layers is much smaller than for the mesa structure (in which air forms the outer cladding).

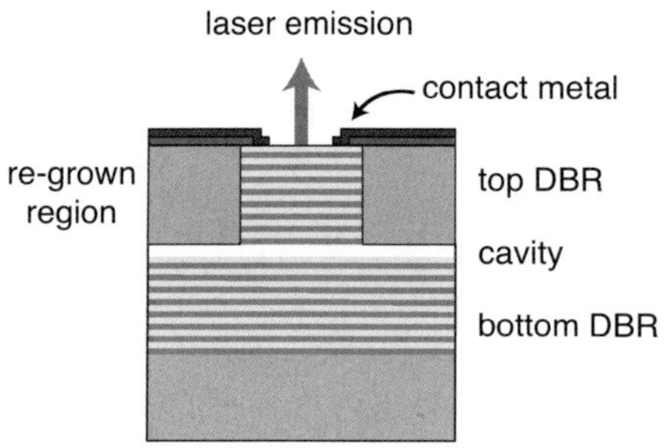

Fig. 4.15. Cross-sectional view of a buried heterostructure VCL. The laser waveguide is etched and refilled with a semiconductor of higher bandgap

The buried heterostructure VCL thus combines good efficiency with single mode operation for larger apertures. Following epitaxial regrowth, the laser surface is once again planar, simplifying contact metallization and subsequent assembly processes. The attractiveness of these features is somewhat offset by the technological effort required to perform the regrowth step.

Dielectric Aperture

One approach for improving VCL performance is the incorporation of a dielectric aperture into the laser cavity (MacDougal et al. 1995; Weigl et al. 1997). Such an aperture, buried near the active region of the laser, may serve to improve current confinement or optical confinement or both. As shown in Fig. 4.16, the aperture is a thin layer directly above the active region, the width of which defines the pumped region in the laser cavity. Usually a non-conducting dielectric layer, the aperture may be combined with a recess etch to allow contacting the n-region of the laser directly at, or close to, the active layer (Scott et al. 1994). Current thus does not need to be injected through the entire upper mirror, significantly decreasing the device resistance.

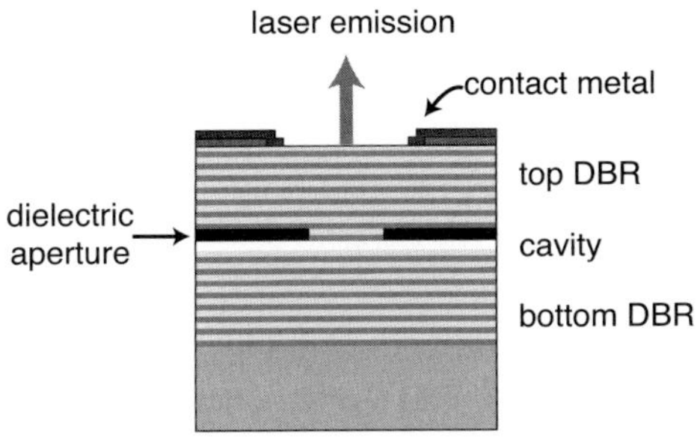

Fig. 4.16. Cross-sectional view of a dielectric aperture VCL. The aperture is placed close to the active region and may improve electrical and optical confinement

The dielectric aperture is typically formed by selective oxidation of a layer expressly grown into the mirror structure during epitaxy. AlGaAs layers with high Al concentration, for example, are easily thermally oxidized in a steam atmosphere; typically AlAs is used, which oxidizes to AlO_x, an electrically insulating material with smaller refractive index than the original AlAs. Oxidation takes place laterally: a large laser mesa is thus etched and the dielectric layer oxidized toward the center, yielding a considerably smaller laser aperture. The aperture will primarily confine the current in the laser, but lateral guiding may also take place if the dielectric material is thick enough to have an optical effect.

Buried dielectric apertures may be combined with any of the VCL structures considered above, giving flexibility in device design. Dielectric aperture VCLs generally have reduced threshold current and improved modulation bandwidth; in addition, the use of dielectric layers which lead to some modicum of lateral guiding also allows control of the mode size. Aperture layers may also be implanted and combined with other structuring techniques, such as relief etching, to independently allow control of the optical and electrical aperture; such approaches have been used to fabricate high-power, single-lateral-mode VCLs (Seurin et al. 2002).

4.5.2 Operating Principle

The VCL, as we have seen, consists of a very short laser cavity, typically one wavelength long, surrounded by two Bragg mirrors for generation of optical feedback.

The operating principle, and to some extent the analysis, is similar to that of the DBR laser we saw in Sect. 4.4; the concept of Bragg reflection (Sect. 4.4.2) of course also applies to understanding the VCL mirrors.

The very short length of the laser cavity, typically not more than several hundred nanometers, changes some of the qualitative aspects of VCL laser behavior when compared to a DBR laser. We recall that the resonance wavelengths of a cavity are a function of its optical length (Eq. 2.121) such that the longitudinal mode spacing (Eq. 2.125) of a very short cavity is quite large. The laser cavity is at the same time the active region of the laser, where the optical gain has a spectral dependence similar to that calculated in Example 3.2 in Sect. 3.2.5. Comparing the longitudinal mode spacing with the spectral width of the gain in a VCL leads to the observation that at most one mode is found within the region of positive gain: the next mode is spaced so far away in wavelength that it is far outside the gain peak. For this reason, VCLs are fundamentally single-longitudinal-mode lasers.

The short length of the cavity has a further implication. A cavity, say, 200 nm long for a VCL with a 5 μm radius aperture, has a total active region volume on the order of 30 μm^3. This volume is all that is available for generation of photons and inversion density and thus achieving lasing. As a result, the finesse of the optical cavity needs to be high to allow the laser to reach threshold such that a high mirror reflectivity is required: typical reflectance values are ≥ 90%. These high reflectance values can only be practically achieved with multi-layer dielectric stacks, such that the Bragg mirrors are the only reflectors which come into consideration for VCLs.

The Bragg mirrors, as we have seen, consist of alternating layers of two materials with different refractive indices; each layer is a quarter wavelength thick. The resultant Bragg resonance gives rise to reflectivity which increases with the number of layer pairs (one high index and one low index) and the index step between the materials. For a multi-layer stack composed of two materials with refractive index n_{low} and n_{high}, the reflectivity of m layer pairs is given by

$$r_{Bragg} = \frac{1 - \prod_{0}^{2m} \frac{n_{low}}{n_{high}}}{1 - \prod_{0}^{2m} \frac{n_{low}}{n_{high}}}, \tag{4.17}$$

where we recall that the intensity reflectance $R = (r_{Bragg})^2$. Rather than evaluate the products, we may also approximate the reflectivity as (Wilmsen, Temkin and Coldren 1999, Chap. 2)

$$r_{Bragg} \approx \tanh\left(2m \frac{(n_{high} - n_{low})}{(n_{high} + n_{low})}\right). \tag{4.18}$$

These simple expressions assume that the entry and exit materials for the mirror layers are the same and have a refractive index between n_{low} and n_{high}.

The plot of Fig. 4.17 shows the reflectance as a function of the number of layer pairs and the refractive index step between them. We see that a medium index step ($\Delta n = 0.3$) results in > 99% reflectance for about 30 layer pairs, which explains the numbers of mirror pairs used in the structural descriptions above. For a top-emitting VCL, the upper mirror usually consists of fewer pairs than the lower mirror, to allow emission of a small fraction (say 2%) of the generated optical field as useful laser emission.

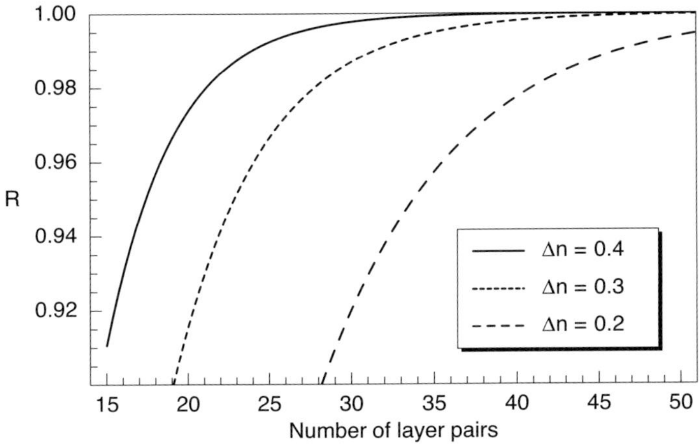

Fig. 4.17. Reflectance of a Bragg mirror as a function of the number of layer pairs for varying refractive index step between layers

As a function of wavelength, a single Bragg mirror with a sufficient number of layer pairs will reflect in a relatively broad, flat stop-band around the resonance wavelength. Two Bragg mirrors separated by a 1-λ cavity will exhibit a transmission resonance at the wavelength defined by the cavity length, as seen in Fig. 4.18. The stop-band due to mirror reflection is about 80 nm wide (810 to 890 nm), but the cavity resonance is clearly seen as a dip at 850 nm. This resonance corresponds to the laser wavelength and is a strong function of cavity length, refractive index and temperature.

For the VCL to lase, the cavity resonance peak must align with the gain peak. Both of these are temperature-dependent and shift to longer wavelengths with increasing temperature, albeit at different rates. Since, as we discussed above, the resonance peaks are spaced very widely in wavelength, the VCL only has one resonance in the mirror stop-band and only one resonance within the spectral range of positive gain. For this reason, temperature drift effects, for example, which lead to a misalignment of the resonance and gain peaks will result in a cessation of lasing.

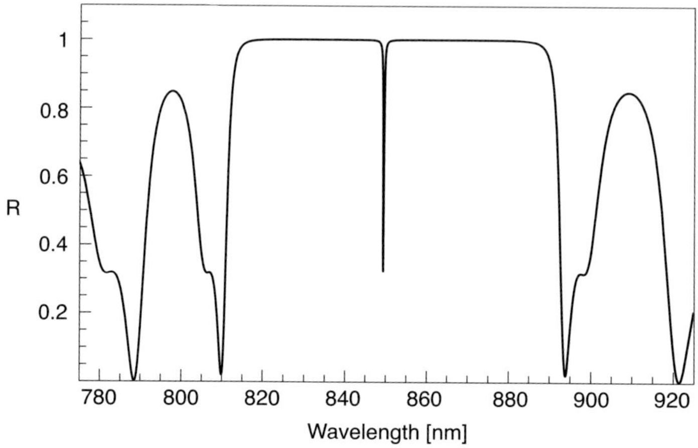

Fig. 4.18. The stop-band and cavity resonance of an 850 nm VCL. The characteristic was calculated for 25 upper and lower $AlAs/Al_{0.38}Ga_{0.62}As$ mirror pairs and a 219 nm thick GaAs cavity

4.5.3 Fabrication Considerations

Fabrication of a VCL involves complex epitaxy but relatively simple subsequent processing. On balance, the overall technological requirements for VCL manufacture as well as the demands on packaging and testing are simpler than for edge-emitting lasers (Chow et al. 1997).

We have seen that a typical VCL structure requires epitaxial growth of between 120 – 180 layers. These must all be grown with nanometer or better thickness accuracy with a variation across the structure of less than 1%. Since the optical thickness is the relevant physical parameter, refractive index and thus composition must also be accurately determined and reproducibly attained. The required quality level has been reached using molecular beam epitaxy (MBE), gas-source MBE or metal-organic vapor phase epitaxy (MOVPE) (Wilmsen, Temkin and Coldren 1999, Chap. 4), where the latter is the technology of choice for volume production of VCLs. A particular layer sequence for a given laser wavelength is usually reproducibly manufacturable following a number of calibration runs; good epitaxialists should be able to hit a new wavelength after one or two growth runs. As a result, VCL epitaxial structures are now grown industrially in large quantities with good reproducibility.

Most standard VCL structures are fabricated solely from semiconductor materials, and most of these in one epitaxial growth step. Exceptions are novel structures which use mirrors based on non-lattice-matched materials (Babic et al. 1995) and buried contact structures (Shau et al. 2000), which may use wafer-bonding techniques to assemble the mirrors. These approaches are generally used for VCL wavelengths for which suitable semiconductors are not available or have insufficient refractive index variation to permit the fabrication of a good-quality Bragg mirror. The GaAs/AlGaAs material system is again the most developed and is the system of choice for VCLs in the 780–980 nm wavelength range. The refractive index steps achievable are large, since AlGaAs may be grown lattice-matched in all compositions from AlAs to GaAs.

Red-emitting VCLs are attractive for emission in the visible range; AlGaInP lasers have been fabricated for wavelengths in the 630–690 nm regime. The III–V nitrides (GaN, AlGaN), under rapid development for blue edge-emitting lasers, are also being investigated for the fabrication of blue VCLs. Long wavelength VCLs are presently the object of considerable research and development effort, as these low-cost lasers may potentially replace the expensive telecommunications lasers for short- and medium-range data communications applications. The InGaAsP, InGaAsN and AlGaAsSb systems have been investigated for the fabrication of VCLs with wavelengths between 1.3 and 1.6 μm. Development of InGaAsP VCLs has been hampered by the very low refractive index steps attainable with this material system (Koeth, Dietrich and Forchel 1998; Rapp et al. 1999); Bragg mirrors thus become very thick, which is unmanageable from the epitaxial growth point of view and undesirable due to the high thermal resistance of most of the relevant quaternary materials. The most successful structures in this wavelength regime are fabricated using hybrid assemblies of mirrors and cavities allowing the combination of high-index-step materials with those required for achieving gain at the correct wavelength; advanced structures also allow tuning the emission wavelength (Vakshoori et al. 1999) and show single-mode behavior even for large apertures (Ohiso et al. 2002).

Following epitaxial growth, the remaining process steps to yield a functional VCL can be quite simple. The etched mesa VCL is fabricated with relatively few demands on process technology and is thus popular for quick processes or the manufacture of lasers with limited demands on lateral mode behavior. Mesa etch is usually by RIE or CAIBE, stops above the active region and is generally followed by the deposition of an insulating and passivating dielectric surface layer; typical mesa heights are $1-2$ μm. Surface metallization is usually patterned by liftoff, where alignment to the top of the etched mesa can be challenging. Buried heterostructure VCLs continue with the re-growth of a high-bandgap material in the etched region; growth technology is typically MBE or MOVPE and extensive surface cleaning is imperative.

Implanted VCLs require an ion implant step to generate the buried damage layer which leads to electrical confinement. Implanted species are typically protons, O^+, N^+ or F^+, and implant depth is to the top edge of the active region, generally about 2.5 μm. The resultant planar surface renders patterning of top-side metallization straightforward (Orenstein et al. 1990).

One very agreeable feature of VCLs is the ability to test them directly after fabrication before chip separation. Since the lasers are surface-emitting, they can be fully characterized using modified wafer probers as employed for testing electrical integrated circuits. The resultant wafer maps may then be used as input for laser packaging, such that only devices with the desired characteristics continue to the back-end of the process; this selection strongly reduces costs and is a primary reason that VCLs are expected, given sufficient market volume, to become the most inexpensive semiconductor lasers.

4.5.4 Microsystem Aspects

Vertical cavity lasers have numerous features which may bring decisive advantages to their use in optical microsystems. Even for the standards of semiconductor lasers, the VCL is compact. Since the long cavity of the edge-emitting laser is not required, the vertical cavity laser size is generally defined by the contact pad, which is considerably larger than the etched or implanted laser itself. Practical considerations usually limit the laser chip size to that which can still be easily handled. As a result, ultra-miniaturized optical microsystems may consider the VCL as a light source with the smallest available dimensions.

From the performance point of view, the VCL is also lucrative for microsystems. Low threshold current implies low overall power consumption, typically an order of magnitude or more below that of a DFB laser, for example. The circular aperture gives rise to a circularly symmetric mode profile, which renders coupling into optical fibers, micro-optical lenses or waveguides straightforward. Mode size and near-field profiles are such that coupling efficiency is typically better than for edge-emitting lasers, whose asymmetric and astigmatic emission characteristics frequently require specialized optical systems to optimize coupling. In addition, the surface aperture of the VCL has permitted the development of micro-lenses which are deposited or patterned directly on the surface as part of the laser fabrication process thereby obviating the need for further alignment and possibly improving stable single-mode operation (Park et al. 2002).

The ability to fabricate VCLs in one- or two-dimensional arrays has implications for the assembly of highly-parallel optical microsystems (Monti di Sopra et al. 2000). For parallel data communications, multi-sensor arrays or custom illumination systems, for example, compact arrays of VCLs may be employed where the spacing of the lasers is usually limited by wiring. Standard fiber array spacing (250 μm) is easily achieved in one-dimensional VCL arrays, such that compact modules based on V-groove mounted fibers may be fabricated with considerably reduced technological demands. Their surface emission properties do make VCLs more difficult to use in planar optical systems; for this case, 45° mirrors or other coupling optics are required to direct the beam into the plane of, for example, a Si or III–V wafer for coupling into a a waveguide. The attendant alignment difficulties make the VCL less desirable than an edge-emitting laser for this particular configuration.

The final consideration is the pecuniary one. As the commercial applications of many microsystems have acute restrictions on system costs, component price often plays a role in the choice of light source. For a variety of implementations, then, the low cost of the VCL may be decisive in evaluating its utility in a microsystem.

Example 4.4

Assignment: *Estimate the number of mirror pairs required to fabricate a VCL Bragg mirror for operation at 1.55* μm *with* $R \geq 99\%$ *using the InGaAsP material system. Assume that* $In_{1-x}Ga_xAs_yP_{1-y}$ *may span the range from InGaP to InGaAs by varying the As mole fraction, y. Approximately how thick would the mirror be?*

Solution: *We may approximate the refractive index of* $In_{1-x}Ga_xAs_yP_{1-y}$ *as a function of y using Eq. 3.65. From this, we may determine that the lowest index is InGaP with* $n \approx 3.40$ *and the highest InGaAs with* $n \approx 3.56$*, yielding* $\Delta n \approx 0.16$.

We may now use Eq. 4.17 to determine the mirror reflectance, recalling that $R = (r_{Bragg})^2$. *Using our high and low index values, we find that about 65 mirror pairs are required to reach* $R \geq 0.99$.

Since the mirror layers are a quarter wavelength thick, we may estimate the thickness by taking an average value for the refractive index ($n_{av} \approx 3.48$) *and estimate the thickness as*

$$t_{mirror} \sim 2\left(\frac{\lambda}{4}\right)\frac{1}{n_{av}} \cdot 65$$

since we have two mirror layers per pair. The total estimated thickness is then about 14.5 μm. *Since two mirrors are required, the total VCL thickness is then on the order of 30* μm*, which, for reasons of electrical and thermal resistance as well as epitaxial growth time, is not feasible. For that reason, most long-wavelength VCLs use alternative materials (AlGaAsSb) or hybridly-bonded mirror systems.*

4.6 Quantum Cascade Lasers

The emission wavelength of the lasers we have considered thus far is primarily determined by the energy gap of the direct bandgap semiconductor from which they are made. The optical cavity, whether horizontal or vertical, serves to select the exact wavelength from the available gain spectrum, but this latter quantity is solely a function of the material. This fact restricts the availability of lasers to wavelength ranges for which an appropriate, direct bandgap laser material exists. Particularly for large parts of the mid- to far-infrared spectrum, either no appropriate semiconductors are known or the ones employed yield lasers with limited performance.

The quantum cascade laser (QCL) gets around this problem by using electronic transitions in a quantum well, as opposed to transitions between the conduction and valence bands, to generate photons (Capasso et al. 2000). The transition wavelength is then a function of the quantum well width, which may be accurately defined during epitaxial growth, and is only indirectly a function of the material. As a result, transition energies may be almost arbitrarily chosen, although getting these wavelengths to lase is another matter. Lasers using the quantum cascade concept have been demonstrated at wavelengths between 3.4 and 67 μm, including a portion of the spectrum particularly relevant for gas sensing.

The QCL structure was proposed in 1971 (Kazarinov and Suris 1971) but not successfully demonstrated until 1994 (Faist et al. 1994). This delay was primarily due to the phenomenal complexity of the epitaxial structure, the technology for which was not sufficiently developed until recently. These lasers have subsequently seen rapid development and have now made a rapid transition from laboratory curiosity to industrial product (Gmachl et al. 2001a); demonstrated wavelengths up to $\lambda = 67$ μm (4.4 THz) (Köhler et al. 2002) show promise for access to an entirely new wavelength range previously unattainable by lasers.

4.6.1 Structure and Operating Principle

The essential difference between the QCL and other semiconductor lasers is in the epitaxial structure of the laser core region; as we will see, the cavities employed are essentially identical to those used for edge-emitting lasers. The double heterostructure waveguide concept is also used for the quantum cascade laser but the laser core region consists of a complex multi-layer structure, a portion of which is shown in Fig. 4.19. The figure shows the energy variation of the conduction band for two stages of a multiple quantum well structure consisting of an injector region and an active region. Electron intra-conduction band transitions in this structure give rise to photon emission; since these are purely electronic transitions taking place solely in the conduction band (and not electron-hole recombination between conduction and valence bands), quantum cascade lasers are referred to as unipolar devices.

The photon emission works as follows: the close proximity of the quantum wells in the injector region, separated by very thin barriers, results in the formation of electron mini-bands. Electrons are injected from these mini-bands across a thicker barrier into the active region, which consists of a thicker quantum well with several discrete energy levels. A subsequent electronic transition between, in our example, levels 3 and 2 results in the emission of a photon. The electron thermalizes to energy level 1 and is transferred via a second quantum well to the mini-band of the following injector. We emphasize again that all transitions are in the conduction band and that no holes are involved.

This process repeats for each stage of the laser core: a typical QCL has 20 to 75 stages and a photon is emitted at each stage. This transfer of an electron from one stage to the next represents the "cascade" of the laser name. The emission energy of the photon is thus a function of the energy difference between the energy levels in the active region quantum well, which is in turn a critical function of the quantum

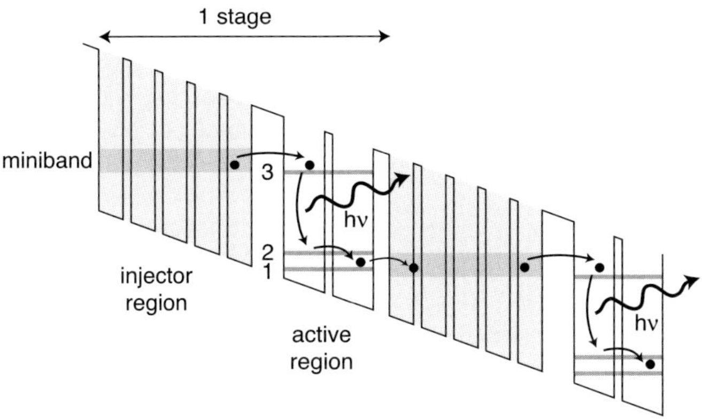

Fig. 4.19. Schematic plot of the conduction band structure for two stages inside the active region of a quantum cascade laser; the quantum wells of each injector region actually get thinner as we move from left to right

well thickness (Eq. 3.67) and not, to first order, of the materials employed. Thus by properly designing and growing quantum well layers of the appropriate thickness, a wide range of wavelengths for which no direct inter-band transitions are available, may be accessed.

This laser core, consisting of up to one thousand layers in all, is the heart of the QCL. The core, as in conventional edge-emitting lasers, is surrounded by top and bottom cladding layers, all of which are n-doped, leading to vertical optical confinement of the generated optical field. Vertical waveguiding becomes increasingly difficult as the wavelength increases, such that specialized waveguide structures need to be used for lasers emitting at wavelengths longer than about 10 μm. Lateral confinement is typically through the use of an etched ridge waveguide. The laser cavity may consist of a Fabry–Perot resonator with cleaved mirrors or use a DFB structure, such that, with exception of the device dimensions and the inner life of the core region, the QCL looks from the outside much like a standard edge-emitting laser.

4.6.2 Fabrication Issues

The fundamental fabrication issue associated with the quantum cascade laser is epitaxial growth of the laser layers. We have seen from the operating principle that emission wavelength is a strong function of layer thickness and that the core consists of repeated quantum well injector and active region stages. With between 20 and 75 stages in the typical QCL, the total layer number varies between 500 and 1 000, each one of which must be grown with sub-nanometer accuracy. As a result, the epitaxy step is critical for successful quantum cascade laser fabrication.

The material system most popular for QCLs is at present InGaAs and InAlAs grown on InP; InGaAs is typically the quantum well material with InAlAs in the barriers. Recently, the GaAs/AlGaAs system has seen rapidly advancing development and it is expected that the antimonides and materials with a large band offset, such as the III–V nitrides (GaN, AlGaN and friends), will become advantageous in the near future. Since the QCL is unipolar and needs only electrons, the layers are typically lightly n-doped; the doping levels are reduced to avoid excess optical losses due to free-carrier absorption, an effect which becomes stronger the longer the wavelength. Total epitaxial layer thickness is between 1.3 and 1.7 μm and total device thickness on the order of 5 μm.

The stripe structure of the QCL is usually defined by an etched ridge waveguide, where the etch depth is usually chosen to strongly confine the current flow as well as the optical field. Gain guiding works in principle for QCLs but is not generally used. Typical waveguide depths are on the order of 5 to 6 μm and widths are usually found in the range 8 to 15 μm. The larger size of the waveguide regions when compared with those we saw in previous sections are in part due to the fact that the emission wavelengths of QCLs are generally at longer wavelengths than standard semiconductor lasers. Some work has been done on the development of buried heterostructure lasers, leading to 300 K CW lasing at $\lambda = 9$ μm, and the use of chalcogenide glasses for burying the laser ridge has been shown to be effective in reducing waveguide losses (Gmachl et al. 2001c). When device fabrication is completed, the lasers are cleaved into bars with typical cavity lengths ranging from 750 to 3 000 μm.

Distributed feedback structures have also been developed using the quantum cascade structure, thereby taking advantage of the single-mode operation and excellent wavelength selectivity of this type of laser (Faist et al. 1997). The laser core region is essentially the same as just discussed, although epitaxy sometimes needs to be completed in two steps, the second following grating formation. The demands on grating fabrication are relaxed with respect to the required periods (the longer wavelengths can use 1 μm periods for a first order grating) but more stringent insofar as achieving an adequate coupling coefficient is concerned.

Post-processing, facet coating and device mounting is in general similar to that for standard edge-emitting lasers. Although room-temperature CW operation of QCLs is now generally possible for $\lambda \leq 9$ μm, most of the accessible wavelength range requires cooled lasers which are generally only operated pulsed. These device configurations require mounting on specialized headers and into cryostats or thermo-electrically cooled headers. However, progress in increasing the operating temperature range and CW operability of QCLs is presently proceeding quite rapidly.

4.6.3 Microsystem Aspects

The most attractive feature of quantum cascade lasers is their range of emission wavelengths, which extend far into the mid-infrared range; both single-mode and ultra-broadband quantum cascade lasers have been demonstrated (Gmachl et al. 2002). As a result, it is expected that QCLs will play a major role in the development of new gas sensing systems, as most atomic and molecular species have fundamental optical absorption spectra in this wavelength regime (Namjou et al. 1998; Gmachl 2001b). The drive to reduce sensor size and cost will likely provide considerable impetus to fabricate these sensors as microsystems with the QCL as a compact light source.

For the most part, the packaging and integration considerations for QCLs are not too different from those of other semiconductor lasers; chip sizes and layout are quite similar. Two aspects may limit microsystem integration in some cases, however: operating temperature and power consumption. As mentioned in the previous section, QCLs originally operated only at low temperatures, whereas room-temperature CW configurations are now standard for a range of wavelengths (Faist et al. 1996). For structures requiring low temperatures, the packaging required may range from mounting on thermo-electric coolers to insertion into a cryostat: whereas the former may be optical-microsystem-compatible, the latter is not. Thermo-electric (Peltier) coolers are available in very compact sizes, in the millimeter range, such that hybrid assembly of QCL-based optical systems whose operating temperature is within a few tens of degrees of 0 °C is feasible.

The issue of power consumption is due to the fact that QCLs are inherently high-power-consumption devices. The generation of one photon per stage through the cascade process implies that optical emission power may be quite high; indeed, pulsed output powers in the Watt range are typical. The requisite electrical input power is then also in this power range, such that wiring and, again, device cooling may present a limitation for very compact microsystems. For applications in which the optical power required is not excessive, packaging and hybrid assembly considerations of QCLs are then similar to those for distributed feedback lasers.

5 Laser Diode Characterization

When an engineer decides to use a semiconductor laser diode as a light source in an optical microsystem, one of her first tasks will be to determine its operating characteristics. Despite availability of data sheets, plots in manufacturer catalogues or vague assertions from colleagues concerning the characteristics, detailed optical and electrical measurement of the individual laser in question is essential for determining its behavior when ultimately installed in a system.

This chapter provides an overview of the measurement techniques required for characterization of a laser diode. Determination of the power/current characteristic is required for evaluation of threshold current, thermal behavior and efficiency. Spectral measurements include emission wavelength, side-mode behavior, polarization, noise and linewidth. All of these characteristics are temperature-dependent, such that the shift with operating temperature needs to be evaluated. Finally, the near-field and far-field intensity distributions are of interest when designing the coupling between the laser diode and the subsequent optical system. We conclude with an overview of the typical types of characteristics which might be expected from the various types of laser diodes we considered in the previous chapter.

5.1 Measurement Setup

The basic characterization of a laser diode can be a rather simple affair whereas measurement of more detailed characteristics rapidly becomes technically more complicated. The most rudimentary measurement setup is shown schematically Fig. 5.1: the laser, packaged on a TO5 header, is placed on a temperature-controlled mount and current is injected into the laser. The optical output is measured by a photodetector or may be coupled into an optical fiber for transmission to a spectrometer or other equipment. The detailed specification of the equipment required to complete a measurement is a function of the data required, as we shall see presently.

5.1.1 Mounting

Common to all measurements is the need to mount the laser being characterized on a test fixture. The type of laser diode mount depends on the packaging of the

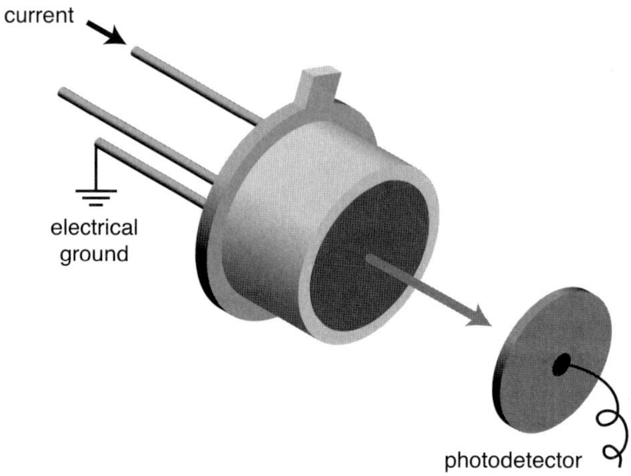

Fig. 5.1. The basic characterization setup required for measurement of optical power output from a laser mounted in a TO5 package

device. The most common mount for discrete lasers is the TO package, shown in Fig. 5.1; the metal TO5 package typically has a diameter of 9 mm and contacts are made to the leads attached to the bottom. The TO mount, which exists in numerous structural variations and sizes, has advantages in that is easily handled, the laser chip is isolated from the environment in its hermetically-sealed package and a photodiode is frequently included in the package to allow monitoring of output power; its size, however, is often not micro-system compatible.

Unpackaged laser chips, such as those which might be hybridly integrated directly into an optical microsystem, require slightly more careful mounting. As shown schematically in Fig. 5.2, the chip is placed onto an electrically-conducting sub-mount which in its simplest form may consist of a machined metal block attached to a thermo-electric cooler. Electrical connection to the laser is through a prober needle, as shown in the photo of Fig. 5.3, or by bond wire to a larger contact section of the sub-mount. For VCLs, the same concept may be used save that the lasers may be characterized at the wafer level and emission is of course from the surface.

5.1.2 Equipment

The equipment required for laser diode measurement varies depending on the characteristics being measured; it ranges from a heat sink, current supply and photodetector to complex interferometric characterization systems.

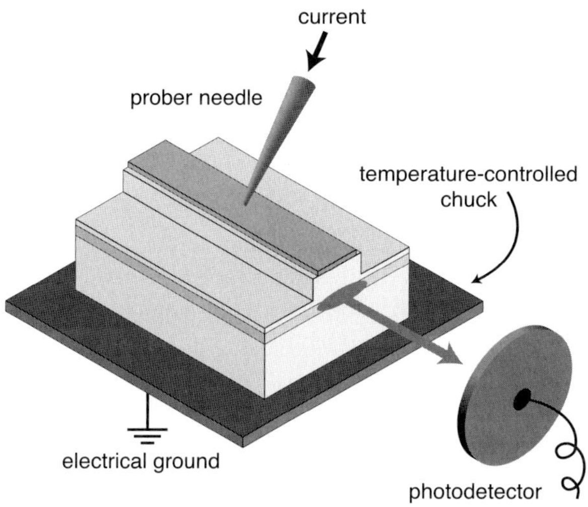

Fig. 5.2. The basic characterization setup required for measurement of optical power output from a laser chip or bar

Fig. 5.3. A stripe laser chip contacted by a prober needle; emission at the facets toward the top and bottom of the photo is seen as bright spots at the edges of the chip

Electrostatic Protection

Laser diodes are very sensitive to electrostatic discharge (ESD). Handling the devices thus needs to be done with care and the workplace as well as the user should be well grounded. Laser measurement setups are frequently built up on optical tables, the metal surface of which is then electrically connected to the user by a conducting bracelet. Current supplies are usually designed with ESD protection in mind, having circuitry to guard against current spikes due to possible fluctuations of the line voltage. In addition, most laser drivers ramp the set current up slowly to avoid excessive laser current, which can also lead to the more-or-less instant demise of the laser.

Current Supply

The laser diode is a current-controlled device, a fact which many first-time experimenters for some reason tend to forget. Electrical power input to the laser is thus via a current source, not a voltage source, and the mistaken use of the latter generally leads to a miniscule smoke cloud and a trip back to the laser supplier. The current range which the supply needs to provide is a function of the laser type being measured: VCLs require a maximum of 1 to 10 mA with good accuracy in the sub-mA range, whereas distributed feedback lasers may require up to 100 mA. Quantum cascade lasers are generally driven with Amperes of current. Thus operating a VCL with a power supply suitable for QCLs is not generally a good idea.

Laser diodes are occasionally operated in pulsed mode, either to avoid temperature effects or because the high-speed modulation behavior is to be measured. The former situation arises in particular for large-area and power lasers. Due to the high level of current injected, the internal laser temperature may rise significantly if the laser is operated continuous wave (CW). Thus pulsed current sources are used, which supply sub-microsecond pulses with a duty cycle of $< 1\%$. As a result, the internal laser temperature quickly returns to ambient levels between pulses and the overall internal laser temperature remains close to the external ambient temperature.

For the case where the high-speed modulation characteristics of the laser are required, quite specialized mounting, probing, excitation and measurement techniques are usually required. “High speed” generally implies modulation bandwidths of > 1 GHz, so that digital modulation in the 10s of femtosecond to nanosecond regime are required. High-speed network analyzers are used for laser stimulation and microwave probing techniques are required to get the signals onto the laser chip; the latter must be mounted so as to minimize parasitic electrical losses. Detectors with a GHz bandwidth are essential for measurement of the emitted output power. This type of measurement, usually restricted to laser diodes with data communications applications, requires considerable expertise and experience to perform and adequately interpret.

Temperature Control

As we will see in Sect. 5.5 below, most of the laser diode characteristics are a function of temperature. For that reason, accurate control of temperature is essential to guarantee realistic and repeatable measurements. Temperature stability is usually achieved by using a thermally massive mount for the laser, either as a chip or in a TO5 package. Using a thermistor as temperature sensor, a thermo-electric (Peltier) controller then maintains a constant sub-mount temperature. For high-power devices or extreme temperature ranges, extensive heat sinking and other heating or cooling mechanisms may be required. In anticipation of the discussion below, we note that the internal operating temperature of the laser may indeed be quite different from the external sub-mount temperature, one of the reasons that pulsed measurements are occasionally preferred.

Detection

Measurement of the characteristics of the light emitted from the laser is the ultimate goal of most laser measurements. For our basic characterization, the optical output power is the parameter of interest. A photodetector is sufficient for this purpose, where a semiconductor photodiode is usually the most practical and inexpensive detector alternative. The photodiode must be absorbing at the laser emission wavelength: silicon works for the visible range to about $\lambda = 1.1$ μm and for the near-infrared wavelength range, to about $\lambda = 1.8$ μm, InGaAs is preferred. Longer wavelengths, as well as measurements which go into the ultraviolet, will require specialized materials: Germanium detectors, for example, are useful into the mid-infrared wavelengths but are operated cooled at 77 K.

Semiconductor photodiodes are available in a wide variety of physical configurations; large-absorption-area devices are available, making optical alignment with the laser unproblematic. Whereas complete calibrated detector systems are available, many users will simply install a photodiode into the measurement system. Two considerations should then be noted: 1) the photodiode signal is a generated photo-current, such that a transimpedance amplifier is required to generate the voltage output required for most measurement equipment; and 2) the absorption efficiency of all semiconductor materials is a function of wavelength, such that calibration of the detector is required at each laser wavelength to be measured yielding a value of mA generated per mW of optical power absorbed at that wavelength.

5.1.3 On-Wafer Testing

One of the decided advantages of surface emitting lasers is that VCLs may be tested in wafer form; since they emit from the surface, cleaving is not required for generation of the optical cavity and the lasers may be characterized before chip separation.

On-wafer testing of VCLs may be accomplished by a slightly modified wafer prober such as that used for electronics testing, as shown in Fig. 5.4. Contact to the laser is through prober needles, where multiple lasers may be contacted simultaneously and measured in sequence; the back-side contact is typically through the wafer chuck, with the exception of high-speed measurements which use two surface contacts. Optical output power is measured using a photodiode placed above the laser and a fiber positioned close to the emitting aperture may be used to couple into a spectrometer for spectral measurements. Using a configuration like this, thousands of VCLs may be characterized per hour in a fully automated fashion.

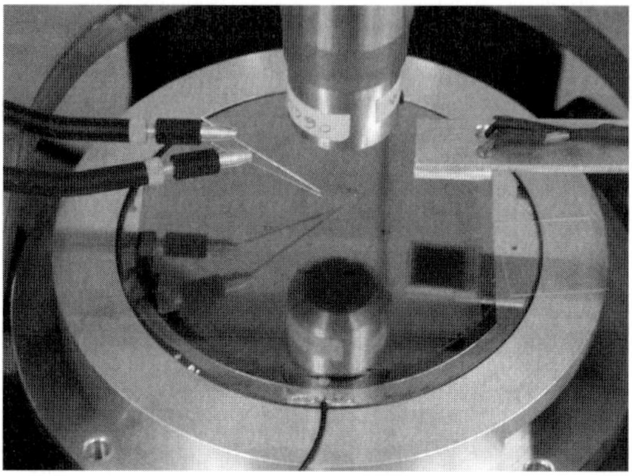

Fig. 5.4. Wafer-scale characterization of VCLs using a wafer prober; laser contact is by the prober needles and the photodetector at the right is subsequently positioned over the laser. Photograph courtesy of U-L-M photonics GmbH

5.2 The PI Characteristic

The first measurement typically done on a laser diode is that of optical output power as a function of input current, yielding the power–current or PI characteristic[1] shown in Fig. 5.5. Power output of the laser is measured either in Watts or dBm; the latter is a logarithmic unit for which 0 dBm is defined as an optical power of 1 mW, so that –10 dBm represents 100 μW, –20 dBm represents 10 μW and so forth[2]. As expected from our theoretical discussion, the optical output starts abruptly at the threshold current, I_{th} [A], and increases linearly as current increases. From the PI curve, or output characteristic, of the laser we may determine I_{th}, the output slope and thus efficiency, the maximum output power and the linearity of the output characteristic.

5.2.1 Threshold Current

The threshold current defines the onset of lasing and is defined by an abrupt increase in optical output power. I_{th} may be determined in one of three ways from

[1] Sometimes referred to as the LI (light–current) characteristic.

[2] dBm should not be confused with dB. Both are logarithmic units but the former defines a power level whereas the latter the ratio between two powers, voltages or currents.

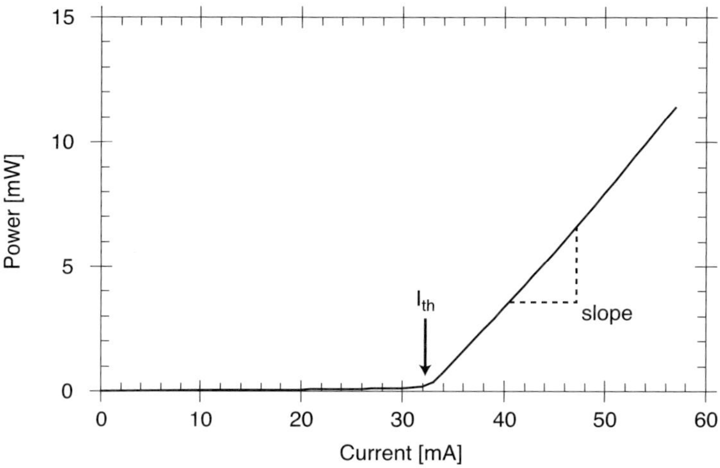

Fig. 5.5. A typical measured PI characteristic of a Fabry–Perot laser; the threshold current and output slope are shown

the PI plot: 1) extrapolation of the linear increase in output power back to the current axis, such that the intercept defines I_{th}; 2) by linearly extrapolating the increase in output power below threshold and taking the intercept with the extrapolated linear increase above threshold; and 3) by taking the second derivative of the PI characteristic and determining the current value at the peak. The last of these (see Fig. 5.7) yields the most accurate and reproducible results but requires the most post-processing of the measured data. Automated measurement systems often use this approach.

Typical threshold current values vary widely according to the type of laser. VCLs, particularly small aperture devices, can have very low threshold currents, in the 100 μA regime; typical values are 1–5 mA. I_{th} for Fabry–Perot lasers are typically found in the range 10–30 mA and distributed feedback lasers 30–60 mA. QCLs, large-area and power lasers will have correspondingly larger thresholds.

Threshold Current Density

Closely related to the threshold current is the threshold current density, J_{th} [A/cm^2], simply defined as

$$J_{th} = \frac{I_{th}}{A} \tag{5.1}$$

for laser active region area A [m^2]. Threshold current density allows a normalized comparison between devices of different size. Typical values range from 100 to 1 000 A/cm^2, where the former represents quite a low figure.

Laser developers will frequently fabricate broad-area lasers and measure the threshold current density. The structure of the broad-area laser implies that the effects of the device edges, which can be significant for stripe lasers, are minimized in comparison to the effects of the laser volume. Thus the laser performance, characterized by J_{th}, is primarily a function of the epitaxial laser layers, allowing these to be optimized independently of any other device effects.

5.2.2 Threshold: A Closer Look[3]

Taking a closer look at what happens at laser threshold will allow us to derive a few useful relationships which may be used to optimize a laser diode design and understand the operating behavior. We recall from Eq. 2.139 that the threshold gain of the laser of cavity length L and with mirror reflectance R is given by

$$\gamma_{th} = \alpha_0 + \frac{1}{L}\ln\left(\frac{1}{R}\right), \tag{5.2}$$

where α_0 represent the background loss; the threshold gain is thus the gain necessary to overcome all the optical losses of the cavity. Since a real laser diode is formed in a waveguide and the relevant portion of the gain is that found in the waveguide active region, we may include the effect of non-ideal mode confinement through the confinement factor, so

$$\Gamma\gamma_{th} = \alpha_0 + \frac{1}{L}\ln\left(\frac{1}{R}\right) \tag{5.3}$$

or

$$\gamma_{th} = \frac{1}{\Gamma}\left(\alpha_0 + \frac{1}{L}\ln\left(\frac{1}{R}\right)\right). \tag{5.4}$$

Thus a poorly confined mode has an overall higher threshold gain since only a portion of it exists in the laser active region.

We saw in Sect. 2.4.2 that the gain is a function of the inversion carrier concentration; Eq. 2.108 gave us the relation

$$\gamma = \frac{\lambda^2}{8\pi\tau}(N_2 - N_1)g(\nu) \tag{5.5}$$

for inversion concentration $(N_2 - N_1)$, carrier lifetime τ and the lineshape function $g(\nu)$.

To provide a useful expression for calculation, we may approximate

---

[3] This section may be skipped without loss of continuity.

$$\frac{N_2 - N_1}{\tau} \approx \frac{J}{qt} \tag{5.6}$$

and

$$g(\nu) \approx \frac{1}{\Delta\nu} . \tag{5.7}$$

Equation 5.6 includes the laser injection current density J and the active layer thickness t, giving an expression for the number of carriers per volume per time pumped into the active region. In Eq. 5.7, we have equated the lineshape function with the reciprocal of the linewidth expressed in frequency, a rough but valid approximation. Thus we may express the gain in the material approximately as

$$\gamma \approx \frac{\lambda^2}{8\pi}\left(\frac{J}{qt}\right)\left(\frac{1}{\Delta\nu}\right) . \tag{5.8}$$

Combining Eq. 5.8 with Eq. 5.4, we may then approximate the threshold current density in the laser as

$$J_{th} \approx \frac{8\pi}{\Gamma\lambda^2} qt\Delta\nu\left(\alpha_0 + \frac{1}{L}\ln\left(\frac{1}{R}\right)\right) . \tag{5.9}$$

This last expression shows us that a small threshold current in a laser is typically achieved for small active layer thickness, low background and mirror losses and high confinement factor.

We may write Eq. 5.9 in the form

$$J_{th} = C_0\alpha_0 + C_0\frac{1}{L}\ln\left(\frac{1}{R}\right) \tag{5.10}$$

for a constant C_0 given by

$$C_0 \approx \frac{8\pi}{\Gamma\lambda^2} qt\Delta\nu . \tag{5.11}$$

Equation 5.10 allows us to evaluate certain laser parameters which are otherwise difficult to measure. If the threshold current density is measured on a set of identical lasers differing only in their cavity length L, one may plot J_{th} as a function of $(1/L)\ln(1/R)$. This plot should be a straight line, with slope C_0 and intercept on the ordinate $C_0\alpha_0$. From Eq. 5.11, we see that a value for C_0 can give us a rough indication of the laser linewidth, assuming we know the active layer thickness and confinement factor. Knowing C_0 and the intercept, we can then evaluate the waveguide background losses, α_0, which are difficult to measure any other way. These type of measurements, however, are typically done by laser designers, rarely by the end user; how a laser user determines linewidth practically is the subject of Sect. 5.3.4.

Example 5.1

Assignment: *Estimate the threshold current density of a $\lambda = 800$ nm GaAs laser given that $L = 500\ \mu m$, $t = 200$ nm, $\Gamma = 0.6$, $\alpha_0 = 10\ cm^{-1}$ and $\Delta\nu = 10^{14}$ Hz.*

Solution: *We may apply the approximation of Eq. 5.9. For a GaAs laser with a cleaved facet, $R = 0.31$ and the remaining parameters are given, so that we find $J_{th} \approx 70\ A/cm^2$; this is quite a low value, but of course ignores any further losses in the cavity.*

From knowledge of J_{th} and Eq. 5.8 we can then also approximate the expected threshold gain to be $\gamma_{th} \approx 56\ cm^{-1}$, a typical value for semiconductors.

Example 5.2

Assignment: *What is the effect of reducing the active layer thickness, t, on threshold current density?*

Solution: *We use the approximation of Eq. 5.9 and find that J_{th} simply decreases linearly with decreasing t. This conclusion seems reasonable, since a smaller active region thickness implies a smaller pumped volume and thus fewer carriers required to reach inversion and thus threshold. Unfortunately this consideration is incomplete.*

The conclusion that J_{th} decreases monotonically with decreasing t is predicated on the assumption that the other parameters in Eq. 5.9 remain constant. This assumption is reasonable save for the confinement factor, Γ. From Eq. 3.81, we saw that the confinement factor is a function of the electric field, the ***y****-dependence of which is in turn a function of the modal solution of the waveguide which is ultimately a function of waveguide thickness. In general, as we saw in Sect. 3.4.3, a thinner waveguide implies poorer confinement. Thus reducing t results in reduced Γ, so that we do not expect J_{th} to continually decrease as the active region thickness decreases.*

We may see this explicitly in the plot of Fig. 5.6 which shows the variation of J_{th} with t for a GaAs laser operating at $\lambda = 850$ nm with $Al_{0.8}Ga_{0.2}As$ cladding layers. Using three core compositions ($Al_{0.5}Ga_{0.5}As$, $Al_{0.4}Ga_{0.6}As$, and $Al_{0.3}Ga_{0.7}As$) giving rise to the three index steps between core and cladding ($\Delta n = 0.15, 0.20, 0.25$), the waveguide equation, Eq. 3.71, was solved for each composition for a range of waveguide thickness values; only the lowest order mode was assumed. The resulting values for z-directed propagation constant β were used to calculate the ***y****-directed propagation constants (Eq. 3.78 – Eq. 3.80) which were in turn used to calculate the electric fields as a function of y. Plugging these into Eq. 3.81 then led to a value for Γ, which we finally put back into Eq. 5.9. The entire process is straightforward if not entirely trivial.*

From the resultant characteristics shown in Fig. 5.6, we see that threshold current density does indeed decrease roughly linearly with t for relatively thick active region layers. As t decreases, the rapid concomitant decrease in *confinement results in a minimum and then increase in J_{th}. Thus we may determine an optimum active layer thickness for minimized threshold current, which is a function of the materials used in the laser core.*

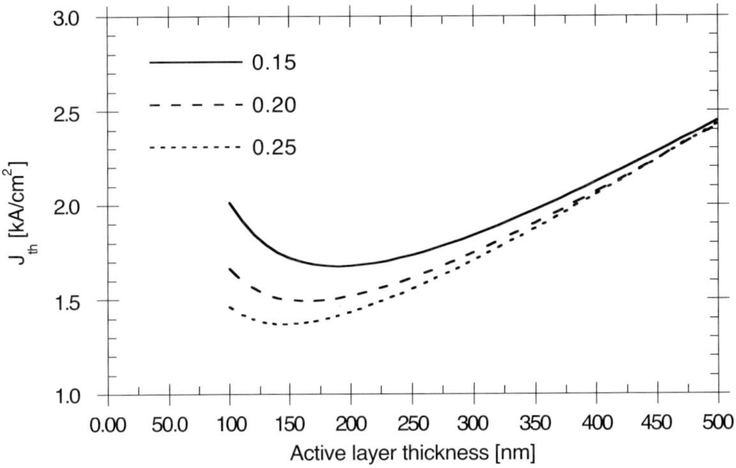

Fig. 5.6. Variation of the threshold current with active layer thickness for a $\lambda = 850$ nm GaAs laser with $Al_{0.8}Ga_{0.2}As$ cladding layers and three core compositions ($Al_{0.5}Ga_{0.5}As$, $Al_{0.4}Ga_{0.6}As$, $Al_{0.3}Ga_{0.7}As$) giving rise to the three index steps ($\Delta n = 0.15, 0.20, 0.25$)

5.2.3 Efficiency

The linear regime for $I > I_{th}$ in the PI characteristic represents the operating region of the laser. The slope gives a measure of the increase in optical output power for a given increase in laser drive current and may thus be characterized by an efficiency.

In the absence of thermal effects and other non-idealities, the slope of the PI curve may be determined by approximating a linear fit to the curve. A more accurate approach is to take the first derivative, dP/dI, as shown in Fig. 5.7 for the PI characteristic of Fig. 5.5. We see from this curve that details in the output characteristic become apparent which are not seen in the PI curve itself: the slope, for example, is not constant but increases slightly with I above threshold and irregularities appear for $I \geq 45$ mA. Fluctuations in the output slope are frequently an indication of subtle (and sometimes not-so-subtle) "kinks" in the output characteristic, often indicative of thermal effects, such as mode-hopping or hole-burning, in the laser. Figure 5.7 also shows the second derivative (d^2I/dP^2) of the PI curve used, as mentioned above, to accurately determine the threshold current.

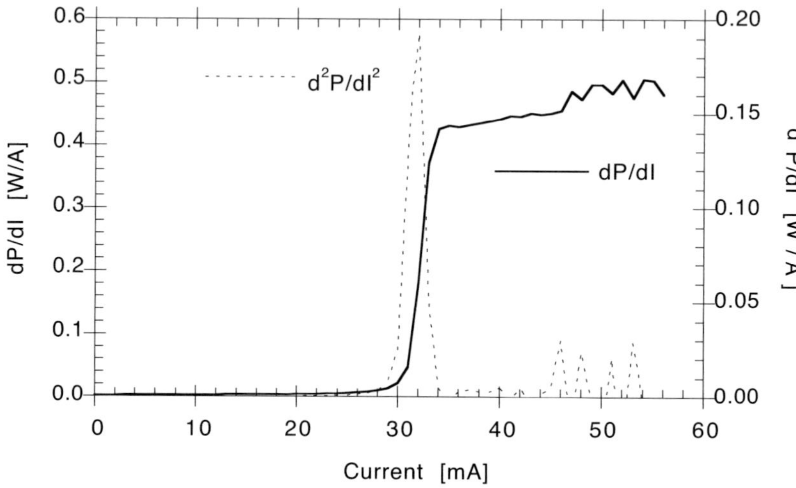

Fig. 5.7. First and second derivatives of the measured PI characteristic of Fig. 5.5

Differential Efficiency

The slope of the output characteristic gives the efficiency of the laser. We may define efficiency by first writing down an empirical expression for the output power of the laser, P_{out} [W], as a function of input current I [A], namely

$$P_{out} = \frac{1}{2}\eta_i\left(\frac{\text{mirror loss}}{\text{mirror loss + background loss}}\right)\frac{h\nu}{q}(I - I_{th}) . \tag{5.12}$$

Going through Eq. 5.12 from left to right: the factor 1/2 is because emission from only one of the two facets of the laser is measured; η_i is the internal quantum efficiency (the number of photons generated per injected electron); the ratio of mirror loss to total loss in the cavity is a measure of the emission efficiency; $h\nu/q$ derives from the energy per photon ($h\nu$) and the charge per electron (q) such that we have an expression giving a ratio of the *numbers* of electrons and photons; and $(I - I_{th})$ represents the current above threshold. The emission efficiency η_e is thus given by

$$\eta_e = \left(\frac{\text{mirror loss}}{\text{mirror loss + background loss}}\right) = \left(\frac{\alpha_m}{\alpha_m + \alpha_0}\right) , \tag{5.13}$$

where α_m is the mirror loss,

$$\alpha_m = \frac{1}{L}\ln\left(\frac{1}{R}\right) . \tag{5.14}$$

We may then re-write the relation of Eq. 5.12 in a simpler form:

$$P_{out} = \frac{1}{2}\eta_i\eta_e\left[\frac{h\nu}{q}(I - I_{th})\right] \tag{5.15}$$

or

$$P_{out} = \frac{1}{2}\eta_d\left[\frac{h\nu}{q}(I - I_{th})\right], \tag{5.16}$$

where we now have three "efficiencies", all dimensionless: η_i is the internal quantum efficiency (typically close to unity); η_e is the emission efficiency; and η_d represents the differential efficiency. The differential efficiency, η_d, is perhaps the most useful of these and the one most often quoted; it ultimately gives a measure of the ratio of the number of photons escaping from the laser to the total number generated. This factor is frequently expressed as differential efficiency *per facet* and is thus related to the slope, or derivative, of the PI curve as

$$\eta_d = \frac{q}{h\nu}\left(\frac{dP}{dI}\right). \tag{5.17}$$

We note that whereas the slope of the PI curve expresses a power as a function of a current, efficiency is a measure of the *number* of photons related to the *number* of electrons, regardless of their energy.

Wall-Plug Efficiency

A further useful efficiency parameter of interest to the engineer who is calculating the power budget of a complete microsystem is the wall-plug efficiency, η_{wp}. Wall-plug efficiency is the ratio of the total optical output power to the total electrical input power, taking all losses into account. Thus

$$\eta_{wp} = \frac{\text{optical power out}}{\text{electrical power in}} \approx \frac{P_{out}}{I^2 R_{laser}} \tag{5.18}$$

for laser diode resistance R_{laser}. The laser is of course a pn junction and thus has a non-linear resistance, but we may approximate the value based on the results of the IV measurement (Sect. 5.2.4) and typically have $R_{laser} \approx 10\ \Omega$. The wall-plug efficiency may be defined and measured for the laser diode itself, the packaged laser or even for an entire module; each step upwards in the hierarchy results in increased electrical losses, typically due to series resistances.

Efficient Summary

When discussing laser "efficiency", we must then carefully consider to which efficiency we are referring; Table 5.1 summarizes what we have described above. The relevant parameter usually depends on the level of design in which one is involved:

an epitaxialist will be concerned with internal quantum efficiency whereas the microsystem designer will likely be interested in the wall-plug efficiency.

Table 5.1. Definition of the various types of "efficiency" in a laser diode

Efficiency	Symbol	Definition	Typical value
internal quantum	η_i	photons / electron	1
emission	η_e	mirror loss / total loss	0.35
differential	η_d	photons escaping / photons generated	0.35
wall-plug	η_{wp}	optical power out / electrical power in	0.2

Example 5.3

Assignment: *The laser whose PI curve and derivatives are given in Fig. 5.5 and Fig. 5.7 is made from GaAs, has a cavity length of 300 μm and emits at* $\lambda = 792$ *nm. Determine* η_d, η_e, *and* η_{wp}, *assuming* $\eta_i = 1$. *Estimate the background loss* α_0.

Solution: *From Fig. 5.7, we see that the slope of the PI curve in the current range 35–45 mA increases slightly; let's take an average value of* $dP/dI = 0.45$ *W/A. From* $\lambda\nu = c$, *we calculate the laser oscillation frequency to be* 3.8×10^{14} *Hz and then from Eq. 5.17 find that* $\eta_d = 29\%$.
Since $\eta_i = 1$, *we can set* $\eta_e = \eta_d$. *From Eq. 5.14, knowledge of the cavity length and the fact that GaAs facets have a reflectance of about* $R = 0.31$, *we can determine the mirror loss to be* $\alpha_m = 39$ cm^{-1}. *Using Eq. 5.13 we thus find that the background losses are* $\alpha_0 = 97$ cm^{-1}.
The wall-plug efficiency can be estimated from the PI curve (Fig. 5.5). Let's take the point at $I = 40$ *mA, where* $P = 3.4$ *mW. Using Eq. 5.18 and the assumption that* $R = 10\ \Omega$, *we find* $\eta_{wp} \approx 21\%$. *This latter is a pretty crude estimate; the calculation is properly done using the instantaneous slope of the IV characteristic to determine the resistance.*

Example 5.4

Assignment: *Assume we have a laser with the same differential efficiency as in Example 5.3 but emitting at* $\lambda = 1.55$ *μm. What is the slope you expect for the linear portion of the PI curve?*

Solution: *Examining Eq. 5.17, we see that the photon energy comes into the calculation. Given that* $\eta_d = 29\%$, *the slope of the PI curve at the longer wavelength would then be* $dP/dI = 0.23$ *W/A, about half the value we measured at* $\lambda = 792$ *nm.*

The reason is that efficiency measures the <u>number</u> of photons emitted per injected electron; however the slope efficiency is a measure of optical power, proportional to the number of photons times their energy. Photons at longer wavelengths have lower energy, so the same efficiency as a short-

wavelength device implies the same number of photons but each with lower energy, and thus lower power output for a long wavelength laser. Thus the output slope of the curve at longer wavelengths is smaller.

5.2.4 IV Measurements

The PI measurement is often accompanied by a voltage–current measurement; in automated measurement systems, this is trivial to implement. The laser diode is of course a diode and has the exponential electrical characteristic we saw in Fig. 3.15; since the PI plot has the current on the abscissa, the IV curve for a laser is then sometimes plotted on its side.

An IV plot has relevance for two aspects. Possible electrical problems, such as high series or contact resistance, may appear in the IV curve as a large offset, distinctly non-exponential increase in current with voltage or a significant linear component; electrical problems with packaging or micro-system mounting may sometimes be tracked down this way. Secondly, the IV curve may be differentiated to yield the instantaneous resistance, useful for calculating a realistic value for wall-plug efficiency. Laser manufacturers also frequently plot (I dV/dI), which allows determination of parameters such as internal resistance and more subtle variations in operating conditions.

5.3 Spectral Measurements

Lasers are frequently used in microsystems because of their excellent spectral properties. Emission at a single, known wavelength with a narrow spectral linewidth is often required for applications, such that the measurement of spectral properties is as relevant as that of output power.

5.3.1 Measurement Setup

Spectral measurements are slightly more involved than power measurements. The laser diode mount, current supply and temperature control may be identical to those discussed in the previous section. As shown schematically in Fig. 5.8, the emitted light must be coupled into a spectrometer which typically uses diffractive optical structures to separate the spectral components of the light.

Spectrometers may either be free-space-optical or fiber-optical components. In the former case, light is collimated and directed through free-space into the entrance slit of the spectrometer. A free-space-optical spectrometer is typically a bulky and sensitive apparatus requiring an optical table and some skill in performing the optical alignment; this type is generally used for very high resolution measurements.

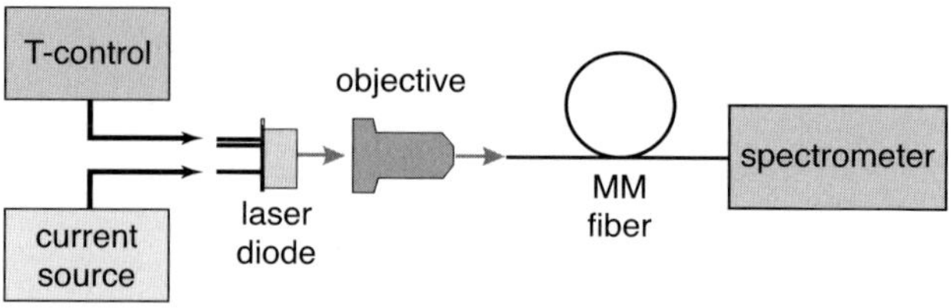

Fig. 5.8. Experimental setup for measurement of laser diode spectral characteristics, assuming a spectrometer with an optical fiber input

Developments in miniaturized and integrated optical technologies have resulted in the availability of a wide range of compact, fiber based spectrometers. These are typically fully self-contained devices with a physical size ranging from table-top equipment to cards which can be inserted into the back of a personal computer. Light is coupled into the spectrometer via an optical fiber, such that the laser emission needs to be focused onto a fiber facet, generally using a microscope objective as shown in Fig. 5.8. A multi-mode fiber is preferred to avoid inadvertently filtering features from the spectrum. The resolution of these compact and miniaturized spectrometers quite literally scales with size, with the smallest versions providing at least an order of magnitude less performance than table-top models, but the ease of coupling and fully self-contained data evaluation is a decided advantage for relatively quick and unproblematic spectral measurements. As a result, most end users of laser diodes will use this type of equipment for laser characterization.

The two essential characteristics of spectrometers are the wavelength range and resolution. The range is typically limited by the diffraction gratings and photodetectors employed and spectrometers which cover the entire visible to near infrared range are available; a wavelength span of 350 to 800 nm is typical. Resolution defines the smallest spectral feature which may be resolved; high-performance table-top models may reach 0.1 to 0.15 nm and miniaturized versions are typically limited to 2 to 5 nm.

5.3.2 Spectrum

A spectral measurement will result in a plot of optical intensity as a function of wavelength. From this spectrum, a number of important laser operating parameters may be determined.

From LED to Laser

The spectrum of a laser is quite different from that of an LED and the transition from one to the other may be used to spectrally define the onset of lasing. This may be seen clearly in Fig. 5.9 in which three spectra are shown for a laser diode in different operating regimes. The spontaneous emission results when the laser is operated well below its lasing threshold; the broad emission peak would correspond to that of an LED. The sub-threshold spectrum was taken a fraction of a milliampere below the threshold current; we see that the spectrum has narrowed noticeably. A fraction of a milliampere above threshold, the lasing curve results; this spectrum is both considerably narrower and about a factor 1 000 more intense.

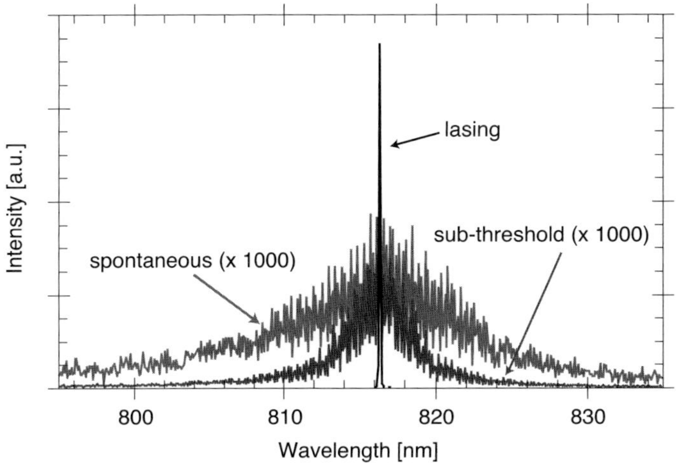

Fig. 5.9. The spectrum of a edge-emitting Fabry–Perot laser measured well below threshold, characterized by the spontaneous emission, directly below threshold (sub-threshold) and directly above threshold (lasing); the former two spectra have been multiplied by a factor 1 000 to make them visible. The vertical scale is linear, not logarithmic

This very abrupt transition from a broad, low-intensity spectrum to one which is very narrow and intense is characteristic of the abrupt laser threshold. All of the gain is restricted to the one mode with the lowest threshold gain and the spontaneous emission is drastically reduced.

Laser Spectrum Parameters

A typical spectrum for a near-infrared Fabry–Perot laser diode is shown in Fig. 5.10; the plot has a logarithmic vertical axis calibrated in dB, such that spectral details with orders of magnitude in intensity variation may be seen[4]. The figure shows the important parameters which may be determined from the spectrum: 1) center wavelength λ_0; 2) side-mode suppression ratio (*SMSR*); and 3) mode spacing $\Delta\lambda_{FP}$. From the given data in the figure, we would assign $\lambda_0 \approx 820$ nm, $SMSR \approx -26$ dB and $\Delta\lambda_{FP} \approx 0.20$ nm.

[4] Recall that dB for optical power or intensity are defined as 10 log (P/P_0).

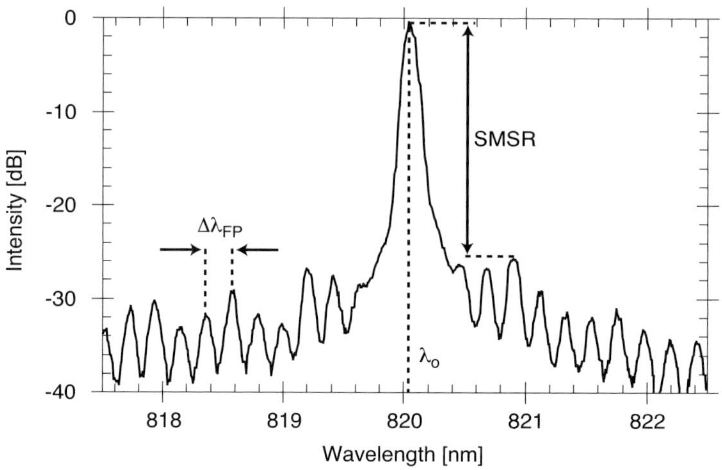

Fig. 5.10. A typical spectrum from an 820 nm Fabry–Perot laser; definition of the most important spectral parameters is shown. The vertical scale is, as it should be, logarithmic

From the spectrum of Fig. 5.10, we would be tempted to state that this is a single-mode laser; indeed, the next highest mode is about 26 dB (almost a factor 400) below the laser peak. Whether this is single-mode or not is a a function of the application: many telecommunications and interferometry applications consider a laser with side-modes higher than –30 dB to be multi-mode.

One parameter which can *not* be determined from the given spectrum is the laser linewidth. We might be tempted to measure the width of the laser peak and regard this as the laser linewidth; what we have measured, however, is usually the resolution limit of the spectrometer. Most good laser diodes have a linewidth in the range of $10^{-3} - 10^{-6}$ nm or less, whereas the spectrometer usually has a resolution of 0.1 nm. Thus the width of the peak as shown in the figure is probably orders of magnitude wider than the true linewidth of the laser; linewidth measurements will be discussed in Sect. 5.3.4.

5.3.3 Mode Spacing

The wavelength spacing between modes in a Fabry–Perot laser, $\Delta\lambda_{FP}$ in Fig. 5.10, allows us to determine, among other things, the optical length of the laser cavity. Conversely, knowing the expected mode spacing may allow the user to predict certain aspects of the laser spectral behavior, such as the occurrence of mode hops.

In Eq. 2.143 of Sect. 2.6.2, we saw that resonant phase matching in a laser cavity is assured for wavelengths λ_0 which satisfy

$$\lambda_0 = \frac{2nL}{i} \quad \text{for } i = 1, 2, 3 \ldots \tag{5.19}$$

for cavity length L and material refractive index n. If we take the 500 μm long GaAs laser emitting at 820 nm whose spectrum is shown in Fig. 5.10, then Eq. 5.19 implies that a typical value for the numeric index is $i \sim 4\,270$. The mode spacing is then the wavelength difference between the mode with, for example, $i \sim 4\,270$ and that with $i \sim 4\,271$.

We can calculate $\Delta\lambda_{\mathrm{FP}}$ by re-writing and taking the differential of both sides of Eq. 5.19:

$$i\lambda_0 = 2nL \Rightarrow d(i\lambda_0) = d(2nL) \tag{5.20}$$

or

$$\lambda_0 di + id\lambda_0 = 2Ldn \tag{5.21}$$

where we have assumed L to be constant[5]. For one mode step, $di = 1$, so that

$$\lambda_0 + id\lambda_0 = 2Ldn \tag{5.22}$$

and thus

$$i + \frac{\lambda_0}{d\lambda_0} = 2L\frac{dn}{\lambda_0}\,. \tag{5.23}$$

We may solve for $d\lambda_0$ to yield

$$d\lambda_0 = \frac{\lambda_0}{2L\frac{dn}{d\lambda_0} - i} \tag{5.24}$$

and since we know from Eq. 5.19 that $i = (2nL)/\lambda_0$, we may thus write

$$d\lambda_0 = \frac{\lambda_0}{2L\frac{dn}{d\lambda_0} - \frac{2nL}{\lambda_0}} = \frac{\lambda_0^2}{n\left(\frac{2L\lambda_0}{n}\frac{dn}{d\lambda_0} - 2L\right)} \tag{5.25}$$

and thus finally the mode spacing in wavelength,

$$d\lambda_0 = \frac{\lambda_0^2}{2Ln - 2L\lambda_0\frac{dn}{d\lambda_0}} \equiv \Delta\lambda_{FP}\,. \tag{5.26}$$

The denominator of the last expression contains the term $dn/d\lambda_0$ which represents the dispersion of the material in the laser core, namely the change of refractive index with wavelength. For accurate calculations, this should be considered in

[5] To avoid a notational mish-mash, we will use $\Delta\lambda_0$ to mean $\Delta\lambda_{FP}$ (thus the mode spacing, not the linewidth) in Eq. 5.20 – Eq. 5.26.

calculating mode spacing, but in many cases the very small changes in wavelength between modes imply that $dn/d\lambda_0 \approx 0$. In this case, Eq. 5.26 simplifies to

$$d\lambda_0 = \Delta\lambda_{FP} = \frac{\lambda_0^2}{2Ln}. \tag{5.27}$$

As a result of the narrow spacing of modes, relative to the laser wavelength, the spacing is often expressed in frequency, which is related to the wavelength value as

$$\Delta\nu_{FP} = \frac{c}{\lambda_0^2}\Delta\lambda_{FP}. \tag{5.28}$$

Example 5.5

Assignment: *Estimate the physical cavity length of the laser whose spectrum is shown in Fig. 5.10; through a secret contact at the laser manufacturer, you know that the laser core is made from* $Al_{0.3}Ga_{0.7}As$.
Solution: *First determine the mode spacing: by inspection, we estimate that* $\Delta\lambda_{FP} \approx 0.2$ *nm. The center wavelength is 820 nm.*
Next we need the refractive index of the core material. Using the approximation of Eq. 3.62 with $x = 0.3$, *we find that* $n \approx 3.39$.
Inverting Eq. 5.27 and entering these values, we find that $L \approx 497$ *μm; taking the inaccuracies of the data into account, we may conclude that the laser is probably designed to be 500 μm long.*

5.3.4 Linewidth

Some applications require knowledge of the laser linewidth. Although even a standard single-mode laser has a linewidth far below the resolution of the spectrometer such as that used to generate Fig. 5.10, the actual spectral width frequently needs to be known for optical data communications and spectroscopy applications (Zappe et al. 2000). The techniques used to determine the linewidth are generally quite involved.

Linewidth Basics

Semiconductor laser linewidth theory is well established (Henry 1982; Arnaud 1996) and is based in part on theoretical work which pre-dates the laser itself. A key expression for the linewidth expressed in frequency, $\Delta\nu$, is given by

$$\Delta\nu = \frac{v_g^2 h\nu\gamma\alpha_m}{8\pi P}F_{sp}(1-\alpha_{lef}^2) \tag{5.29}$$

which is very similar to a relationship given by Arthur Schawlow (1921–1999) and Charles Townes (b. 1915) (Schawlow and Townes 1958) and is thus sometimes referred to as the modified Schawlow-Townes formula. In Eq. 5.29, v_g [m/s] represents the group velocity of the mode moving in the laser (typically c/n_g); $h\nu$ the photon energy [J]; γ [m^{-1}] the gain; α_m [m^{-1}] the mirror loss, given by Eq. 5.14; P [W] the laser output power; and F_{sp} [ ] a spontaneous emission factor. The latter quantity may be found from

$$F_{sp} = \frac{\alpha_a + \gamma}{\gamma}, \tag{5.30}$$

where γ represents the gain and α_a [m^{-1}] is the absorption coefficient at the laser wavelength. A typical value for the spontaneous emission factor is $F_{sp} \approx 2.6$.

The final term in Eq. 5.29 is α_{lef} [ ], known as the linewidth enhancement factor. This factor relates the changes in real and imaginary refractive indices as a function of carrier concentration N [cm^{-3}], namely (Kuksenkov et al. 1995)

$$\alpha_{lef} = \frac{\frac{dn_R}{dN}}{\frac{dn_I}{dN}}\left(\frac{4\pi}{\lambda}\right) \tag{5.31}$$

for real part of the refractive index n_R and imaginary part of the refractive index n_I (recall the definition of Eq. 2.69 in Sect. 2.2.8 and the discussion in Sect. 2.6.1); the factor $4\pi/\lambda$ is traditional and occasionally omitted in some definitions of α_{lef}.

The linewidth enhancement factor is important in determining the spectral properties of a laser and can be determined experimentally for a laser diode (Krüger and Krüger 1995). α_{lef} gives a measure of the change in real refractive index when the imaginary refractive index, which is identical to the gain, is changed by increasing or decreasing pumping. Since the resonance wavelength of a laser is a function of the optical cavity length, essentially the product ($n_R L$), changing the pump level when, for example, modulating the laser output intensity will result in a variation of the laser wavelength, a phenomenon known as chirp. A laser with a large value for linewidth enhancement factor exhibits a large chirp and thus a large variation of emission wavelength with change in pump level, in addition to a large value for linewidth. Typical values for α_{lef} are in the range 3.5–4.0, although narrow linewidth laser can have values as low as 1.7 (Monti di Sopra et al. 1999)

Returning to Eq. 5.29, we see that a narrow linewidth may be achieved by assuring a small cavity mirror loss, small α_{lef}, small absorption losses and reduced spontaneous emission; these are parameters which can, to some extent, be considered in laser design. $\Delta\nu$ is also inversely proportional to laser output power, P, such that for a given laser, linewidth deceases as output power increases. One often plots measured $\Delta\nu$ as a function of $1/P$, which gives a linear characteristic with an intercept on the $\Delta\nu$ axis representing the residual linewidth (a lower limit on linewidth) and the slope a value for the linewidth-power product for that laser.

Linewidth Measurement and Optical Feedback

Due to the wavelength resolution limits of most standard spectrometers, measurement of linewidth usually requires specialized techniques; we will look at two below. We digress briefly beforehand to discuss the measurement setup, particularly the mounting and control of the laser diode, which are both more involved for these measurements than for those discussed previously.

The linewidth of a laser diode is very sensitive to the operating conditions of the laser, particularly current and temperature. Even the smallest fluctuations in drive current will lead to a measurable variation in the linewidth, such that linewidth measurements typically require ultra-low-noise current sources; current noise levels below 0.5 nA/√Hz are typical. The optimal current source is frequently a battery with simple current stabilization electronics but without connection to the power grid.

Temperature deviations likewise cause the linewidth to change and stabilization in the milli-Kelvin regime is usually necessary. Fluctuations in device temperature due to air currents above the optical table are easily seen in the linewidth measurement such that the experimenter's physical movement can noticeably affect the result; physical isolation of the laser diode is to be recommended.

One of the most drastic effects on laser linewidth is optical feedback. Feedback of the emitted laser light back into the cavity due to reflection from an external surface can strongly affect the laser behavior: the linewidth may be increased or decreased, depending on the feedback conditions (Petermann 1995; Hamel, van Exter and Woerdman 1992), the noise behavior is strongly affected (Hong et al. 2003) and in extreme cases, feedback leads to coherence collapse in the laser cavity (Lenstra, Verbeek and den Boef 1985). Feedback levels as low as 10^{-8} can noticeably affect laser behavior and the effects are seen in Fabry–Perot as well as distributed feedback (Favre 1987) and vertical cavity (Law and Agrawal 1997) lasers.

Optical feedback effects may be reduced by the use of an optical isolator. Usually based on crystals exhibiting the Faraday effect, named in honor of Michael Faraday (1791-1867), and a strong magnetic field (see Sect. 6.1.5), the optical isolator essentially transmits polarized light in only one direction; reflected light moving back to the laser is reduced, typically by 40 dB for bulk isolators. The optical isolator is generally large, heavy and quite expensive, although fiber-based models are more compact and more affordable. Their use in a linewidth measurement setup is, however, essential.

Fabry–Perot Interferometer

A relatively straightforward technique for linewidth measurement is through the use of a scanning, high-Q resonant Fabry–Perot cavity. The resonance wavelength of this etalon may be tuned by physically changing the cavity length through the use of piezo translators; the etalon length is swept through this small length range at a rate of several hundred Hertz so that the cavity acts as a very high resolution, variable wavelength filter. The scanning Fabry–Perot is usually quite a compact structure; commercial versions are often fiber-coupled, so that the user does not have to concern herself with cavity alignment. The typical measurement setup is shown in Fig. 5.11.

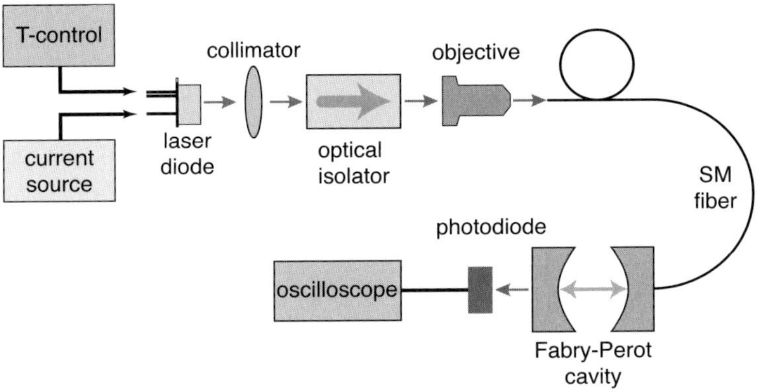

Fig. 5.11. A typical setup for measurement of laser linewidth using a scanning Fabry–Perot etalon

Scanning Fabry–Perot etalons for linewidth measurements typically have very high values for finesse; $F > 12\,000$ is typical. When coupled with a free spectral range of about 6 GHz, we see from the relationship of Eq. 2.127 that the expected resolution (given by the full width of the resonance peak) is about 500 kHz. From Eq. 5.28, we see that this corresponds to a wavelength resolution of about 1 fm, considerably below the 0.1 nm achievable with a standard spectrometer.

Recalling the laser mounting considerations discussed above, the laser emission is typically collimated and coupled through an optical isolator into a single-mode fiber via a microscope objective; the fiber transmits the light to the scanning Fabry–Perot etalon. A photodetector at the output of the etalon is then connected through a transimpedance amplifier to an oscilloscope whose sweep is synchronized with the scan of the cavity. The resulting output, an example of which is shown in Fig. 5.12, is the transmission though the etalon which is then directly the laser line-shape, assuming that the linewidth is larger than the etalon resolution. The numerical value for linewidth is then typically taken as *FWHM* of the characteristic.

Self-Homodyne Techniques

An alternative approach for linewidth measurement is through the use of homodyne (Baney and Sorin 1990; Ryu and Yamamoto 1986) or heterodyne (Schmid et al. 1996) techniques. In these methods, the light from the laser is split into two parts in an interferometer; these are de-correlated from one another using a delay

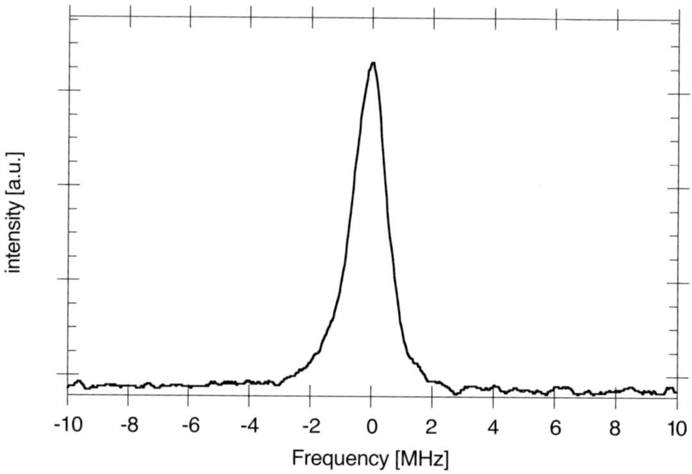

Fig. 5.12. Linewidth of a laser diode measured using a scanning Fabry–Perot etalon with a resolution of 600 kHz; the linewidth value (*FWHM*) is about 1.8 MHz. Measurement is of a near-infrared VCL emitting at 852 nm

line and the resultant interference between the two portions results in a peak whose width is twice the laser linewidth.

We may see this more clearly by considering the self-homodyne measurement configuration of Fig. 5.13. The same laser mounting concerns are relevant here; the setup is identical to the point where the light is coupled into a single-mode fiber. The light is then coupled into an unbalanced Mach-Zehnder interferometer; one arm of the interferometer has a delay line consisting of a long fiber loop (typical lengths are 2–5 km) and the other, shorter arm, may have mechanical polarization controllers to permit optimal polarization matching of the two signals. The length of the delay loop, which defines the extent of decorrelation between the signals and thus ultimately the spectral resolution of the setup, is typically limited by optical losses in the fiber. The output of the interferometer is directed onto a photodetector attached to an electrical spectrum analyzer.

If the delay line, the fiber loop in this case, is considerably longer than the coherence length of the laser (typically $L \geq 5L_c$), the optical signals in the two arms of the interferometer are decorrelated. When recombined at the output, the signals interfere to yield the auto-correlation function (convolution) of the two signals, namely

$$\boldsymbol{E}_T = \boldsymbol{E}(t) \otimes \boldsymbol{E}(t-\tau) \tag{5.32}$$

where τ represents the delay of the fiber delay line; we have used electric field as the signal variable. Since they are decorrelated, the two fields which interfere have independent phases; their random phase variations are due primarily to spontaneous photon generation which has an exponential time dependence, $\exp(-t/\tau_{spon})$,

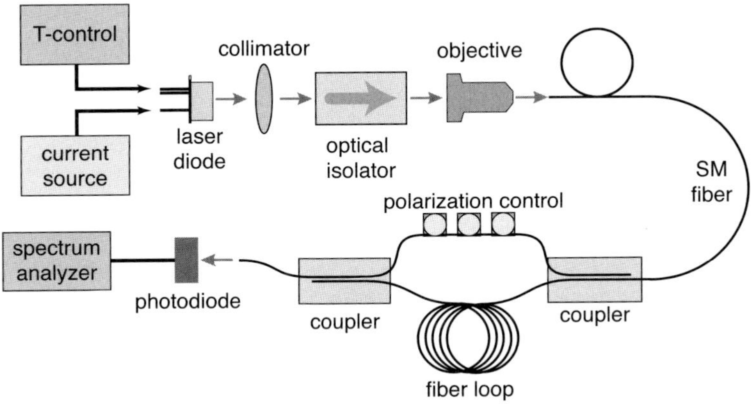

Fig. 5.13. A typical setup for measurement of laser linewidth using the self-homodyne approach

characterized by the spontaneous carrier lifetime. This random phase variation is what leads to a finite, non-zero value for $\Delta\nu$.

It can be shown that the convolution of the two de-correlated fields leads to a Lorentzian lineshape, such that the power spectral density as a function of frequency ν, $S(\nu)$ [W^2/Hz], at the output is given by

$$S(\nu) \propto \frac{\Delta\nu}{(\nu - \nu_0)^2 + \Delta\nu^2} \tag{5.33}$$

for laser oscillation frequency ν_0 [Hz] and linewidth $\Delta\nu$. The self-homodyne configuration transforms the laser oscillation frequency to $\nu_0 = 0$, so that the spectral line of the laser is given by a spectrum centered at zero and with a width equal to the linewidth. Thus $\Delta\nu$ is given by the *HWHM* (half-width half-maximum) of the resultant characteristic shown in Fig. 5.14; in this case, the measured spectrum for $\nu > 0$ (a "single-sided" spectrum) was duplicated and folded back for $\nu < 0$ to facilitate curve fitting.

The self-heterodyne approach is similar, save that an acousto-optic modulator in one arm of the interferometer modulates the signal at a frequency ν_{mod}; the output spectrum is then a complete Lorentzian (a "double-sided" spectrum) centered at ν_{mod}. The self-heterodyne method then translates the laser frequency back to the much smaller ν_{mod}, whereby the entire laser spectral line (both sides of the peak wavelength) is measured directly.

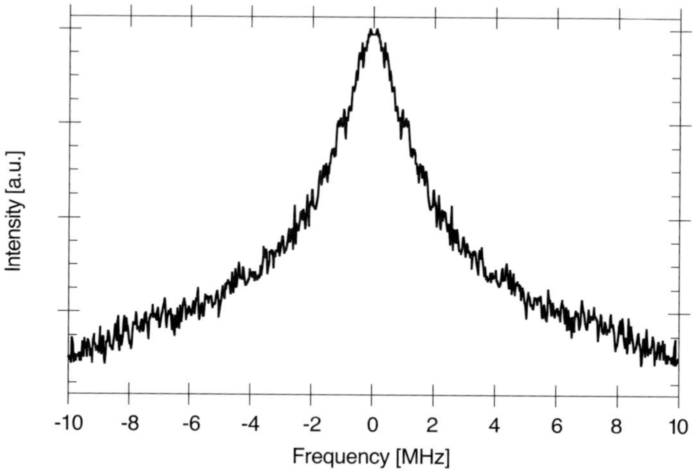

Fig. 5.14. Linewidth of a laser diode measured using a self-homodyne setup with a resolution of 100 kHz; the linewidth value (*HWHM*) is about 3.2 MHz. Measurement is of a near-infrared VCL emitting at 762 nm

5.4 Advanced Characterization

Whereas the power output and spectral behavior of a laser diode are of central importance, many specialized microsystem applications require knowledge of more detailed aspects of laser behavior. Some of these, such as polarization, are easy to measure while others, such as modulation bandwidth, require considerable experience and equipment investment. We discuss a few aspects of the relevant characterization procedures and expected results.

5.4.1 Polarization

Laser light is typically polarized. The polarization direction is relevant for applications which employ polarization-sensitive optics, such as polarizing beam splitters or diffractive optical structures. Polarization direction is defined as the direction in which the electric field points.

Characterization is quite simple: a polarization filter is placed between laser output and the photodetector and a comparison of the intensity for two orthogonal filter directions gives a measure of the polarization direction of the laser emission. Sheet polarizers, fabricated from dichroic sheets of stretched organic molecules, are applicable for low-power laser diodes; however, since they filter the unwanted polarization direction out by absorption in the sheet itself, high-power lasers can easily damage this type of polarizer. Alternatively, bulk optical polarizers made from birefringent materials (such as $CaCO_3$ or quartz) may be used. The Glan-type polarizer, in the Glan-Thompson or Glan-Taylor configuration (see Sect. 6.1.5), consists of two prisms, either closely spaced or cemented together, and directs one of the polarizations out of the optical axis onto an external absorber; this type of polarizer is generally used for high-end and high-power measurements. For all cases, the measurement resolution is a function of the filtering capability of the polarization filter expressed as an extinction ratio; sheet polarizers typically have an extinction ratio on the order of 10^{-4}.

Edge-emitting laser diodes are generally linearly polarized with the polarization direction in the plane of the laser surface. Near threshold, the preponderance of spontaneous emission, which is unpolarized, results in a relatively poor polarization ratio, but for laser drive currents significantly above threshold, the polarization ratio is on the order of $1{:}10^3$. Edge-emitters are consequently considered polarization-stable.

VCLs, on the other hand, do not have *a priori* a defined polarization direction. Due to the circular aperture, the physical structure does not lead to a preferred polarization direction and, in theory, all polarization directions are degenerate. In practice, anisotropies in the structure as well as birefringence and dichroism effects result in pinning the polarization direction parallel to the crystal planes of the substrate (Martin-Regalado et al. 1997). Unfortunately this still leads to two degenerate orthogonal polarization directions and the output of the VCL may often "flip" from one polarization direction to the other as drive current is varied (van Exter, Willemsen and Woerdman 1998) or if thermal lensing effects become active (Panajotov et al. 1998).

Polarization-stable VCLs are still an unsolved problem at the time of writing. Numerous approaches to pin the polarization direction have been proposed and demonstrated, with varying success: the use of anisotropic (311) substrates (Mizutani et al. 1998), anisotropic mesa shapes (Yoshikawa et al. 1998) and anisotropic mechanical deformation (Monti di Sopra, Brunner and Hövel 2002; Kallweit et al. 2002) have been investigated. For the user of garden-variety VCLs, complete polarization characterization of each individual VCL at all drive currents and temperatures of interest is essential before installing the device into a microsystem with polarization-sensitive optics.

5.4.2 Noise

Laser diode noise is a complex but well-studied phenomenon (Petermann 1991); noise characteristics are closely related to modulation bandwidth as well as spectral linewidth. For semiconductor lasers used in data communications systems, high-resolution sensors and spectroscopy, characterization of noise is required to optimize system performance.

Basic Noise

Noise is essentially a measure of the fluctuation of laser parameters with time. The time-dependent output power, *P(t)*, for example, may be expressed as

$$P(t) = \langle P \rangle + \delta P(t) \tag{5.34}$$

in which $\langle P \rangle$ represents the mean value of power, averaged over time, and $\delta P(t)$ the fluctuation about this mean value; we may write analogous expressions for wavelength and any other laser parameter.

The total noise-based fluctuation of a parameter may be characterized by the relative intensity noise, *RIN*, measured in [dB] or [dB/Hz]. *RIN* measures the ratio of parameter fluctuations to the mean value of that parameter and is thus defined as

$$RIN = \frac{\langle \delta P(t)^2 \rangle}{\langle P(t)^2 \rangle} . \tag{5.35}$$

The numerator gives the mean square of the noise distribution where we recall that the mean square of a parameter $\Delta\xi$ is given by

$$\langle \Delta\xi^2 \rangle = \overline{(\xi - \bar{\xi})^2} \tag{5.36}$$

in which $\bar{\xi}$ denotes the average value of the parameter ξ. *RIN* describes the fluctuation of intensity about a mean value; these fluctuations can have a number of origins including thermal effects, the discrete character of electrons and photons as well as the statistical nature of current and light transmission.

The magnitude of parameter fluctuation, and thus the noise level, may be characterized by the spectral noise density Ξ_N [W^2/Hz] such that the relative intensity noise may be expressed as

$$RIN = \frac{2\Delta f \Xi_N}{\langle P^2 \rangle} . \tag{5.37}$$

This relation is often written in a slightly different form, namely

$$\frac{RIN}{\Delta f} = \frac{2\Xi_N}{\langle P^2 \rangle} \tag{5.38}$$

thus giving a measure for the noise per unit bandwidth ($\Delta f = 1$ Hz). These last two relations explain the units generally used for *RIN*: in Eq. 5.37 it is given as the ratio between two powers, thus expressed as dB, and in Eq. 5.38, the noise is given as the power ratio per unit bandwidth, thus dB/Hz. The latter is more common.

RIN is a measure of intensity noise, for example the variation of the output intensity of individual modes in a laser diode. We see from Eq. 5.37 or Eq. 5.38 that intensity noise varies with P^{-2}, so that increasing the output power will effectively decrease *RIN*. The noise level is typically highest near laser threshold and decreases as drive current increases. We also note that increasing the laser bandwidth (see section Sect. 5.4.3) increases the value for *RIN*, since more spectral components contribute to the total noise. Spectrally, *RIN* increases with decreasing frequency as $1/f$, which represents the ubiquitous if poorly understood phenomenon of $1/f$ noise.

Numerous other forms of noise contribute to fluctuations in the semiconductor laser output parameters; we will outline only a few of these here. Mode partition noise derives from the variable distribution of total output power between various laser modes. The power of individual modes may fluctuate but the total output power integrated over all modes may be relatively constant such that overall *RIN* is small. Mode partition noise is worst for single-mode lasers such that a high *SMSR* (typically 60 dB or more) is required to reduce this noise figure. This noise source is especially acute for VCLs (Kuchta et al. 1993) such that multi-mode lasers are often preferred for short-range communications applications in which the noise performance is a crucial determinant of system performance.

Closely related to mode partition noise is mode-hopping noise. In the vicinity of a mode-hop (see Sect. 5.5), due to current or thermal tuning of the semiconductor laser, the total output power may dynamically switch between the two or more modes close to lasing. The relative intensity noise is thus increased in the vicinity of a mode hop. In VCLs, it has also been see that *RIN* increases in the vicinity of a polarization flip.

Spectral stability of laser output is characterized by frequency or phase noise. This noise component defines the variation of output wavelength or phase with time. The white frequency noise spectrum is the primary source of finite laser linewidth.

Noise Measurement

Whereas the measurement setup for noise characterization can be relatively simple, considerable experience is required to adequately interpret and make sense of the results. As shown in Fig. 5.15, the light from an appropriately mounted and temperature-controlled laser diode is directed through an optical isolator onto a photodetector. The electrical output of this detector is connected to an electrical spectrum analyzer which directly gives the noise intensity as a function of frequency.

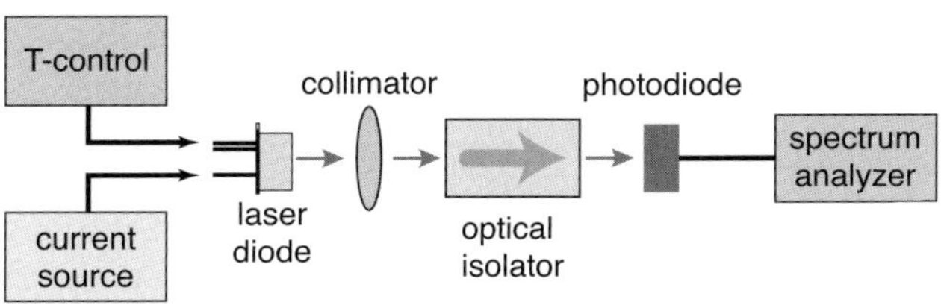

Fig. 5.15. Characterization setup used for measuring laser diode noise

Special care must be taken in the choice of components; the same considerations as described for linewidth measurements in Sect. 5.3.4 apply here. Low noise current sources, photodiodes and amplifiers are essential if the noise measurement is to be of the laser output as opposed to the measurement equipment. The electrical spectrum analyzer directly plots the energy in a given spectral interval (typically 1 Hz) as a function of frequency; the bandwidth of the spectrum analyzer is chosen to include all the frequencies of interest, and typically reaches into the GHz range. Note that this is not the same piece of equipment as the spectrometer we used to measure the optical spectrum in Sect. 5.3.2, also sometimes called an "optical spectrum analyzer" which measures optical frequencies, in the 100 THz range.

The direct measurement of *RIN* yields a plot such as that shown in Fig. 5.16; this characteristic is measured at frequencies in the range 5 to 20 kHz and we clearly see the decrease of *RIN* as current is increased above threshold. The noise units are [Hz^{-1}] in this plot; [dB/Hz] would use a linear vertical axis.

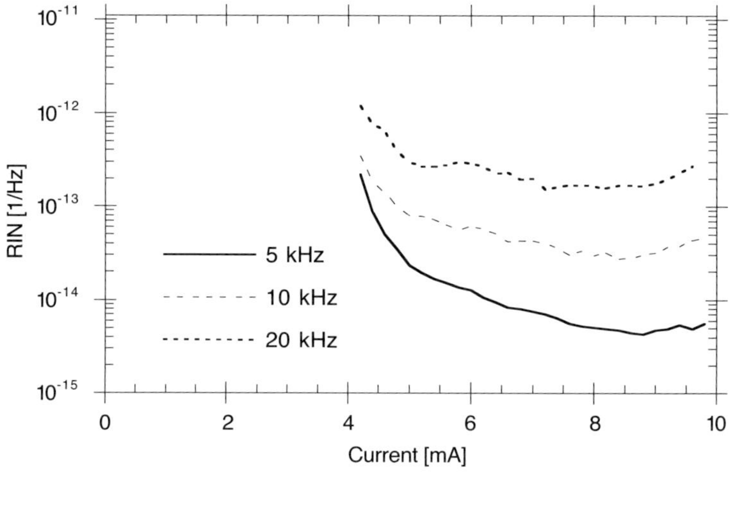

Fig. 5.16. Plot of *RIN* as a function of current for a laser with I_{th} = 4 mA measured at three frequencies

Frequency noise of laser emission is more difficult to measure as the phase or frequency variations require a stable reference wavelength. One typically uses a strong atomic or molecular absorption line, such as gaseous Rb at 780 nm or Cs at 852 nm (Ikegami, Sudo and Sakai 1995) and biases the laser so that the mean emission wavelength is on the steep flank of the absorption peak. Variations in the wavelength of the laser then result in moving back and forth (in wavelength) on this flank which gives rise to a large intensity change of the optical signal transmitted through the gas. Thus frequency noise is translated into intensity fluctuations allowing calculation of the former in units of [GHz/Hz] or [nm/Hz].

5.4.3 Modulation Bandwidth

Operation of lasers in microsystems which require high-speed modulation of the laser output, such as for optical data transmission, requires that the modulation bandwidth of the laser be measured. Analysis of high-speed laser modulation is well established (Tucker 1985; Lau and Yariv 1985; Lau 1989) and modulation speeds approaching 50 GHz have been demonstrated. The measurements themselves require advanced equipment and considerable experience to perform and interpret, the more so the higher the modulation speed.

Laser Dynamics

The dynamics of laser modulation are based on the time-dependent interaction of carrier and photon densities in the laser cavity, N_e [cm^{-3}] and N_p [cm^{-3}] respectively. These are related by the laser rate equations (Coldren and Corzine 1995, Chap. 5; Nagarajan and Bowers 1999) which are given by

$$\frac{dN_e}{dt} = \frac{\eta_i}{qV}I - (R_{sp} - R_{nr}) - v_g \gamma N_p \tag{5.39}$$

and

$$\frac{dN_e}{dt} = \left(\Gamma v_g \gamma - \frac{1}{\tau_p}\right) N_p + \Gamma R'_{sp} \; . \tag{5.40}$$

In the previous two expressions, η_i is the internal quantum efficiency; V [m^3] the active region volume; I [A] the laser current; R_{sp} and R_{nr} [cm^{-3}s^{-1}] the spontaneous and non-radiative recombination rates, respectively; R'_{sp} [cm^{-3}s^{-1}] the spontaneous emission into the mode of interest; v_g [m/s] the group velocity; γ [cm^{-1}] the gain; Γ [ ] the confinement factor and τ_p [s] the photon lifetime.

For steady-state laser emission, the time derivatives are zero and the rate equations directly yield time-invariant values for N_e and N_p. From these, the power output of the laser may be expressed as a function of the photon density N_p simply as

$$P = \eta_e h \nu \frac{N_p}{\tau_p} \Gamma V \; . \tag{5.41}$$

Small-Signal Response

Time-varying current leads to time-varying carrier concentration and photon density. Assuming that these parameters may be represented by the sum of constant (N_{p0}) plus sinusoidally time-varying (N'_p) components[6], thus taking the form (using photon density as an example)

$$N_p(t) = N_{p0} + N'_p e^{j\omega t} , \tag{5.42}$$

the rate equations may be solved to yield the small-signal response of the laser. The solutions lead to a transfer function, *H(ω)*, as a function of modulation frequency ω, which has the form

$$N_p(\omega) = N_{p0}H(\omega) = N_{p0}\frac{\omega_R^2}{\omega_R^2 - \omega^2 + j\omega\zeta} \tag{5.43}$$

where ω_R [rad/s] is a resonant frequency and ζ [rad/s] represents a damping term. The photon density, and thus output power of the laser, thus exhibits a classical resonance behavior, as seen in Fig. 5.17; the transfer function peaks at the resonant frequency ω_R where the height and width of the resonance is defined by the damping ζ. For low frequencies, *H(ω)* is constant and usually normalized to unity; far above the resonant frequency, *H(ω)* decreases with 40 dB/decade.

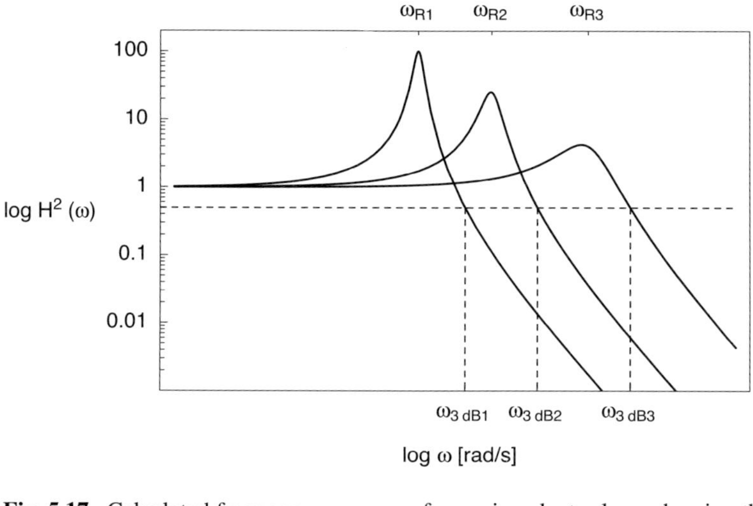

Fig. 5.17. Calculated frequency response of a semiconductor laser showing the relaxation oscillations for three values resonance frequency, ω_{R1}, ω_{R2} and ω_{R3} with variable damping terms: $\zeta_1 = 0.1\omega_1$, $\zeta_2 = 0.2\omega_2$ and $\zeta_3 = 0.5\omega_3$. The modulation bandwidths, defined as 3 dB below the low-frequency value of *H*(ω), are given by ω_{3dB1}, ω_{3dB2} and ω_{3dB3}

The resonance in a semiconductor laser, the relaxation oscillation, defines the modulation bandwidth of the laser. Above threshold, the resonance frequency may be approximated as (Coldren and Corzine 1995, Chap. 5.3.1)

[6] Note that we distinguish between ω [rad/s], the electrical modulation frequency of the laser, and ν [Hz], the optical oscillation frequency of the laser; the former ranges from Hz to GHz, the latter is in the 100 THz range.

$$\omega_R^2 \approx v_g \frac{N_p}{\tau_p} \gamma_d \tag{5.44}$$

or

$$\omega_R^2 \approx \frac{\Gamma v_g \gamma_d}{qV} \eta_i (I - I_{th}) \tag{5.45}$$

where γ_d [m^2] represents the change in gain with carrier concentration,

$$\gamma_d = \frac{\partial \gamma}{\partial N_e} \tag{5.46}$$

and is known as the differential gain. In addition, the damping term may be approximated as

$$\zeta = \zeta_R \left(\frac{\omega_R}{2\pi}\right)^2 + \zeta_0 \tag{5.47}$$

where $\omega_R/(2\pi)$ represents the resonance frequency in Hertz; ζ_R and ζ_0 are usually found from a fit to the measurement data.

We see from Eq. 5.44 and Eq. 5.45 that the resonance frequency increases with N_p so that increasing the output power by increasing the laser current increases the resonance frequency. The laser modulation bandwidth is defined as the frequency for which $H(\omega)$ decreases by 3 dB below its low-frequency value and, for $\zeta \ll \omega_R$ (i.e., low damping), is approximately given by

$$\omega_{3dB} \approx \left(\sqrt{1 + \sqrt{2}}\right)\omega_R \,. \tag{5.48}$$

Increasing the resonance wavelength is therefore a design criterion for high-speed laser diodes.

Large-Signal Response

The calculation of small-signal response was predicated on the assumption that the laser output varies about a bias point above laser threshold. Large-signal modulation results when the laser is turned on and off, thus varying the current from below to above threshold, and becomes an issue for digital modulation of the laser output. For this case, an analytical solution of the rate equations is not possible and numerical approaches must be employed. Resulting parameters of interest include, for example, the turn-on delay, which in turn defines the maximum modulation speed.

Measurement

As we mentioned at the outset, high-speed modulation measurements of laser diodes are not particularly simple. Part of the problem is that, at high modulation frequencies, the parasitic response of virtually all components of the measurement system plays a role and may even dominate the behavior. The measured transfer function, $H_{meas}(\omega)$, is then given by

$$H_{meas}(\omega) = H_{laser}(\omega) \cdot H_{external}(\omega) \tag{5.49}$$

where $H_{laser}(\omega)$ is the intrinsic transfer function of the laser, given in Eq. 5.43, and $H_{external}(\omega)$ the transfer function of the measurement system, including the contacts, bond pads, wiring, probers, connectors and external circuitry.

The laser may be characterized while mounted in a specialized high-speed package or on the chip level; contact to the laser is typically through coaxial probers with which both signal and ground are brought to the laser under test. Either 50 Ω impedance-matched cables and connectors or, for very high frequencies, microwave waveguides are used. Laser power supply is through a microwave signal generator or network analyzer which, for small signal measurements, applies a high-speed modulation on a dc bias. The measurement treats the laser as a two-port and the measurement system typically yields values for the s-parameters; of particular relevance are the reflection, s_{11}, and transmission, s_{21} from which the small-signal transfer function, $H(\omega)$, may be determined.

The large-signal behavior is typically characterized by an eye diagram, such as that shown in Fig. 5.18 (Czotscher et al. 1999). These characteristics are generated by sending a pseudo-random bit sequence to the laser by a network analyzer; the abscissa is a time axis and the ordinate gives the laser output response. A laser which may be modulated at the given large-signal pulse rate will lead to "open eyes", as seen in the figure; as the frequency is increased beyond the bandwidth of the laser, the "eye" will close.

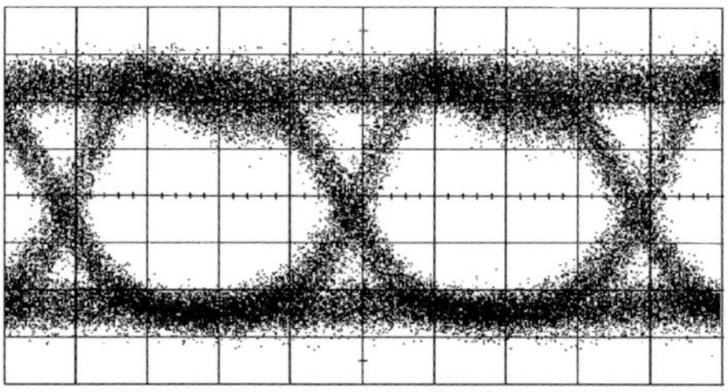

Fig. 5.18. Eye-diagram resulting from the large-signal measurement of a 1.55 μm, InP-based DFB laser modulated at 10 Gb/s; the time axis is 25 ps/div. Characteristic courtesy of the Fraunhofer Institute for Applied Solid State Physics

5.4.4 Reliability

A user who incorporates a semiconductor laser diode into an optical microsystem would like it to operate for a longer time with stable characteristics. In particular, the laser should provide repeatable output power and wavelength for a given current and temperature with little long-term drift and should certainly not suddenly stop functioning altogether. Measurement and optimization of laser diode reliability and lifetime is a topic which has concerned semiconductor laser developers since the early history of the technology and represents a fundamental issue in the acceptance of diode lasers for use commercially-viable optical microsystems (Fukuda 1991).

The degradation mechanisms which lead to a long-term drift in the laser output characteristics are manifold (Takeshita et al. 2002). Degradation of the crystal structure, including migration of defects and "dark lines", localized regions of enhanced non-radiative recombination due to clusters of defects, represent fundamental material problems. The emission facets are subject to oxidation, defect migration or degradation, such that locally-enhanced heating can lead to catastrophic mirror damage. The laser package is also an issue, since metal diffusion, solder degradation and the reliability of bond wires may all contribute to limiting the laser lifetime.

The effect of these degradation mechanisms may take the form of diverse changes in operating characteristics. For a given current and temperature, the output power generally decreases with time, the wavelength may drift, the noise will increase and the near- and far-field patterns may shift. In addition, the occurrence of catastrophic optical damage (COD) will result in a sudden cessation of lasing which, at least in edge-emitting lasers, is typically characterized by extreme facet damage.

Reliability and lifetime measurements are generally done by laser manufacturers when developing new laser designs and to monitor production of existing designs; high-end laser diode users may perform measurement of specialized reliability parameters themselves. To assure long laser lifetime and thus a long mean-time-to-failure (*MTTF*), manufacturers will typically operate large numbers (hundreds to thousands) of lasers continuously at a variety of elevated temperatures, typically while periodically monitoring threshold current and output power. The "failure" of a laser is subject to definition: it clearly fails if it stops working altogether due to COD and a long-term reduction in output power by 30% or 50% is also considered a failure. Based on the *MTTF* as a function of temperature, which typically follows an Arrhenius distribution with an activation energy characteristic for each failure mechanism (Moser and Latta 1992), extrapolation to room-temperature *MTTF* can be made. Using this approach, expected lifetimes exceeding 10^6 hours (> 100 years) have been determined (Meier and Graf 1996).

Spectroscopic applications not only require stable output power but also stable emission wavelength. Characterization of spectral aging is slightly more involved as a stable reference wavelength must be employed. A strong atomic absorption peak may be used (Ikegami, Sudo and Sakai 1995) and the laser locked to the absorption wavelength; the current required to reach this wavelength (at constant temperature) is measured and the drift monitored as a function of time. The extent of spectral aging varies strongly between devices and even within groups of nominally identical devices; values of about 50 GHz/year have been determined for VCLs optimized for spectroscopy (Monti di Sopra et al. 2001).

5.5 Temperature Effects

All of the previously described laser characteristics are a function of temperature. It is for this reason that the point of temperature stabilization of the laser under test was belabored. The temperature sensitivity also results in the necessity for temperature-stabilization features when laser diodes are used in applications where performance stability is required yet ambient temperature may vary. Since temperature-stabilization components represent an expensive part of a complete optical microsystem, the shifts in laser behavior as a function of temperature need to be characterized to determine the acceptable operating temperature range of the system.

The effective temperature of a laser may shift due a change in ambient temperature, chip temperature and internal junction temperature. The junction temperature is a function of not only ambient temperature but also of the laser drive current; changing the output power of the laser by varying the laser current therefore also changes its junction temperature, even for constant ambient. Thus shifts in the operating characteristics of laser diodes due to changes in drive current are also ultimately temperature effects.

In general, laser diode performance gets worse with increasing temperature. As we will see in greater detail below, the threshold current (or current density) increases, efficiency decreases and noise increases. The emission wavelength shifts to longer values, in and of itself not a degradation, but the change in wavelength must be considered for applications dependent on wavelength stability. On the other hand, temperature and current tuning of emission wavelength may prove to be an attractive feature in some systems.

5.5.1 PI vs. Temperature

The power-current characteristic changes quantitatively and qualitatively with temperature. The threshold current shifts to larger values with increasing temperature and the slope of the curve will generally decrease. In addition, the PI curve may exhibit kinks and will usually become non-linear with increasing temperature, causing the output characteristic to flatten; this latter effect is manifested as a decrease in efficiency as temperature and drive current are increased. We may refer forward to the example characteristics of Sect. 5.7 to see the behavior explicitly: the PI curves for a DFB laser shown in Fig. 5.39, for example, clearly show the shift in threshold current and the change in the slope of the curve as temperature is increased.

The variation of I_{th} with temperature is given approximately by the empirical relationship (Pankove 1968; Seki et al. 1996)

$$I_{th}(T) = I_{th0}e^{T/T_0} \tag{5.50}$$

where T [°C] is the temperature, I_{th0} [A] the threshold current at $T = 0$°C and T_0 [°C] is a fit parameter; T_0 does not correspond to any physical temperature in the laser and is merely an empirical parameter derived from an analysis of the temperature behavior by means of a curve fit. Equation 5.50 is a simplified approximation and relationships which provide a better fit have been proposed (Prosyk, Simmons and Evans 1998).

T_0 provides a measure for the temperature sensitivity of the laser threshold; smaller values imply a greater sensitivity to temperature. The sensitivity is to first order a function of the laser material; typical values for AlGaAs range from $T_0 = 100$–200°C whereas for InGaAsP values on the order of $T_0 = 50$°C are more typical.

A plot of I_{th} and η as a function of temperature for constant drive current, as shown for example for the DFB laser in Fig. 5.40, clearly shows the increase of the former and decrease of the latter with increasing temperature. The two effects are both related to the same physical processes in the laser. As temperature increases, the internal efficiency decreases, non-radiative recombination, particularly Auger recombination, increases and thermal carrier leakage from the heterojunction increases. These processes result in an overall poorer device performance. Since the InGaAsP material system, that from which telecommunications lasers are predominantly made, suffers from significant Auger recombination, the temperature dependence of this recombination processes explains why these lasers generally have low values for T_0 and thus high temperature sensitivity (Li and Bradford 1995; Pan and Chyi 1996).

Quantum well-based lasers generally have improved temperature performance when compared to non-quantum well devices, due to confinement effects in the quantum well. Vertical cavity lasers also show distinct temperature performance (Taylor and Evaldsson 1994), particularly since the spectral position of the gain peak and the cavity resonance have different dependencies on temperature. As a result, the gain and cavity peaks will rapidly become misaligned as temperature is varied, resulting in saturation and subsequent decrease in output power with increase in current, evidenced by the classic flattening of the VCL PI curve with increasing current.

5.5.2 Spectrum vs. Temperature

Changing laser temperature also results in a shift of the laser wavelength; usually, increasing T leads to increasing λ. In edge-emitting Fabry–Perot lasers, the wavelength shift is predominantly due to a change in the refractive index of the laser material with temperature: the optical length of the cavity increases with increasing temperature so that the resonant wavelength also increases. The physical length of the cavity also changes with temperature, but this is a higher order effect and generally negligible. The shift in wavelength is material dependent and is in the range of 0.25 nm/°C for GaAs-based materials and 0.5 nm/°C for the InGaAsP system.

An example of the wavelength shift with temperature is shown in Fig. 5.19, measured for a GaInP-based Fabry–Perot laser emitting in the red. The drive current was held constant and the device temperature increased externally. The red-shift in wavelength of almost 10 nm is clearly seen, as is the decrease in output intensity. For the highest temperatures, the increase in threshold current is such that the laser barely operates above threshold and the emission approaches that of an LED.

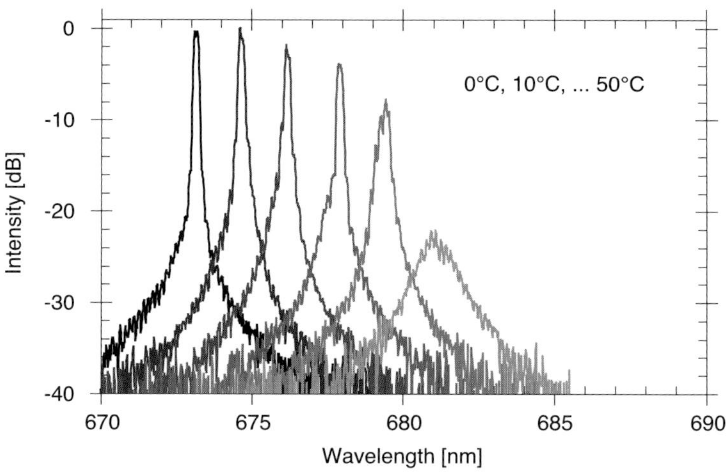

Fig. 5.19. Variation of the emission spectrum of a 670 nm edge-emitting Fabry–Perot laser as a function of temperature for constant drive current; temperature varies from left to right from 0°C to 50°C in 10°C steps. The increase in wavelength and decrease of output intensity is clearly seen

For lasers which use a distributed feedback structure to form the laser cavity, such as DFB, DBR or VCL lasers, the temperature tuning of wavelength is more complicated. The change in material refractive index with temperature results in a change of the resonance behavior of the grating reflector, such that the complete temperature-dependent device behavior is a combination of shifts of the refractive index as well as of the grating reflection wavelength. In a DBR laser, for example, the gain peak moves at 0.25 nm/°C and the cavity resonance peak at 0.07 nm/°C; the resultant modal shift in temperature results from a complex interplay of these shifts and is on the order of 0.06 – 0.09 nm/°C (Hoftsetter and Zappe 1997). In addition, non-uniform heating effects may result in anomalous mode-hop behavior, giving rise to jumps which decrease in wavelength as overall device temperature is increased.

For DFB lasers, the gain peak (for an InGaAsP device) may move at 0.4 nm/°C whereas the grating resonance peak shifts with 0.05 – 0.1 nm/°C, also due to a change in the material refractive index. We recall that lasing occurs at a wavelength at which the gain has a peak and the cavity has a resonance (recall the discussion which led to Fig. 2.21); if the gain and resonance peaks move with temperature at different rates, the laser will degrade if the overlap between the two becomes small. In a DFB laser, the resonance and gain peaks move apart in this manner with temperature and it may occur that the DFB laser will begin to lase in a Fabry–Perot mode rather than a distributed feedback mode for large temperature increases. The same type of relative shift in gain and cavity peaks is seen in VCLs, where increases in temperature can give rise to sufficient misalignment between the peaks so that the VCL stops lasing altogether.

The temperature effects described in this and the last section may occasionally be used, for example, to tune the laser emission wavelength. In general, however, wavelength-sensitive applications require stable emission at a desired frequency with no shift, even if ambient temperature changes. Laser emission stability may then be assured through the use of temperature-stabilization techniques, such as Peltier coolers and the associated sensors and control electronics. Including such a system in every laser module often leads to unacceptable system costs, such that considerable effort is being expended in the design and fabrication of temperature-insensitive laser diodes (Lu, Blaauw and Makino 1996; Silver et al. 2002; Tansu and Mawst 2002).

5.5.3 Current Tuning

As mentioned in the introductory section above, the relevant temperature for determination of the operating characteristics of a semiconductor laser is that of the junction and cavity. This temperature may be varied by changing the temperature of the laser and its package as a whole but also changes if the internal power dissipation is modulated due to an increase or decrease in laser drive current. Thus the laser operating conditions, predominantly wavelength, vary with laser drive current but the ultimate physical effect is a change in internal device temperature.

The current-dependence of laser wavelength is a strong function of the device structure. Depending on the current distribution in the active region, the bulk and contact resistance, heat-sinking and related factors, a given current change may lead to greater or lesser internal temperature change and thus shift in wavelength. Typical values which might be expected for current-dependent wavelength shift are on the order of ~0.03 nm/mA for Fabry–Perot edge-emitting lasers and ~0.5 nm/mA for VCLs.

Since increasing current leads to increasing power dissipation and thus increasing temperature, wavelength generally increases with a rise in drive current. In contrast to the roughly linear shift in wavelength with changing temperature, wavelength varies as I^2 since power is proportional to the square of current. As we see in Fig. 5.20, the wavelength shift may either be continuous or subject to discrete jumps, depending on the laser in use. The jumps in wavelength, known as mode-hops, may be understood from the schematic of Fig. 2.21: for a relatively dense spectrum of resonance peaks, a shift in the gain spectrum with temperature will result in the resonance at the next higher wavelength in eventually having a higher gain than the currently lasing wavelength. The laser then "hops" to that next peak, accompanied by a sudden jump in wavelength. This effect is particularly acute in Fabry–Perot lasers but is never seen in VCLs; due to the short cavity, the VCL has only one resonant cavity peak in the vicinity of the gain peak (the next possible cavity peak is very far away in wavelength), so that there is effectively no other mode to "hop" to.

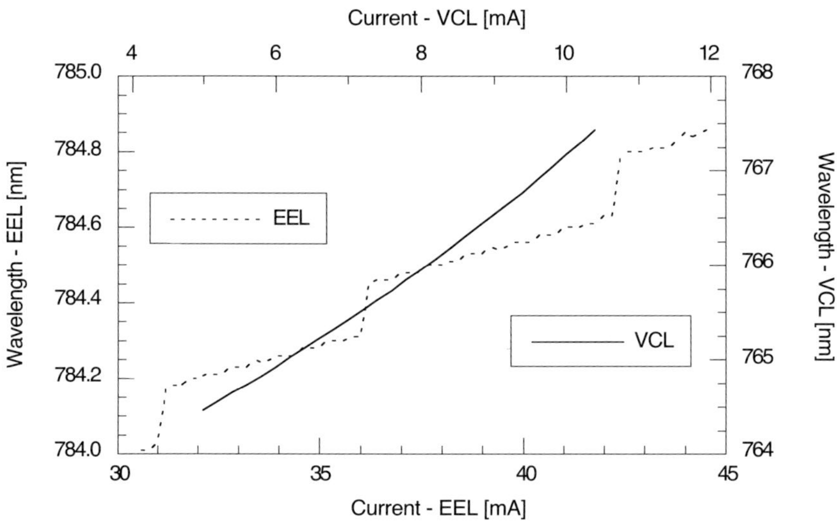

Fig. 5.20. Wavelength tuning by current for an edge emitting laser (EEL) and a vertical cavity laser (VCL); note the difference in horizontal and vertical scales for the two lasers. External temperature was held constant. The EEL exhibits three mode-hops whereas the VCL is mode-hop-free

Current wavelength tuning represents a useful tool for reaching an accurately desired wavelength; some laser types, such as VCLs, may be continuously tuned over several nanometers. The effect is quite fast, so current variations may be used to scan the emission wavelength over a certain range, such as for spectroscopy applications in which the wavelength is scanned over an absorption peak (Zappe et al. 2000). In other laser types, the occurrence of mode hops may limit the wavelength tuning range.

5.6 Intensity Profiles

The light from a laser diode or LED built into an optical microsystem will ultimately need to be coupled into some optical element: lenses, fibers, waveguides, beam-splitters and so forth. Optimization of optical coupling will generally result in system performance improvement and accomplishing this requires knowledge of the spatial intensity distribution of the optical field emitted from the laser.

Measurement of spatial intensity profiles take two forms: the near-field and the far-field. The two are distinguished by the measurement distance from the emitting facet. The near-field profile is the intensity distribution directly at the emitting facet of the laser, known as the Fresnel region. As the point of observation is moved away from the facet, there is a gradual transition to the Fraunhofer region, which then corresponds to the far-field intensity distribution. For a laser emitting at a wavelength λ_0 with an aperture size a [m], the far-field regime is reached for distance from the aperture d [m] given by

$$d > \frac{a^2}{\lambda_0} . \tag{5.51}$$

For typical laser diode apertures ($a \sim 5$ μm) and wavelengths (say down to $\lambda_0 \sim 500$ nm), we find that $d \sim 50$ μm; thus measurement of intensity a few millimeters from the laser facet, a practical value, is quite definitely in the far-field region.

The near- and far-fields are related by the spatial Fourier transform. Emission from a small, well-defined facet will lead to a broad far-field distribution, due predominantly to diffraction effects. A large aperture will result in tighter far-field profile and a Gaussian emission profile (such as from a VCL) will give rise to a Gaussian far-field.

5.6.1 Near-Field Measurement

The near-field measurement is an imaging measurement. As shown schematically in Fig. 5.21, the laser is mounted, temperature-stabilized and driven at a constant current. The optical emission is imaged through a lens, in practice frequently a microscope objective with power 20x to 40x, onto an imager. The latter is often simply a CCD allowing monitoring the near-field intensity distribution at video frame rates. For wavelengths in the visible to near-infrared, standard CCDs can be used, but longer wavelengths will require specialized imagers sensitive at those wavelengths. Neutral density filters are often required to reduce the optical intensity on the imager to prevent signal saturation, blooming or, in some cases, damage to the device.

A typical near-field intensity distribution of an edge-emitting laser diode is shown in Fig. 5.22. The rectangular emission profile shows the shape of the laser

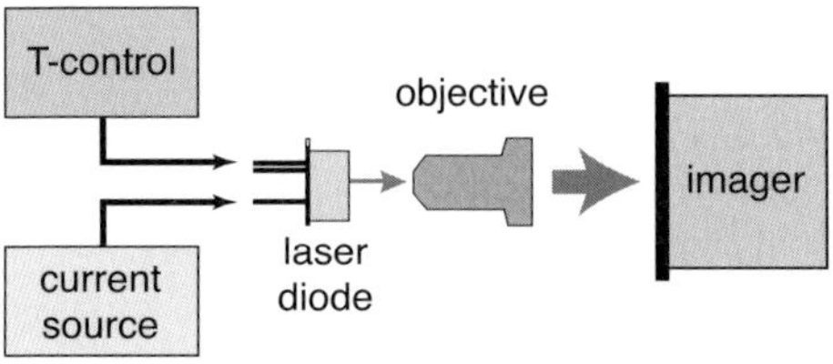

Fig. 5.21. Experimental setup for measurement of laser diode near-field characteristics; emission from the facet is imaged through an objective lens onto an imager, such as a CCD

aperture, in this case about 5 μm wide and ≤ 1 μm thick. The near-field profile of a vertical cavity laser is, in contrast, typically circularly symmetric and is well approximated by a Gaussian intensity distribution. The near-field of an LED, with a large and usually ill-defined aperture, can be quite complex, as the example of Fig. 5.27 shows; the chip, elements of the package and the bond wires are visible in the near-field image.

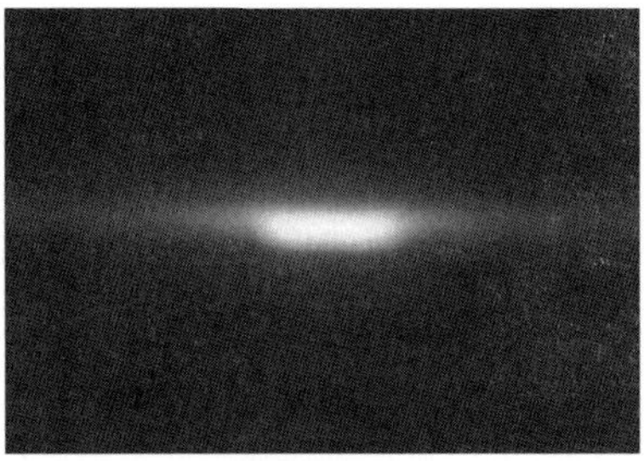

Fig. 5.22. Near-field pattern of an edge-emitting laser, taken with a CCD camera, a magnifying objective and a neutral density filter; note the asymmetric structure. The bright image has a width of about 5 μm

A further aspect of laser diode emission which is relevant for near-field measurements is the astigmatism. Astigmatism, which is discussed in general for optical components in Sect. 6.1.2, derives from the fact that the lateral and transverse foci are found at different positions in the laser. Whereas the focus of the transverse beam component (normal to the junction plane) is typically at the laser facet, the lateral component (parallel to the junction plane) can have a focus behind the laser facet, a significant distance inside the laser cavity. Index-guided edge-emitting lasers generally have little astigmatism, typically less than 5 μm focus distance difference, but the lateral focus of gain-guided lasers can be over 50 μm behind the facet. Characterization of astigmatism requires relatively involved measurements (Sun 1997) and the effect should not be confused with beam ellipticity, which we discuss in the following section.

5.6.2 Far-Field Measurement

When an optical microsystem is assembled, light is coupled from a laser diode into another optical element; this element, with some exception, is typically placed some distance from the laser facet, such that the illumination distribution is in most cases the far-field intensity profile. For that reason, measurement of the far-field is generally of fundamental importance in the characterization of a complete system. Measured far-field distributions are also frequently used in the design of a system, since typical intensity distribution values need to be known to allow optimization of size and placement of lenses, fibers and other optical components.

Measurement of the far-field distribution is mechanically a bit more involved than that required for a near-field measurement since the optical intensity needs to measured as a function of angular position about a complete hemisphere. Two approaches may be used to accomplish this: 1) the laser may be rotated about two normal axes centered at the emission facet, as shown schematically in Fig. 5.23, and the intensity measured by a fixed photodetector; or 2) the laser may be fixed in position and the photodetector scanned in two dimensions across an entire hemisphere. In both cases, the center of rotation must be at the emission facet of the laser. To assure measurement of the far-field characteristic, the spacing between laser and detector must satisfy Eq. 5.51; for high angular resolution, spacings of centimeters to 10s of centimeters are typically used.

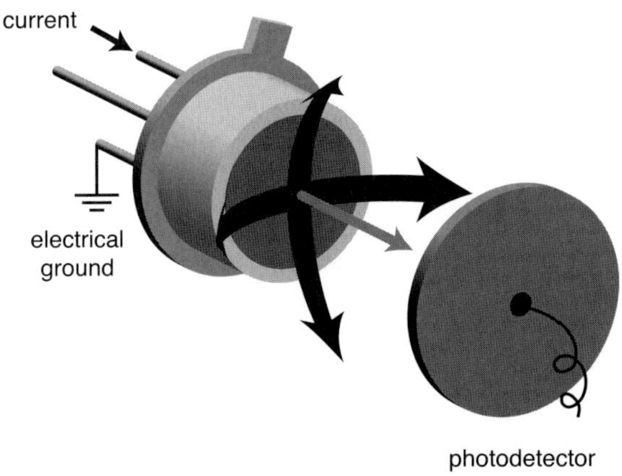

Fig. 5.23. Measurement of the far-field characteristic of a laser diode, shown here in a TO package. The package is rotated in two axes centered about the emission point of the laser and the photodetector remains fixed

The results of the far-field measurement are then typically plotted in the form of intensity as a function of spherical angle, as shown for two dimensions in the far-field plot of an LED in Fig. 5.24, or as a one-dimensional slice through this characteristic. As the tails of the distribution toward low intensities typically have an exponential type of decay for a Gaussian intensity distribution, the width of the far-field profile is typically given by the *FWHM* of the characteristic[7]. For an LED or a circularly-symmetric VCL, the *FWHM* may be on the order of 5° – 12° and is also circularly symmetric. For an edge-emitting laser, with an asymmetric aperture typically 200–500 nm thick (in the transverse direction) and 2–5 μm wide (in the lateral direction), the far-field intensity distribution is then highly elliptical. Typical *FWHM* values in the transverse direction (normal to the chip surface) are 30°–50° and in the lateral direction 10°–20°. The elliptical far-field distribution combined with the astigmatism of edge-emitting lasers are the source of the considerable difficulty encountered when coupling the light from these devices into other optical components.

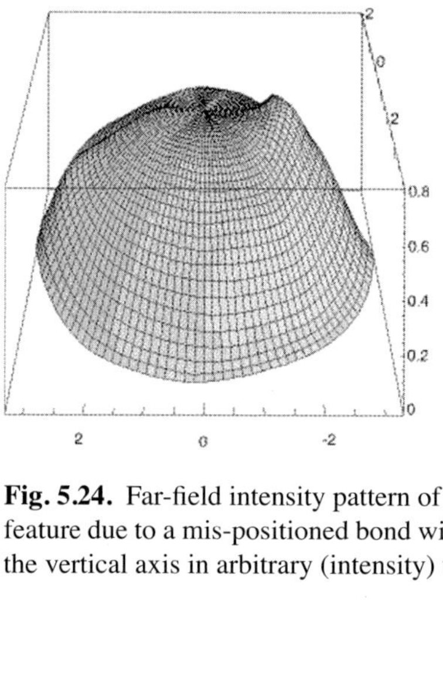

Fig. 5.24. Far-field intensity pattern of an LED; the bump at the right is an anomalous feature due to a mis-positioned bond wire. The lateral and transverse axes are in degrees and the vertical axis in arbitrary (intensity) units

[7] For an assumed Gaussian intensity distribution, the $1/e^2$ width is sometimes used, which has a width $\sqrt{2/\ln(2)} \approx 1.699$ times the *FWHM* value.

5.7 Typical Characteristics

This final section acts somewhat as an appendix to the previous discussion. We have collected a few measurements from the types of lasers of utility for optical microsystems; the results are designed to give the reader a hint as to the type of characteristics she might expect when measuring a laser of various types. The results presented are only intended to represent *typical* values; your mileage may vary.

5.7.1 LEDs

Light-emitting diodes are available at a host of wavelengths and powers; we have measured red, green, blue and white ones here. The white LED is basically a blue semiconductor emitter with a portion of the spectrum converted to yellow using the approach outlined in Sect. 4.1.3. The devices were all mounted in a standard lensed 3 mm plastic package; chip or surface-mounted-device (SMD) structures may have slightly higher output powers but poorer far-field characteristics. Measurements were at room temperature.

The power output versus input current for the four devices is shown in Fig. 5.25, the resultant spectra at room temperature in Fig. 5.26 and the near-field image of a blue LED in Fig. 5.27. The latter clearly shows features of the chip, bond wires and package. The far-field intensity distribution of a different diode, shown in Fig. 5.28, shows the intensity as a function of angle from the emission axis of the LED; we see that the intensity distribution is uniform, circular and without any undesirable features.

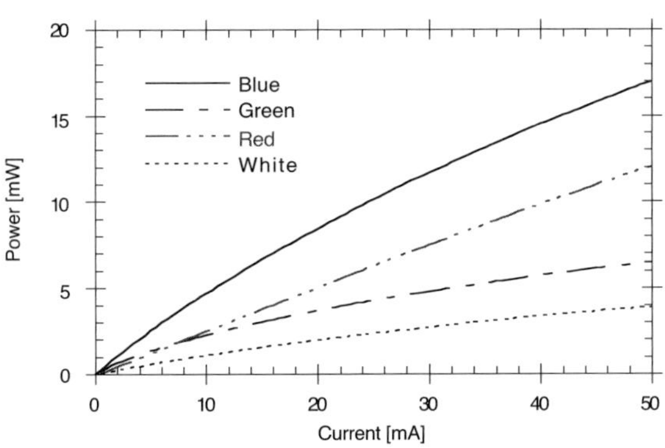

Fig. 5.25. PI curve for blue, green, red and white LEDs

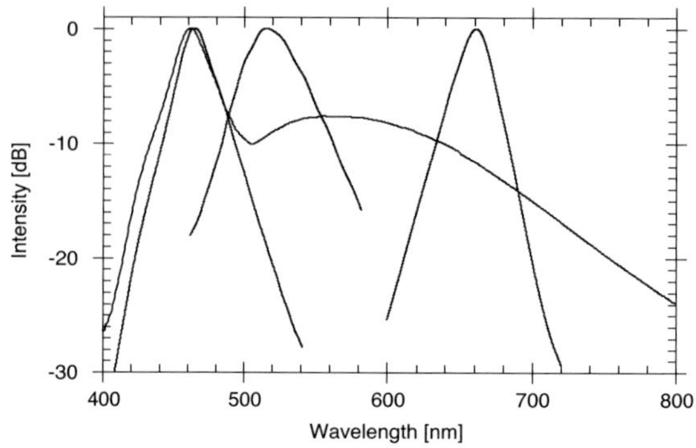

Fig. 5.26. Spectra for blue, green, red and white LEDs; note that the white LED has a peak in the blue (the pump) and around 550 nm in the yellow (fluorescence converted)

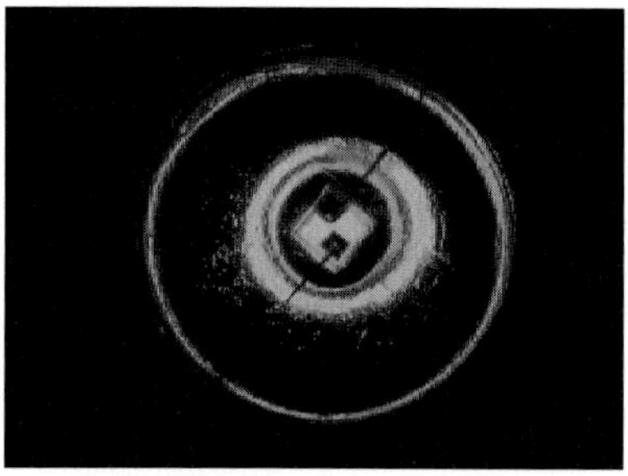

Fig. 5.27. Near-field image of the emission from a blue LED; the chip, bond wires and features of the package are clearly visible

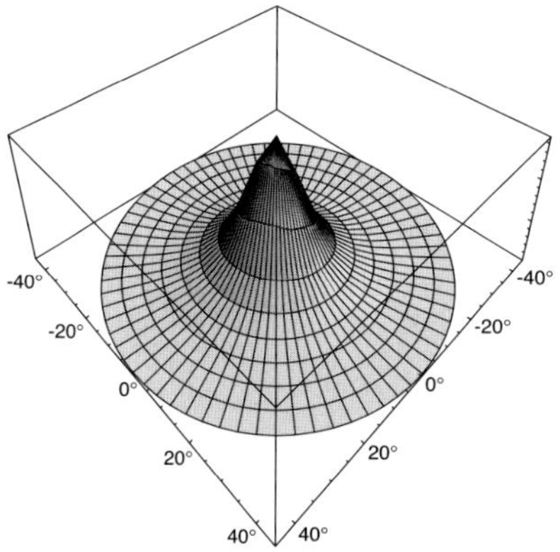

Fig. 5.28. Three-dimensional intensity plot of the far-field of a white LED; the vertical direction represents the intensity and the planar axes the angular position from emission axis

5.7.2 Fabry–Perot Lasers

As an example for an edge-emitting Fabry–Perot laser, we chose a red-emitting laser such as that found in laser pointers or scanners. The device was mounted in a TO package with a glass window and no lensing.

The PI characteristics as a function of temperature are shown in Fig. 5.29; the output power is linear over the measurement range and the characteristics shift to the right with increasing temperature. The plot of threshold current and efficiency in Fig. 5.30 clearly shows the increase of the former and decrease of the latter with increasing temperature. The emission spectrum at room temperature is shown in Fig. 5.31; we see that emission is distinctly multi-mode with a clearly discernible modulation on the spectrum; this is expected for a Fabry–Perot laser. The variation of emission wavelength with current for a range of temperatures, shown in Fig. 5.32, demonstrates the current and temperature tunability; the sudden jumps in wavelength are mode-hops.

The far-field intensity distribution of the edge-emitting laser is shown in Fig. 5.33; in this three-dimensional representation, the vertical axis represents the intensity and whereas the lateral and transverse axes give the position, in units of degrees. The ellipticity of the emission is clearly seen, as are two intensity peaks, suggesting that two modes of roughly equal intensity are lasing.

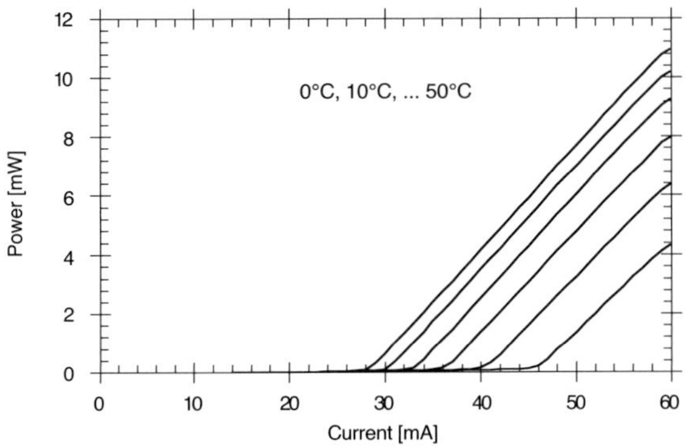

Fig. 5.29. PI curves for a 670 nm edge-emitting Fabry–Perot laser as a function of temperature; temperature varies from left to right from 0°C to 50°C in 10°C steps

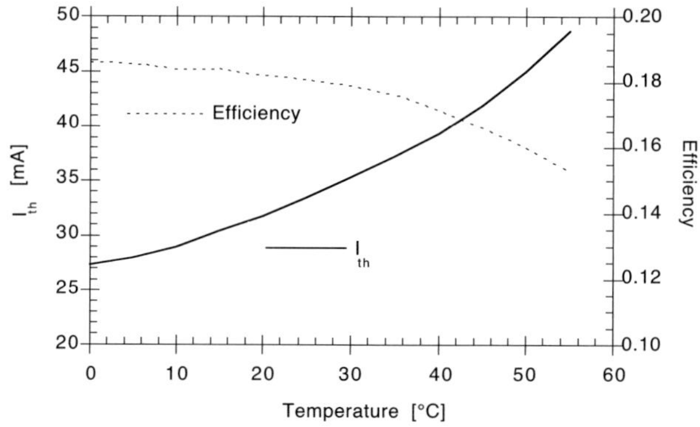

Fig. 5.30. Threshold current and efficiency for a 670 nm edge-emitting Fabry–Perot laser as a function of temperature. This laser has a measured $T_0 = 94.95°C$

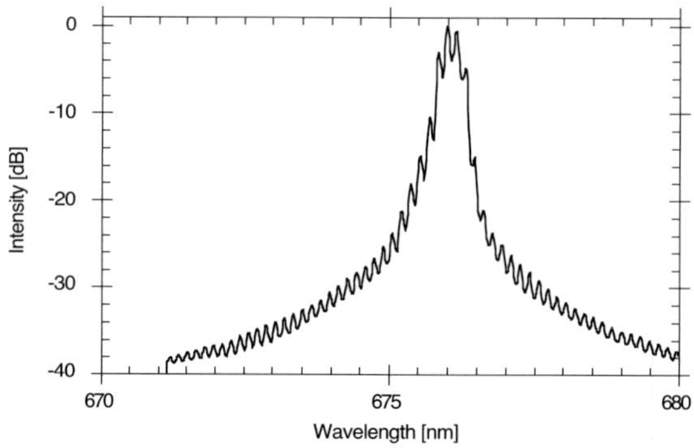

Fig. 5.31. Emission spectrum for a 670 nm edge-emitting Fabry–Perot laser at 20°C

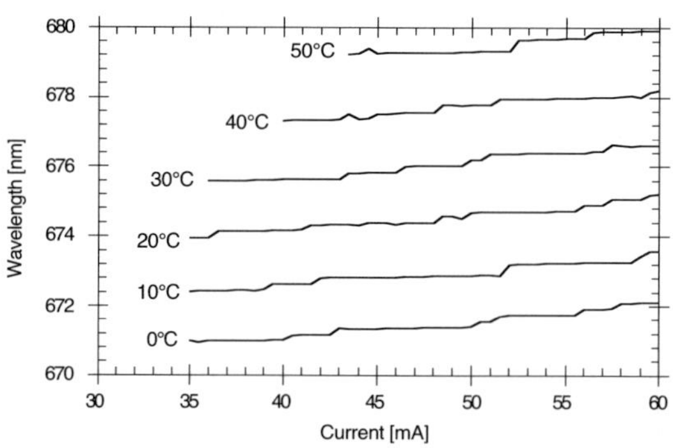

Fig. 5.32. Emission wavelength for a 670 nm edge-emitting Fabry–Perot laser as a function of current for a range of temperatures; mode-hops are clearly seen

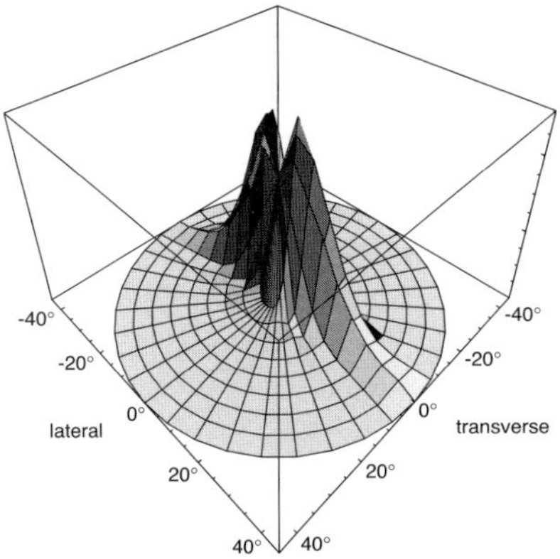

Fig. 5.33. Three-dimensional intensity plot of the far-field of a 670 nm edge-emitting laser; the lateral direction is parallel to the junction plane and the transverse normal to it

5.7.3 VCLs

A standard 850 nm VCL, such as those used for short-range data communication, was chosen for the vertical cavity laser example. The laser was again mounted in a TO package and was unlensed.

The PI curves as a function of temperature shown in Fig. 5.34 show the typical VCL behavior: low threshold current and a marked flattening of the characteristic with increasing current. The maximum output power is less than 1 mW. The spectrum of Fig. 5.35 shows fundamentally single-longitudinal-mode behavior, with no indication of any lateral modes. The tuning behavior (of a different diode) shown in Fig. 5.36 demonstrates the strong wavelength dependence on temperature, but also an absence of any mode hops. The far-field intensity distribution shown in Fig. 5.37 is in this case a two-dimensional representation; the lighter areas represent higher intensity, as a function of angular position from 0 to 2π. We note the "doughnut"-like emission peak at some distance from the device center. Such a "doughnut" mode is typical for VCLs with a larger aperture driven at relatively high currents; the classical Gaussian intensity profile is only observed for low drive currents, before higher-order lateral modes begin to lase.

As mentioned in Sect. 5.1.3, one of the agreeable features of VCLs is that they may be tested on-wafer; if an automated wafer prober is used, wafer maps of the laser characteristics may be made. An example of such a map is given in Fig. 5.38, which shows the distribution of emission wavelength across a 2-inch wafer. Inhomogenieties in the epitaxial growth or processing clearly manifest themselves as regions of varying wavelength. For a spectrally highly-demanding application, such a wavelength map may be used, for example, to pre-select those VCLs which emit exactly at the desired wavelength.

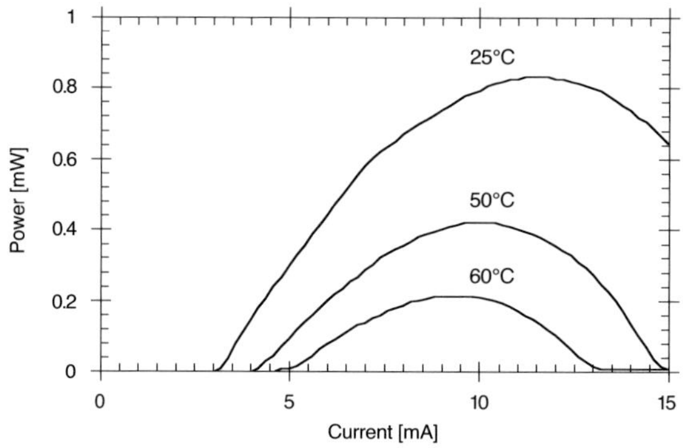

Fig. 5.34. PI curves for a VCL as a function of temperature; device temperatures of 25 °C, 50°C and 60°C are shown. The rollover behavior is typical for VCLs but operation below the peak output power is recommended

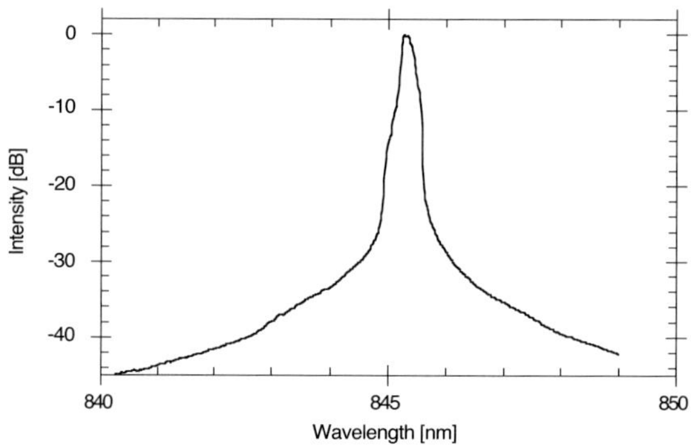

Fig. 5.35. Emission spectrum for an 850 nm VCL at 20°C

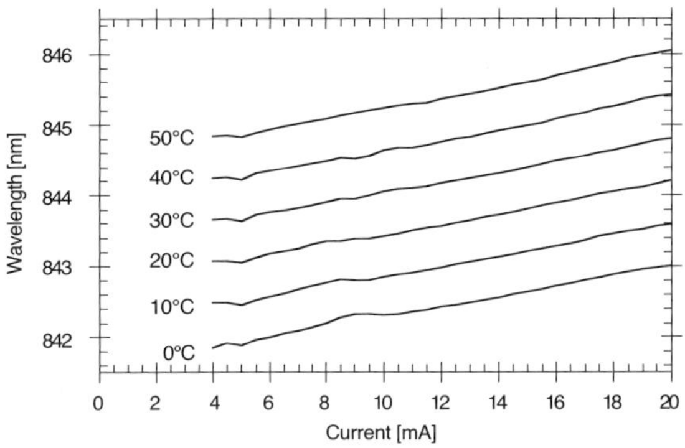

Fig. 5.36. Emission wavelength for an 850 nm VCL as a function of current for a range of temperatures

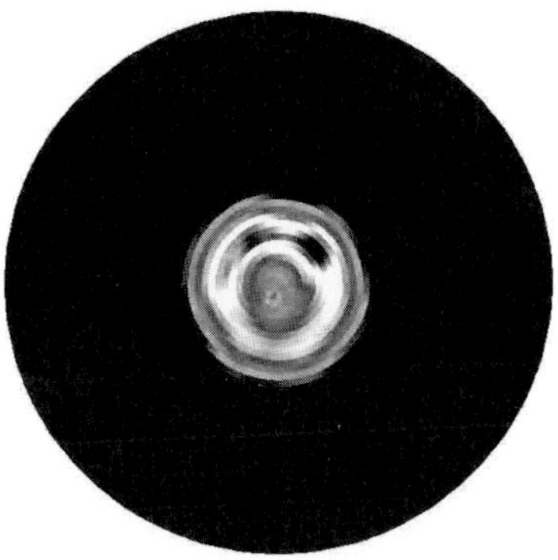

Fig. 5.37. Two-dimensional representation of the far-field intensity distribution of a VCL; the lighter areas correspond to higher emission intensity as a function of angle (0 to 2π)

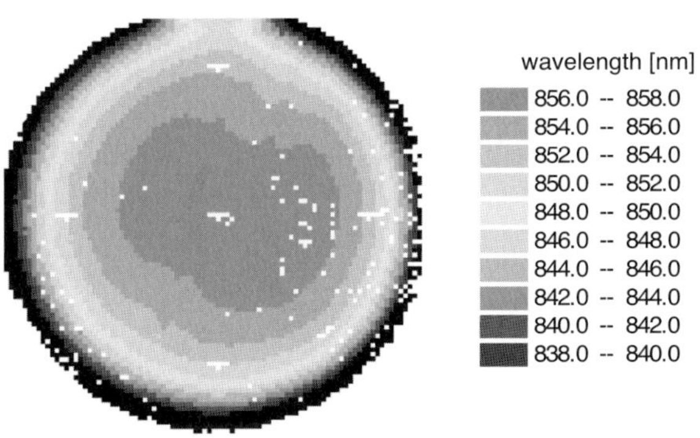

Fig. 5.38. Map of the emission wavelength of about 7 000 VCLs fabricated on a 2-inch wafer; each pixel represents one VCL

5.7.4 DFB Lasers

Distributed feedback lasers are rarely available in chip form commercially, save perhaps to telecommunications component or subsystem manufacturers. As a result, the measurements on the DFB laser shown below are from a device mounted in a butterfly package with a fiber pigtail; laser emission was measured directly at the fiber.

We note from the PI characteristics as a function of temperature, shown in Fig. 5.39, that the threshold current, total current range and maximum optical output power are all higher than for the edge emitting laser and certainly than for the VCL. The shift in threshold current and the decrease in efficiency with temperature, shown in Fig. 5.40, is larger than for the 670 nm laser; we already see in Fig. 5.39 that the slope of the PI curves decreases markedly with increasing temperature. The spectrum of Fig. 5.41 shows the excellent spectral performance expected of DFB lasers: single-mode emission and a *SMSR* in excess of 50 dB. Likewise, the tuning curves of Fig. 5.42 show a uniform increase in wavelength with current, with a small slope but complete absence of mode hops. Using temperature and current in combination would thus allow continuous access to any wavelength in a band of about 5 nm for this laser.

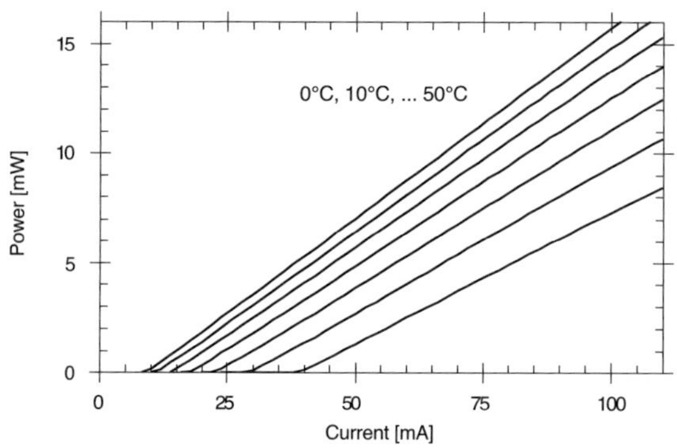

Fig. 5.39. PI curves for a 1550 nm edge-emitting, fiber pigtailed DFB laser as a function of temperature; temperature varies from left to right from 0°C to 50°C in 10°C steps

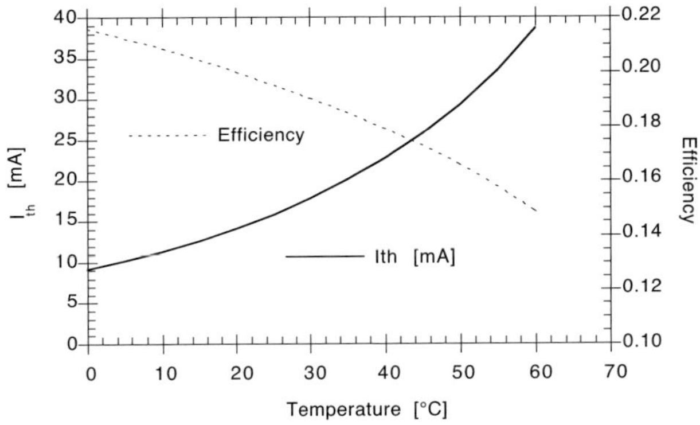

Fig. 5.40. Threshold current and efficiency for a 1550 nm edge-emitting, fiber pigtailed DFB laser as a function of temperature. This laser has a measured $T_0 = 42.19°C$

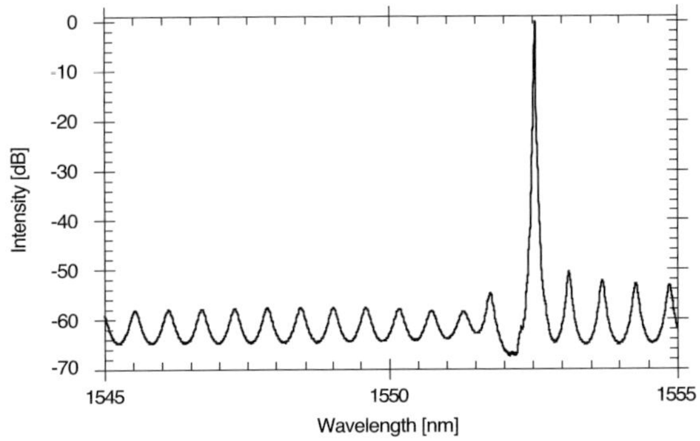

Fig. 5.41. Emission spectrum for a 1550 nm edge-emitting, fiber pigtailed DFB laser at 20°C

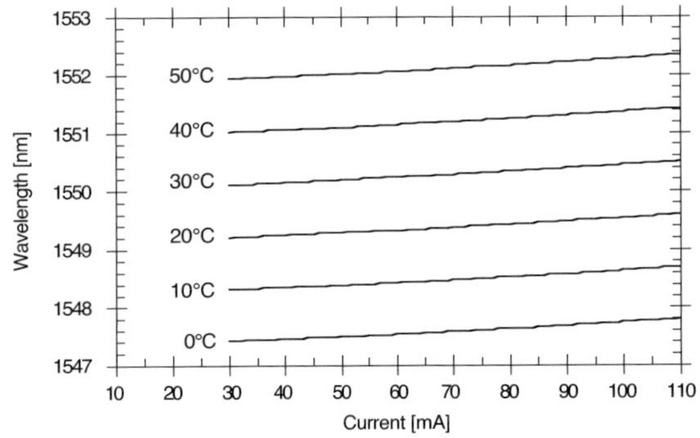

Fig. 5.42. Emission wavelength for a 1550 nm edge-emitting, fiber pigtailed DFB laser as a function of current for a range of temperatures

6 Micro-Optical Components

When used in microsystems, laser diodes are rarely found alone. The emitted light must typically be focused or collimated and directed through other optical components to perform the function for which the optical system is designed. In particular, components are required for laser beam shaping, spectral and spatial filtering, intensity and phase modulation, guiding, splitting and finally detection. In keeping with the considerations of a microsystem, the optical components which perform these functions are typically of small physical size and the discipline which considers their design, fabrication, assembly and characterization is thus termed micro-optics.

An example of the micro-optical components which may be required for the fabrication of a useful laser diode module is shown in the schematic drawing of Fig. 6.1. The light emitted from the laser diode is focused using a ball lens, directed through an optical isolator and a modulator whereupon it is subsequently focused onto the facet of an optical fiber. Such an optical arrangement is typical for laser diode modules used in optical communications systems.

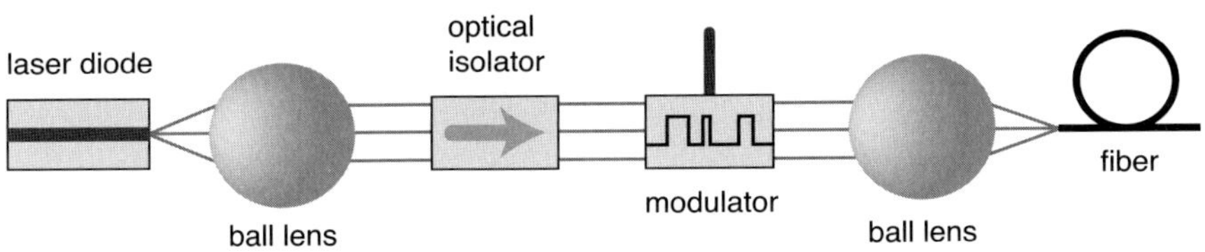

Fig. 6.1. Schematic drawing of the micro-optical components which may be found in a laser diode module: the laser emission is collimated by a ball lens, directed through an optical isolator and intensity modulator and focused by a second ball lens onto the facet of an optical fiber

The field of micro-optics and optical microsystems is a large and growing one (Sinzinger and Jahns 2002; Herzig 1997; Zappe 2003). In this chapter we consider a number of the components essential for the fabrication of complete laser diode-based optical microsystems; in Chapter 7 we will then discuss the techniques required for their alignment and assembly.

6.1 Refractive Microlenses

Refractive lenses use refraction at a surface to change the direction of a light beam and are most popularly known for their role in macroscopic imaging systems; microscopic lens structures are even known in nature (Aizenberg et al. 2001) and utilize the same effects. Whereas laser diode-based optical microsystems can employ miniaturized versions of classic refractive lenses, the need for very compact, robust and cheaply manufacturable components will often exclude the common types of glass lenses. Micro-optical refractive lenses are playing an increasingly important role in these applications and we will discuss a few of the relevant structures below.

The optical physics behind classic lenses, whether large or small, and micro-optical refractive lenses is identical. Calculation of the displacement of optical rays through a lens of a given shape does not vary with size, at least at length scales considerably longer than the wavelength of light, but diffraction effects become more important as the lens aperture decreases. We will thus discuss classic refractive lens theory as it applies to all sizes of lenses and refer in Sect. 6.1.3 to those factors relevant for micro-optical lenses.

6.1.1 Basic Geometric Optics

Geometric optics refers to the study of light refraction through a lensing surface and is one of the oldest and most established physical disciplines (Hecht 1987, Smith 1990). Refraction describes the bending of a light ray at an interface between two materials with differing optical refractive indices, such as the surface of a glass lens in air. For a light ray incident onto the interface with an angle θ_i, measured from the normal to the surface, the angle of the transmitted beam, θ_t, is given by Snell's law

$$n_i \sin\theta_i = n_t \sin\theta_t \tag{6.1}$$

named after Willebrord van Roijen Snell (1580-1626). In Eq. 6.1, n_i refers to the optical refractive material on the incident side of the interface and n_t the optical refractive index on the transmission side of the interface.

When a plane wave is incident onto an elliptically curved convex surface, it can be shown that the initially parallel rays are focused to a point behind the surface, as shown in Fig. 6.2; the lens is said to be converging. For a concave surface, also shown in the figure, initially parallel rays diverge and for normal incidence onto a planar surface, the rays are undeflected. Most refractive lenses are made from a combination of these three surfaces.

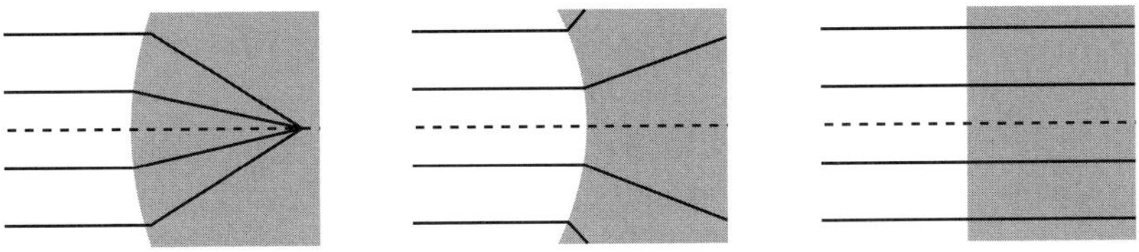

Fig. 6.2. Refraction of the parallel rays of a plane wave at convex *left*, concave *center* and planar *right* surfaces

Function of a Lens

Most lens systems are known to non-specialists for their role in imaging systems. In such a system, the light emitted from an object a distance s_o [m] from the lens is directed through the lens system and generates an image of that object at a distance s_i [m] from the lens. For example, a camera objective generates a real image on the plane of the film or CCD. Lens systems may also generate virtual images of the object, which need a further lens systems (such as a human eye) to subsequently generate a real image (on the retina in this case).

For laser diode-based systems, lenses are relatively infrequently used for imaging. Instead, lenses are generally used for collimation or focusing a laser beam. A collimated beam is essentially a plane wave in which the rays propagate parallel to each other; the width of the beam changes little with propagation distance. Alternatively, in a focused optical beam, the rays converge at a single point. A popular application is the transformation of the emitted beam from a laser diode facet, approximately a point source, into a collimated beam which is then transmitted through the remainder of an optical microsystem.

Focus

Whereas the focus to a point for a converging lens is theoretically ideal for an elliptically curved surface, most refractive lenses are spherical. This fact is due to the classic manufacturing techniques for lenses, which typically use a grinding and polishing process which automatically lead to a spherical profile. Exceptions to this rule are aspheric lenses, which may be designed with almost arbitrary profiles; these are complicated to manufacture and, with the exception of replicated lenses, typically very expensive.

Given a spherical convex lens, the light rays from a plane wave converge to a focus at an image focal distance f from the lens surface, as shown in Fig. 6.3. For a lens with radius of curvature R made from a material with optical refractive index n_L, the focal length may be determined from

$$f = \frac{Rn_L}{n_L - n_0} \tag{6.2}$$

where n_0 is the refractive index of the surrounding medium, typically air. Equation 6.2 is valid for the so-called paraxial approximation, namely for rays close to the optical axis. We thus see that surfaces with larger curvature (smaller R) and materials with larger refractive index values (n_L) lead to lenses with shorter focal lengths.

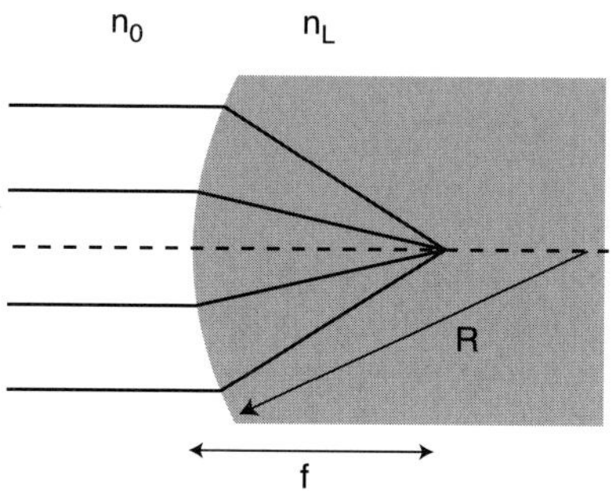

Fig. 6.3. Definition of the focal length f of a spherical convex lens with radius R; the external material has refractive index n_0 and the lens n_L. The optical axis is given by the dotted line through the lens center

Lens Types

Practical lenses are made by combining convex, concave and planar surfaces. Figure 6.4 shows a selection of lens types with either two curved surfaces or one curved and one planar surface. The bi-convex and plano-convex are converging lenses and the bi-concave and plano-concave are diverging lenses. The radii are given for the left surface (R_1) and right surface (R_2) and the meaning of the polarity will be discussed below.

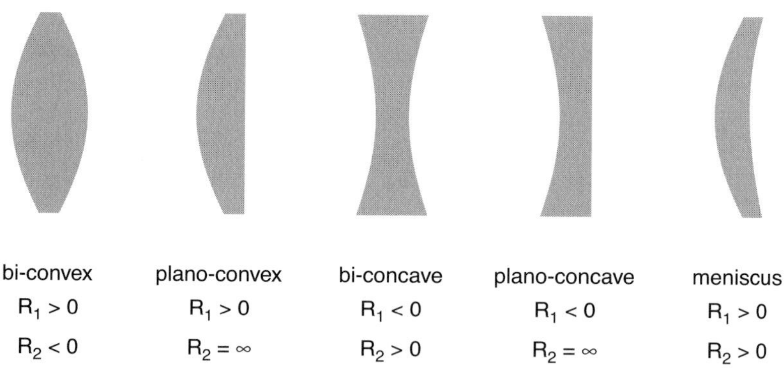

Fig. 6.4. Typical thin lenses combining convex, concave and planar surfaces. The radius of the left surface is given by R_1 and the right surface by R_2

In practice, accurate calculation of lens parameters for an optical system design is done via numerical simulation using ray-tracing techniques (Shannon 1997); however, good approximations are in many cases possible using simplified representations of the lens. Although the lenses clearly have a finite, non-zero thickness, one useful simplification in optical analysis is possible by making the "thin-lens approximation". In this case, only the refractive effect of the two curved surfaces is considered and the accuracy of the approximation is sufficient in most applications. For the case of the thin lens, the distance between the lens and an object to be imaged, s_o, and the distance between the lens and the image of that object, s_i, are related to the lens parameters by the Lensmaker's formula, given by

$$\frac{1}{s_o} + \frac{1}{s_i} = \left(\frac{n_L}{n_0} - 1\right)\left[\frac{1}{R_1} - \frac{1}{R_2}\right] \tag{6.3}$$

for the refractive indices and lens radii as defined above.

For the case in which either the image or the object is infinitely far from the lens, the former corresponding the case of a collimated beam, we see that either $s_i \to \infty$ which from Eq. 6.3 implies that $s_o \to f$, or that $s_o \to \infty$ implying that $s_i \to f$. From these considerations, we may simplify the above expression to

$$\frac{1}{f} = \left(\frac{n_L}{n_0} - 1\right)\left[\frac{1}{R_1} - \frac{1}{R_2}\right] \tag{6.4}$$

and realize that

$$\frac{1}{s_o} + \frac{1}{s_i} = \frac{1}{f}. \tag{6.5}$$

The previous equation is referred to as the Gaussian lens formula and represents a useful expression for rapid calculation of relative object and image distances.

Sign Conventions

As we saw in the definitions of Fig. 6.4, lens radii as well as image and object distances may be positive or negative; the sign conventions for these parameters allow us to define the curvature of the surfaces and the positions of the generated images as we move through a lens system. The diagram of Fig. 6.5 allows us to define the polarity of these parameters with respect to their position relative to the lens surface. The point V represents the vertex of a lens with radius R centered at C. If C is to the *right* of V, the radius $R > 0$; conversely, if C is to the *left* of V, $R < 0$. In addition, if the object is to the *left* of V, $s_o > 0$ and if the image is to the *right* of V, $s_i > 0$.

Calculations using Eq. 6.3 – Eq. 6.5 should adhere to these sign conventions. These expressions allow a calculation of the lens radii of curvature and spacings required, for example, to collimate a laser beam using a microlens. The object distance is that between the laser and the lens and the image distance is, in this case,

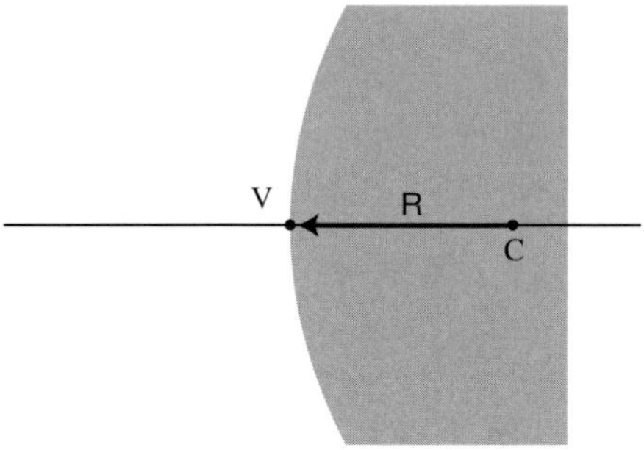

Fig. 6.5. Diagram for the definitions of the sign conventions for lens radius

set to infinity. The relations allow a quick estimate of the total system and lens dimensions required.

Example 6.1

Assignment: *Design a plano-convex microlens such as that shown in Fig. 6.6 to collimate a laser beam placed 500 μm from the planar lens surface; assume you have BK7*[1] *glass with* $n_L = 1.52$*. Determine the radius of curvature, maximum aperture diameter and the resultant lens thickness. Is this a "thin lens"? What about a lens with aperture size half the maximum value? What would the object and image distances using this lens be if we wanted to focus the laser emission to a point?*

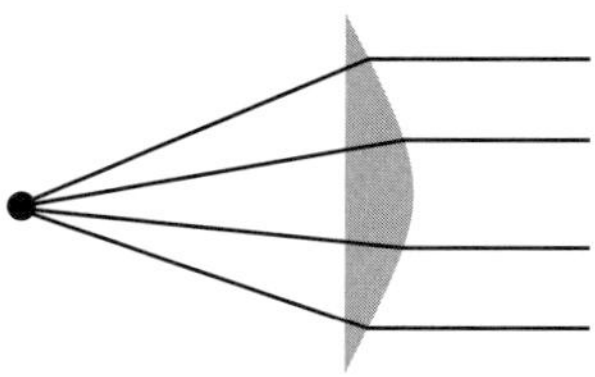

Fig. 6.6. A plano-convex lens for collimation of a laser beam

Solution: *We approximate the design of such a lens by hand. For collimation of a laser beam, assume that the laser is a point source; a 500 μm lens/laser spacing implies that* $f = 500\ \mu m$*.*
We may use Eq. 6.4 to calculate the radius. We assume that the lens is operated in air, so that $n_0 = 1.0$ *and we are given* $n_L = 1.52$*. For a plano-convex lens,* $R_1 = \infty$ *and we note from the sign conventions that* $R_2 < 0$*. Thus Eq. 6.4 becomes*

---

[1] BK7 is a catalogue designation from the glass manufacturer Schott and represents a very popular glass for optical applications

$$R_2 = f\left(\frac{n_L}{n_0} - 1\right)$$

and we calculate $R_2 = 260\ \mu m$. *That is quite a small radius (i.e., quite a large curvature).*

The maximum aperture diameter of the lens is twice the curvature radius; in that limit, the lens consists of a half-sphere. Thus we may write $D_{max} = 520\ \mu m$. *The lens height is then equal to the radius of curvature, so that* $h = 260\ \mu m$. *Such a structure is by no means a "thin lens".*

If we restricted the aperture to half the maximum, then $D = D_{max} = 260\ \mu m$. *Calculation of the lens height requires a bit of simple geometry (or the intrepid reader may look ahead to Eq. 6.14) which in this case yields* $h \approx 35\ \mu m$. *Such a structure comes closer to our definition of a "thin lens". Figure 6.7 compares these two lenses using a numerical ray tracing program; we see that the "fat" lens has a completely different laser/lens spacing than the thin lens, not to mention the significant aberration which the former would yield for portions of the beam at the lens edges.*

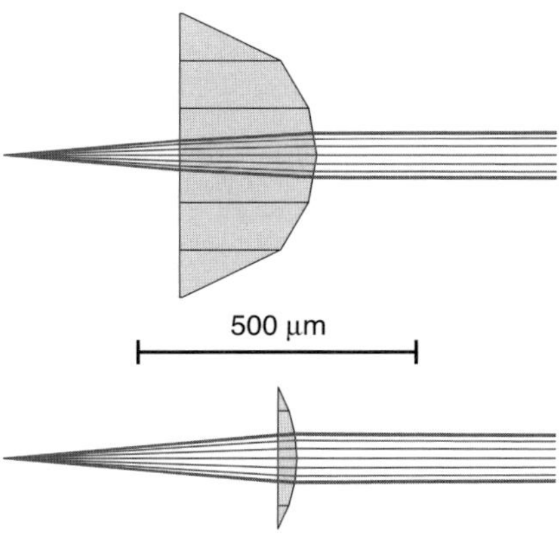

Fig. 6.7. Comparison of ray tracing results for the plano convex lens with maximum aperture *top* and half the maximum aperture *bottom*; the apparent facetting of the lens is due to the graphics rendering of the simulation

Using the lens as a collimator implies that the lens is placed a focal distance (500 μm) away from the source. Using the Gaussian lens formula of Eq. 6.5, we see that focusing the emission from a laser to a point would require that both object and image are at twice the focal distance ($s_i = s_o = 2f$) *such that the laser and its focused spot are both* $2f = 1$ *mm from the lens.*

Numerical Aperture

Lenses typically have a finite diameter or aperture. As a result, only a limited cone of rays from a uniformly emitting point source will traverse the lens: the numerical aperture (*NA*) of a lens defines how much light the lens can gather. The schematic drawing of Fig. 6.8 shows the lens aperture *D* defined by the lens mount (or, alternatively, simply the lens diameter); the lens is illuminated from a point source from which rays with a maximum angle θ_{max} still enter the lens. From these parameters, we define the numerical aperture as

$$NA = n_0 \sin\theta_{max} \tag{6.6}$$

for refractive index of the external medium (usually but not always air) n_0. From simple geometry, we see that the maximum angle is related to the aperture *D* and focal length *f* of the lens as

$$\theta_{max} = \tan^{-1}\left(\frac{D}{2f}\right) \tag{6.7}$$

such that

$$NA = n_0 \sin\left[\tan^{-1}\left(\frac{D}{2f}\right)\right] \tag{6.8}$$

which for lenses with small *NA* (i.e., $f \gg D$) may be approximated as

$$NA \approx n_0 \frac{D}{2f}. \tag{6.9}$$

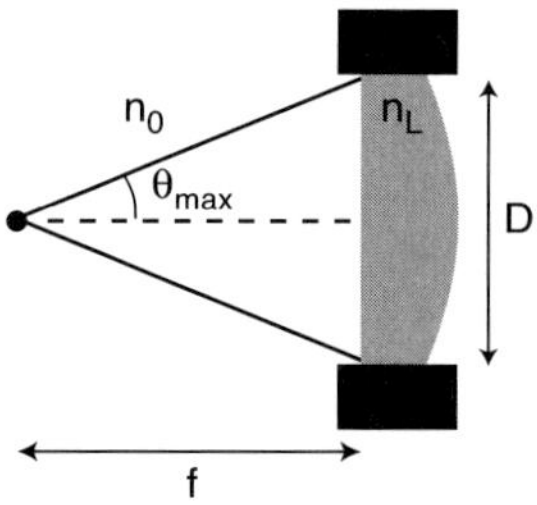

Fig. 6.8. A mounted lens showing the aperture diameter *D* and the maximum input angle θ_{max} for illumination from a point source

A larger value for *NA* thus implies greater light gathering power of the lens; a larger aperture will generally lead to a larger numerical aperture. We see that for lenses in air ($n_0 \approx 1$), the numerical aperture is always $NA \le 1$. Values greater than unity may be achieved using an external medium with $n_0 > 1$ (typically an oil) but such techniques are typically limited to specialized applications in microscopy and are not generally practical for compact optical microsystems.

An alternative description of the light gathering power of a lens is via the "f-number" or *f/#*, defined as

$$f/\# = \frac{f}{D} = \frac{1}{2NA}. \tag{6.10}$$

The *f/#*, used predominantly as a descriptor for photographic objectives, is smaller for lenses or lens systems with larger aperture or larger *NA*.

For lenses used to focus the emission from a laser, the numerical aperture also defines the smallest spot size of the focused beam. High-quality lenses are typically diffraction limited, meaning that diffraction effects due to the finite value of *D* dominate any other aberration effects. The focus spot takes the form of a circular Airy function (named for George Airy (1801-1892)) with alternating light and dark rings; the width of the central spot, *d*, focused by a lens with numerical aperture *NA* is approximately given by

$$d \approx 1.22\frac{\lambda}{NA}. \tag{6.11}$$

for illumination wavelength λ. Smaller spot sizes are thus achieved with shorter wavelengths, insofar wavelength may be varied in a given system application, and by using a larger *NA*. This latter fact explains why optically-demanding systems often use lenses with a large diameter.

Example 6.2

Assignment: *Due to the large width of the emitted beam from a semiconductor laser, a typical laser diode collimator requires NA ~ 0.7 – 0.8. Does the "thin" version of the lens (the one with half the maximum aperture) calculated in Example 6.1 fulfill this requirement? What are NA, θ_{max} and f/# for this lens? Can we improve matters by just increasing D?*

Solution: *The "thin" lens from Example 6.1 had $D = 260\ \mu m$ and $f = 500\ \mu m$; from Eq. 6.7, this yields $\theta_{max} = 14.6°$ such that NA = 0.25 and f/# = 1.99. Clearly this lens does not come close to the required numerical aperture.*

By increasing the aperture D for constant focal length f, NA increases. For $NA \geq 0.7$, we can calculate from Eq. 6.8 that we would require $D \geq 1.96 f$, However, considering the fact that a spherical lens cannot have an aperture larger than twice its radius of curvature (at which point the lens is a half-sphere), we can determine from the equation in Example 6.1 that we are limited to $D \leq 1.04 f$ for the refractive indices given. This last restriction limits achievable NA with a lens of this type to $NA \leq 0.45$. Thus increasing the aperture will only help matters up to a point. The design of a a high NA system typically requires compound lenses and is not trivial.

6.1.2 Aberrations

We mentioned in the previous section that most classically fabricated lenses as well as most microlenses have a spherical cross-section. The accuracy with which the manufacturing process achieves the shape of the sphere defines the quality of the lens and thus deviations from the theoretically ideal behavior we assumed in Sect. 6.1.1. In addition, diffraction effects as well as the fact that many lens properties are wavelength-dependent lead to non-ideal lens behavior and thus aberrations in the lens characteristics (Born and Wolf 1980; Hecht 1987).

However, aberrations are present even for theoretically ideal lenses operated at a single wavelength far from the diffraction limit. Known collectively as the Seidel aberrations, in honor of Philipp Ludwig von Seidel (1821–1896), these five aberrations represent deviations from the first-order lens theory as presented in the previous section. The paraxial approximation was predicated on the assumption that the light rays were sufficiently parallel to the optical axis to allow us to write $\sin\theta \sim \theta$. Since the series expansion for $\sin\theta$ is given by

$$\sin\theta = \theta - \frac{\theta^3}{3!} + \frac{\theta^5}{5!} - \ldots \tag{6.12}$$

values of ray angle sufficiently greater than zero require consideration of the higher order terms. The Seidel aberrations include those of third order ($\propto \theta^3$); higher order aberrations include further terms in the series but will not be considered here (Shannon 1997).

Spherical Aberration

Ideally, a lens would have an elliptical cross-section. Since manufacturable lenses are generally spherical, the focusing characteristics change as the input ray (from the image) moves away from the optical axis. In particular, as shown in Fig. 6.9, the focus position moves toward the lens as the distance between the ray and the optical axis, h, increases. This effect represents longitudinal spherical aberration. For the focus of a collimated beam onto a focal plane, the spherical aberration will result in a diffuse spot, since those portions of the beam further from the optical axis will come to a focus at a different position on the plane than those close to the axis.

Correcting spherical aberration is generally done by using an aspheric lens. In this case the lens profile is optimized to reduce or eliminate spherical aberration, at the expense of considerably higher manufacturing costs. In macroscopic systems, a cemented doublet, two lenses of differing optical characteristics glued together, is also often used, also a more complex and expensive component than a normal spherical lens.

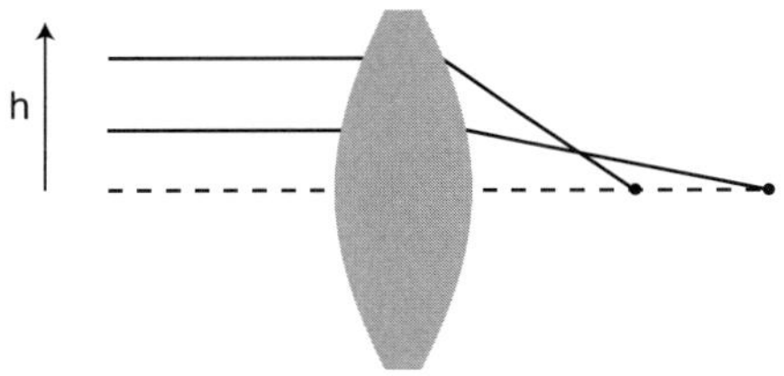

Fig. 6.9. Longitudinal spherical aberration: the focus position is a function of distance from the optical axis, *h*

Example 6.3

Assignment: *We would like to use a plano-convex lens for collimation of a diode laser beam. Which way should it be oriented: flat or curved side toward the laser?*

Solution: *Spherical aberration plays a central role in deciding upon the lens orientation. Using a ray tracing simulation, we can examine the focus of a plano-convex lens when illuminated over its full aperture with a perfectly collimated beam. The result is shown in Fig. 6.10: the collimated beam is incident from the left and focused ($f = 500\ \mu m$) on the right of the lens. We see that the upper example, in which the curved side of the lens faces the collimated beam, has noticeably less spherical aberration than the lower example. For the case in which the flat side of the lens faces the collimated beam, those rays incident at the edges of the lens focus at a point closer to the lens. Thus for beam collimation using a single plano-convex lens, the curved side should face the collimated side and the flat side should face the focus (i.e., the laser diode).*

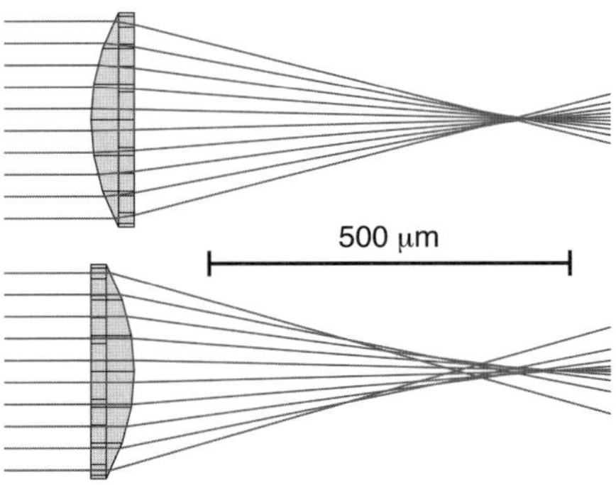

Fig. 6.10. Ray tracing simulation of a plano-convex lens used to focus a collimated laser beam; the configuration with the planar side of the lens turned toward the focus *top* shows less spherical aberration than with the curved side turned toward the focus *bottom*

In general, lenses with differing curvatures on the two sides, such as the meniscus lens of Fig. 6.4, should be oriented so that the flatter side (i.e., that with the largest radius of curvature) faces the point source.
As an aside, the simulation should more realistically have been done with a point emitter at the focus of the lens and we should have examined the collimated side; unfortunately, the effect of spherical aberration is more difficult to see in that simulation and thus serves less well as an example.

Coma

For illumination at an oblique angle to the optical axis, the focus is likewise subject to severe aberration. The magnification of an obliquely incident beam is a function of the distance from the optical axis, such that a collimated input does not focus to a point but rather to a comet-like image. This aberration is thus termed coma.

Coma is relatively difficult to correct; macroscopic lens systems typically use cleverly positioned field stops to reduce the size of the lens aperture and thereby mitigate the effects of coma by eliminating certain oblique rays. For laser beam collimation or focusing, where illumination is typically parallel to the optical axis, coma is less of an issue but for imaging system it represents a persistent aberration.

Astigmatism

Rays incident on a lens at an oblique angle to the optical axis are also subject to astigmatism. This aberration results from the difference in focal length for two different components of the incident beam as shown in Fig. 6.11. As a result, the image of a spot is always defocused vertically or horizontally.

Understanding astigmatism requires that we look in the form of the incident optical beam in slightly greater detail. Referring to the figure, we see that illumination from a point source leads to a cone of rays and includes one principle ray, namely the center of the beam which traverses the center of the lens. The cone of rays may be decomposed into its components on two orthogonal planes. The first is the plane formed by the optical axis and the principle ray and is termed the tangential plane. The second is normal to the first, also includes the principle ray and is referred to as the sagittal plane. Both planes contain a fan of rays defined as a projection of the components of the three-dimensional cone of rays.

Astigmatism results when we consider that the focus distances of the rays in the tangential and sagittal planes differ, as we see in Fig. 6.11. In this drawing, the tangential rays focus more closely to the lens leading to a tangential image; this image is not a point, since the sagittal rays focus at a longer distance from the lens, and thus takes the form of a focal line rather than a focal point. Analogously, at the sagittal focus, the tangential components are no longer in focus, such that a sagittal image, also a focal line, is formed with an orientation normal to the tangential image. Somewhere between the tangential and sagittal foci lies a point where both components have a minimum focus spot size, yielding the colorfully named "circle

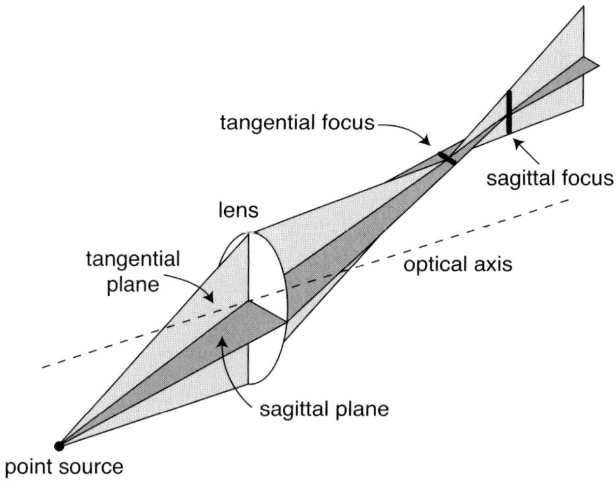

Fig. 6.11. Astigmatism of an optical beam under oblique incidence; the foci for the sagittal rays (dark gray plane) and tangential rays (light gray plane) are focal lines at different positions

of least confusion". This circle represents the best focus that can be achieved with an astigmatic lens.

As an aside, we note the astigmatism of a laser diode, as mentioned in Sect. 5.6.2, is only slightly related to astigmatism of a lens as we have described it. In a laser diode, the two orthogonal components of the emitted beam (the transverse and lateral components) appear to come from two different "point sources" behind the laser facet; the positions of these point sources depend on the laser, but are typically a few to a few tens of microns behind the facet, inside the laser cavity. As a result, the transverse and lateral components of laser diode emission appear to have different foci and this effect leads to laser diode astigmatism. The ultimate effect in an optical microsystem is similar to that of a lens as just discussed but the origins are physically quite distinct.

Field Curvature and Distortion

The two final lens aberrations in the group of third-order Seidel aberrations are important for imaging applications but of limited relevance for laser diode-based systems. Field curvature, also known as Petzval curvature for Józeph Petzval (1807–1891), derives from the fact that for an ideal imaging system, both the object and the image should be curved with the radius of curvature of the lens. Since these are instead generally planar, images are generally curved toward the lens; the projection of this curved image onto a plane then leads to aberrations, particularly far from the optical axis. Field curvature cannot be corrected using a single lens, but compound systems using both convex and concave lenses may be designed to mitigate the effect.

Finally, since focal length and magnification vary with the distance between the object and the optical axis, the image of an extended (i.e., two-dimensional) object will be distorted, again particularly at the image edges from the optical axis. Depending in the lens, the image of a square object will either have concave edges ("pincushion" distortion, usually found for convex lenses) or convex edges ("barrel" distortion, usually found for concave lenses). Distortion effects are strongly reduced for thin lenses and may be corrected using aperture stops in compound lens systems.

Chromatic Aberration

The five Seidel aberrations which we have just discussed are characteristic of all spherical lenses, even if illuminated with a single wavelength. Chromatic aberration is of a different nature altogether. From Eq. 6.4, we saw that the focal length of a refractive lens is a function of the refractive index of the lens material, n_L. Refractive index is in general wavelength-dependent, so that the focal length of a lens also varies with wavelength leading to chromatic aberration.

In general, $n_L(\lambda)$ decreases as wavelength λ increases, in the visible wavelength range. As a result, the focal length of a convex lens, for example, will increase as the wavelength of the incident light increases. For an object which emits a broad spectrum of wavelengths (such as a white source), the resulting image will then have a spread of spectral components about the focus position.

For the single-wavelength emission of laser sources, chromatic aberration tends to be a minor issue in the specification of an optical system. If various wavelengths are used from a variety of sources, or a broad-band source (such as an LED, particularly white LEDs) is employed, then the chromatic behavior of the lens system should be considered for an optimal design.

6.1.3 Refractive Microlens Structures

Refractive microlenses, as mentioned at the outset of this section, may be considered as miniaturized macroscopic lenses: until the size becomes so small that diffraction effects become significant, the optical behavior may be calculated using classic analysis techniques. Fabrication is another matter; at sub-millimeter size scales, conventional grinding techniques are no longer practical and new approaches for lens fabrication have been developed.

For the fabrication of *small* refractive optical components, with sizes into the millimeter range, conventional glass polishing techniques may still be applied. Such mini-lenses, while commercially available, tend to be expensive, particularly for inclusion in large-volume microsystems. An alternative approach is the use of replicated lenses, fabricated in optical plastics using embossing or molding techniques and a high-quality master. Replicated lenses may in some cases not achieve the performance levels of their glass cousins, due in part to the limited types of plastic materials available for fabrication; on the other hand, since the master structure only needs to be fabricated with high precision once, aspheric or other complex profiles may be cheaply fabricated using these techniques. Some disposal camera lenses, for example, have a relatively complex optical structure.

Fabrication of Refractive Microlenses

For refractive microlenses with aperture sizes significantly below 1 mm, alternative techniques have been developed. Most approaches are based on microfabrication technology using many of the same processing techniques as those used for microelectronics or semiconductor laser fabrication.

A popular and well-established technique is the use of thermal reflow of photoresist (Hutley 1998). Cylindrical posts are defined in photoresist on a suitable substrate, typically glass or silicon; the post diameter is chosen to be that of the ultimate aperture of the lens. The substrate is then heated above the melting point of the photoresist (typically 120° – 140°C) whereupon the posts melt and take on a spherical surface curvature. This curvature is defined by the surface tension of the molten material and is usually a very good approximation to a sphere. Variations on this technique include the use of custom base layers to influence the numerical aperture (Haselbeck et al. 1993) and advanced exposure techniques to achieve non-spherical lens profiles (Grunwald et al. 1998; Grunwald, Mischke and Rehak 1999) The use of microfabrication technologies, particularly standard photolithography techniques to define the lens posts, implies that these lenses may be easily fabricated in large two dimensional arrays, which represents a significant advantage for this fabrication approach.

The thus structured photoresist may itself be used as a lens for wavelengths at which the material is sufficiently transparent; however, photoresist lenses are typically not terribly robust, are easily subject to physical damage and have a limited useful temperature range. Considering these limitations, the photoresist lens structures are most often etched into the glass or semiconductor substrate using well-defined dry etch techniques. If properly done, the exact shape of the photoresist lens is transferred into the substrate such that glass or silicon is then used as the lens material; an example of a Si lens fabricated by this method is shown in Fig. 6.12. Alternatively, varying the gas composition during dry-etch transfer of the lens profile into the semiconductor may be used to generate an aspheric profile from a spherical photoresist lens. The photoresist lens (or the etched substrate) may also be used as a master for generation of a mold from which plastic-based lenses may be fabricated using replication technologies.

For a microlens with radius of curvature R, aperture radius r, and made from a material with optical refractive index n_L used in a medium (usually air) with refractive index n_0, the focal length f is given by

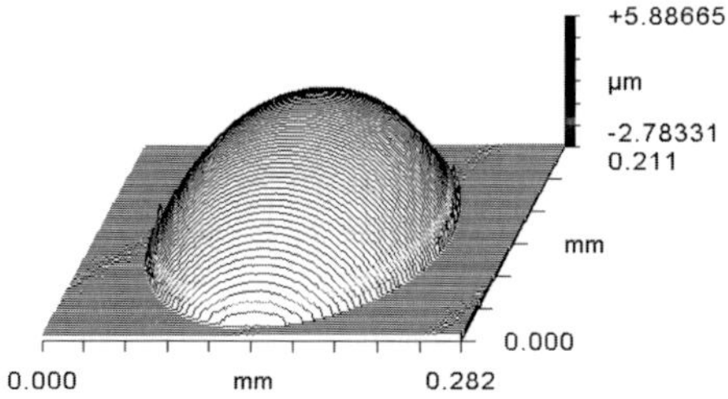

Fig. 6.12. Interferometric surface profiler image of a microlens with 200 µm diameter and a height of about 7.5 µm fabricated by photoresist reflow and etched into silicon

$$f = \frac{Rn_0}{n_L - n_0} . \tag{6.13}$$

The height of the lens above the substrate, h, may easily be found from

$$h = R - \sqrt{R^2 - r^2} \tag{6.14}$$

where lenses with very small curvature radius (and thus large height for a small aperture) may be difficult or impossible to fabricate using the techniques just described. Typical values for lens aperture are $15 \le r \le 500$ µm and lens heights are usually restricted to $h \le 100$ µm, often limited by available photoresist thicknesses. Lens structures of this type generally have a numerical aperture in the range $0.15 \le NA \le 0.45$.

Example 6.4

Assignment: *We would like to fabricate the smaller of the microlenses (the one with D = 260 µm) designed in Example 6.1 using the microfabrication techniques just discussed; recall that a working distance (object–lens distance) of 500 µm was required to collimate a laser beam. The lens will now be fabricated as part of a 500 µm thick glass substrate (BK7 with* $n_L = 1.52$*) as shown in Fig. 6.13. Use a simple ray-tracing model to discuss the design considerations of the lens.*

Solution: *The lens will be illuminated through the glass substrate, as schematically shown in Fig. 6.13. Due to refraction at the bottom surface, the rays emitted by the laser source will tedious to calculate by hand; so we will use a ray-tracing model to optimize our design.*

We begin by just trying the lens as we designed it in Example 6.1: radius of curvature 260 µm, aperture 260 µm and height 35 µm. The ray-tracing result is shown in the top figure of Fig. 6.14. We see two problems right away: the working distance (i.e., spacing between the laser diode point source and the flat bottom of the glass substrate) is only 100 µm, not

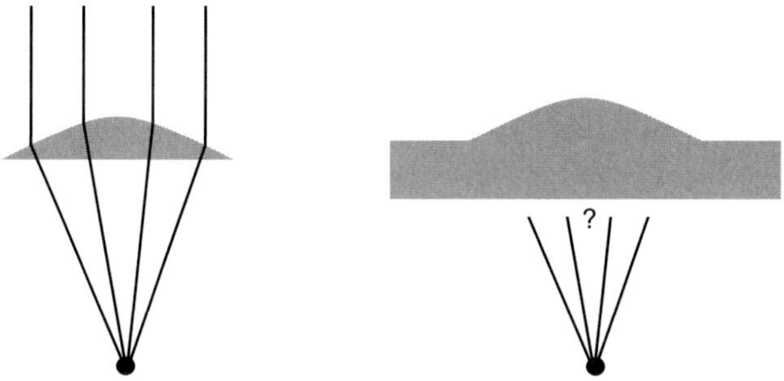

Fig. 6.13. A plano-convex microlens *left* will have a more complex beam propagation pattern when fabricated as part of a glass substrate *right*

500 μm as specified. Secondly, the maximum angle of the beam is about 15°, implying NA ~ 0.26. Let's see if we can do better.

To increase the working distance, we should increase the focal length of the lens. A few tries shows that a curvature radius of 489 μm leads to a focal length of the lens (in air) of 940 μm and, when traced through the glass substrate, leads to the required working distance of 500 μm. Lens aperture is 500 μm and the NA of the lens is the same as the first design. However, another problem may have arisen. The height of this new lens is 69 μm; that considerably thicker than most photoresist systems used to define a microlens of this type.

Assuming that we have photoresist available with a maximum thickness of 20 μm, we can stay with our new lens design and decrease the aperture to reduce the maximum thickness. The result is shown in the bottom of Fig. 6.14, where we see that D = 280 μm is all that can be achieved. the numerical aperture of this lens is only about NA ~ 0.13 implying a maximum acceptance angle of less than 8°.

These calculations show some of the compromises which must be considered when designing a microlens; maximum aperture, feature height, focus length and NA may not all be optimal for a particular application.

Fresnel Lenses

The feature height of microlenses is often limited by aspects of the fabrication technology involved; if photoresists are used to define the lenses and structured using re-flow or other techniques, the thickness of the resist will define the maximum possible feature height of the lens, as we just saw in the previous example. Most common resists are limited in thickness to less than 80-100 μm. Other aspects of the manufacturing process may also limit the feature height as may, in some cases, the dimensions of the micro-optical system.

One approach to reduce the maximum feature height is the use of a Fresnel lens. Developed by Augustin Jean Fresnel (1788-1827), originally as a means to reduce

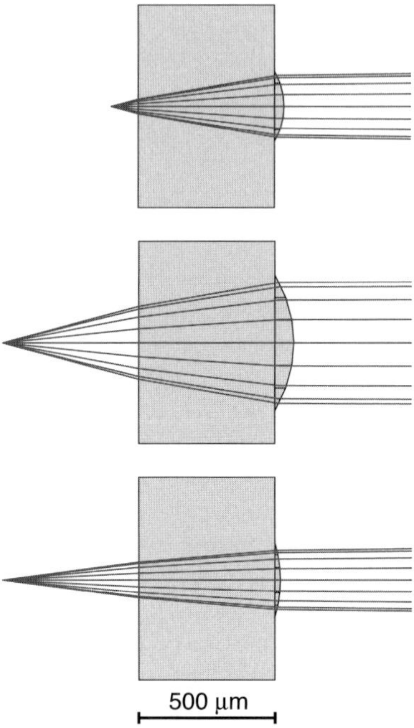

Fig. 6.14. Ray tracing simulations of microlenses fabricated on a 500 μm thick BK7 glass substrate: R = 260 μm, D = 260 μm and h = 35 μm *top*; R = 489 μm, D = 500 μm and h = 69 μm *center*; R = 489 μm, D = 280μm and h = 20 μm *bottom*

the size and weight of lighthouse lenses, the Fresnel lens divides the curved surface of the lens into zones which are separated from each other by abrupt steps. We can imagine that the original lens profile is folded back onto itself at regular intervals such that for a lens with N zones, a refractive lens with an original height of H will have a Fresnel equivalent with maximum feature height H/N. An example of such a Fresnel microlens is shown in Fig. 6.15; the lens with 2.7 mm radius consists of 12 zones with a maximum feature height of 88 μm.

The Fresnel lens as described here (and as it is classically used) is a purely refractive optical element; the zone sizes and vertical steps are much larger than the wavelength so that the Fresnel lens does *not* rely on diffractive effects. We will examine a Fresnel-like diffractive optical element in Sect. 6.2.3 which looks much like a Fresnel lens and is sometimes mistakenly called such but functions on a different principle.

6.1.4 Ball and GRIN Lenses

The designer of an optical microsystem has a number of other options when considering the types of lens structures required to perform a particular optical func-

Fig. 6.15. A micro-optical Fresnel lens fabricated in PMMA; lens radius is 2.7 mm and the maximum feature height 88 μm

tion. Ball and GRIN lenses are refractive optical structures with sizes in the millimeter to hundreds of micrometer range; hybrid assembly considerations are often such that sizes of this order are not a disadvantage. However, integration into compact arrays is generally difficult or impossible using these relatively macroscopic components.

Ball Lenses

A ball lens or spherical lens is essentially a glass sphere with accurately specified radius and typically made from a high-grade optical glass with precisely known refractive index and other optical properties. The spherically-curved surface provides the lensing function, as shown in Fig. 6.16, whereby a ball lens may no longer be considered a "thin" lens in the sense of Sect. 6.1.1. Ball lenses are mass-produced using glass polishing techniques and are usually sold in lots of hundreds or thousands; in these quantities, with the exception of specialty products, they tend to be relatively inexpensive.

Ball lenses are available in diameters ranging from 110 μm to several millimeters; standard sizes include 150 μm, 225 μm and 400 μm. The glass used is typically of high optical quality with refractive indices ranging from $n_L = 1.5$ to $n_L = 2$ for specialty glasses; some types are also available with an anti-reflection coating. Typical focal lengths for the smaller lenses are 100 – 250 μm.

The advantage of using a ball lens in a micro-optical system is the nicely symmetric nature of the sphere. Since the sphere has the same curvature in all directions and thus has no optical axis, positioning and aligning the ball lens may be considerably easier than a conventional lens; the orientation of the lens is unimportant. On the other hand, the optical performance is limited and handling may be an issue for smaller ball lenses; unlike microlenses or microlens arrays fabricated on a larger glass or semiconductor substrate, the small size of the lens requires mechanical positioning equipment or a vacuum pencil and a steady hand.

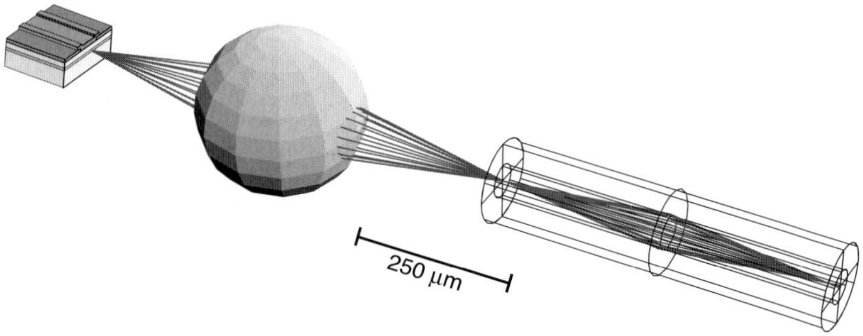

Fig. 6.16. Ray-tracing simulation of a 225 μm ball lens used for coupling a laser beam into a multi-mode fiber with a 50 μm core; the laser chip is shown schematically and is not to scale

GRIN Lenses

The gradient-index or GRIN lens works on a different principle than standard microlenses or ball lenses. The lens consists of a glass cylinder in which the refractive index has a well-defined radial profile. This refractive index profile typically has a parabolic form and thus decreases from the center as

$$n(r) = n_{max}\left[1 - \frac{n_{max} - n_{min}}{n_{max}}\left(\frac{r}{a}\right)^2\right] \tag{6.15}$$

for lens radius a, refractive index maximum n_{max} and minimum n_{min}; the radial parameter r varies from 0 at the lens center to $r = a$. The typical GRIN lens is 0.5 – 2 mm in diameter and 1.25 – 10 mm long.

As shown schematically in Fig. 6.17, the effect of the parabolic refractive index profile is to bend rays incident at an oblique angle to the optical axis back toward the optical axis; rays thus propagate sinusoidally through the lens. The length of one period of such a sinusoid is termed the lens pitch such that different lengths of the GRIN material give rise to different optical effects. As we see in the figure, a 0.23 pitch lens will focus a collimated beam at a point slightly behind the lens; 0.25 pitch focusses a collimated beam at the lens surface; 0.29 pitch is used to connect focus to focus; and 0.50 pitch will connect a collimated beam to a collimated beam.

A GRIN lens, such as the 0.23 pitch example, may be used to collimate the beam from a laser diode. The emission facet is positioned at the focus, at a finite distance from the lens surface. The advantages of GRIN lenses include that they may have a very short working distances (down to zero in the case of the 0.25 pitch example) and thus placed arbitrarily close to a semiconductor laser facet. Typical numerical aperture values are about 0.5.

GRIN lenses are also used as cylindrical lenses, with light incident onto the long side of a block of gradient index material. In this configuration, they are useful for collimation of one axis of the asymmetrical laser diode output.

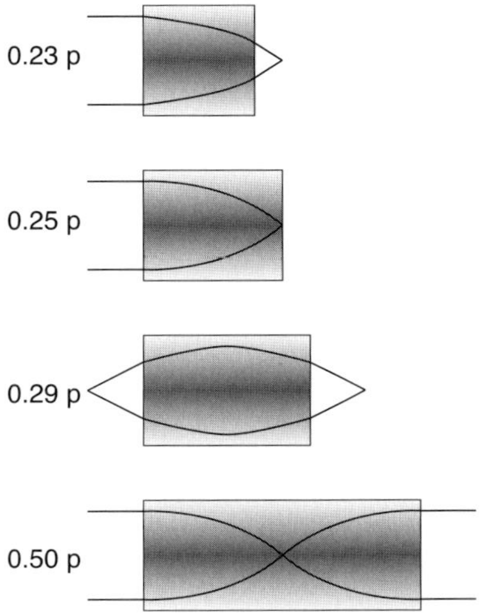

Fig. 6.17. Beam propagation through GRIN lenses of various lengths; *top to bottom* 0.23 pitch; 0.25 pitch; 0.29 pitch; 0.50 pitch

6.1.5 Other Micro-Optical Components

Lenses may be the most critical passive components in most laser diode-based optical microsystems and are perhaps also those parts requiring the most careful design. In general, a variety of other optical elements may be required for the construction of a complete system and we will provide a brief overview of a few of these here. With regard to their size, most of these components would not necessarily be considered "micro-optical" but many are available in dimensions which would allow incorporation into an optical microsystem.

Mirrors

Mirrors are used for beam deflection, focusing or for the fabrication of an optical cavity. High-quality mirrors typically consist of a high-reflection coating deposited on a glass substrate and are anything but trivial components. The substrate is usually chosen to have a low thermal coefficient of expansion; popular materials include fused silica, a low-expansion borosilicate glass (such as Pyrex®) or a high-tech composite such as Zerodur®. The latter is a glass ceramic with virtually zero thermal expansion and of great interest for telescope mirrors which, with diameters often exceeding 8 m, are not particularly suitable for micro-optical systems.

The reflective coatings are often metal films, whose composition must be chosen to provide high reflectivity at the wavelength of interest and exhibit a high long-term stability. Aluminum is probably the most popular reflective film, but generally has reflectance below 0.95 and, as it oxidizes readily, must be coated with a passivating film. Silver provides higher reflectance but also oxidizes. Gold may be used to achieve reflectance values exceeding 0.99 and may be used in a broad wavelength range, up to about $\lambda = 16\ \mu m$. Unfortunately, Au is quite soft and thus needs to be coated with a scratch-resistant film. Multi-layer Bragg structures (such as those used for VCL mirrors, as discussed in Sect. 4.5) may also be used as discrete mirror components; these are of course highly wavelength- and input angle-dependent.

Surface quality is a further factor in determining the performance of a mirror. Flatness, usually measured as a fraction of the wavelength of interest, will determine the optical properties of the reflected beam. For operating wavelength λ, mirrors with flatness $\lambda/4$ are of standard quality, surfaces with $\lambda/10$ are used for high-quality imaging systems and a flatness of $\lambda/20$ is required for interferometric or multi-path applications.

The mirror shape depends on the intended optical function. Planar mirrors are used for deflection, ideally without altering the shape of the beam. Concave or spherical mirrors are used for focusing a laser beam and are often used in applications or in wavelength ranges where refractive lenses are impractical or do not exist. Finally, retro-reflectors, also called "corner cubes" or "cat eyes" are formed by a mutually orthogonal arrangement of three planar mirrors and have the property that an incident beam is reflected back in the direction from which it came, even for non-normal incidence. Retro-reflectors are commonly used for beam alignment and positioning optics, as well as in distance measurement systems and for optical delay lines.

The sizes available for mirrors tend to be relatively large; for micro-optical applications, the user is often best advised to fabricate the lens herself using available evaporation techniques and metals using, for example, a microscope cover glass as a substrate. This structure may then be cleaved or sawn to sizes below 1 mm. An alternative is the use of etched facets in a semiconductor substrate. Wet etches, which stop on a particular crystal plane, can yield mirror facets of optical quality and very specialized dry etching techniques have been used to achieve the same. The etched facets are typically coated with a high-reflectance metal using angle evaporation. The approach is useful if the system technology allows the use of the semiconductor substrate in the ultimate design; a silicon-optical-bench approach (see Sect. 7.2.5) is one example.

Prisms

Light may also be deflected using refractive, as opposed to reflective, effects; prisms perform this kind of function, often with greater functionality than that which is achievable using mirrors. A prism is typically a glass block, made from the same type of materials as classic refractive lenses, with an arrangement of normal and angled surfaces. A light beam entering a prism may then be subject to deviation, rotation or displacement; dispersion effects also allow the separation of a multi-wavelength input beam into its spectral components (probably the most classic demonstration in basic optics) or into its polarization components.

An arrangement of prisms is very useful for circularizing the intrinsically elliptical beam emitted from a laser diode. The use of an anamorphic[2] prism pair, or Brewster telescope[3], as shown in Fig. 6.18, allows the beam to be expanded in one direction only, such that any degree of ellipticity may be corrected. The magnitude of the beam expansion may be changed by rotation of the prisms, allowing considerable flexibility but requiring accurate and stable positioning for use in a standard system. Such an arrangement is popular in macroscopic optical systems for collimating the emission of a laser diode.

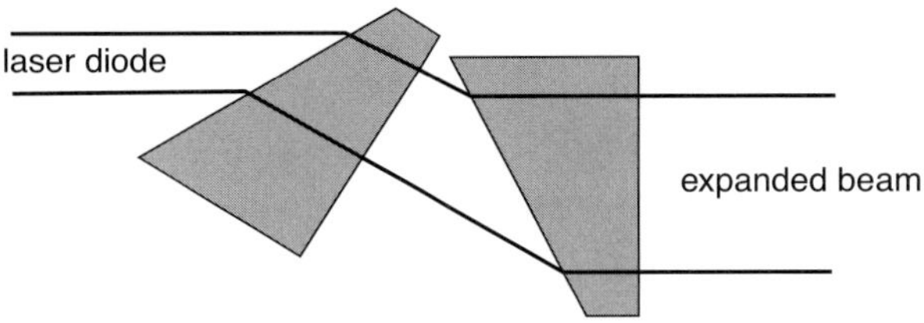

Fig. 6.18. An anamorphic prism pair used for expanding the elliptical beam for a laser diode in one dimension only, allowing circularization of the beam

Most prisms are relatively macroscopic structures with sizes typically in the millimeter range, but some manufacturers offer components with dimensions down to $500 \times 500 \times 500\ \mu m^3$. Fabrication is by classic optical grinding and polishing techniques which limits the sizes available.

Beam Splitters

Occasionally, an application will require that the emission from a laser is divided into two separate beams; a beam splitter accomplishes exactly this. The simplest structure, a cube beam splitter, consists of two right-angle prisms cemented together at their hypotenuse, with a partially-reflecting coating at the interface as shown in Fig. 6.19. This type of beam splitter has the advantage that it may be used in a relatively wide wavelength range, that the polarization state of the beam is maintained and that the splitting ratio may be arbitrarily chosen during fabrication. Polarizing beam splitters are also available in which the materials are selected so that one polarization state is transmitted and the other reflected. Sizes are generally 3 mm on a side or larger.

---

[2] Anamorphic optical components are those with differing magnification in two primary axes normal to the optical axis.

[3] Named for David Brewster (1781-1868), better known to the general public as the inventor of the kaleidoscope.

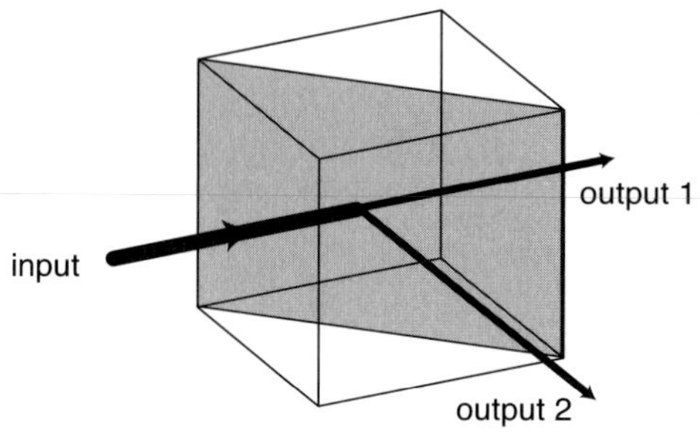

Fig. 6.19. A cube beam splitter based on two right-angled prisms cemented together

An alternative structure is the plate beam splitter which uses precisely defined optical coatings on a thin substrate to define the splitting ratio; values from 50:50 to 10:90 are achievable. By varying the coating thickness or composition across the plate an adjustable beam splitter may be made. This type of structure occasionally suffers from ghost images due to reflection from the plate boundaries.

Polarizers

Laser beam polarization is relevant for numerous applications; although many laser diodes emit radiation in a well-defined linear polarization direction, some, such as VCLs, do not. Polarizers are used to transmit one desired polarization direction.

The best known polarizers are the sheet polarizers, also commonly known as Polaroid films. These consist of dichroic sheets made from long organic molecules stretched into alignment and thus exhibiting birefringence. The filters remove the unwanted polarization direction by absorption, thereby limiting the powers which may usefully be transmitted: high-power laser diodes will easily damage the polarizer. The extinction ratio (measured using two crossed sheet polarizers) is generally about 10^{-4} and this type of component allows large incidence angles and is thus robust with respect to positioning. The sheets are available in large sizes and may be trimmed to the size required; sub-millimeter dimensions, however, are difficult to achieve.

Reflection polarizers use multi-layer stacks, typically oriented at the Brewster angle, and are thus often used as polarizing beam splitters. Such polarizers are typically made by evaporation of a sequence of optical materials and may use small glass substrates; cleaving or sawing may be used to achieve sub-millimeter sizes.

Bulk polarizers are typically macroscopic components, very useful for laser-based optical systems, but difficult if not impossible to miniaturize. Made from birefringent materials, such as calcite ($CaCO_3$) or quartz, these function by dividing the unpolarized input beam into two orthogonal polarizations which propagate in two different directions, one of which may be absorbed. The structure is usually similar to that of a polarizing beam splitter: the Glan-Thompson polarizer, for example, uses two prisms cemented together to deflect one polarization into an absorber. The Glan-Taylor polarizer uses two prisms separated by an air gap, but is otherwise virtually identical. The Wollaston prism, finally, transmits both polarizations but at an angle of typically 20°, allowing access to both components if necessary. These bulk polarizers are generally of high quality and allow high laser powers, but remain macroscopic components such that they are more useful for characterization of laser diodes than in a semiconductor laser microsystem.

Retarders and Isolators

The retarder is a phase-shifting element which typically rotates the polarization of an input field. The two most common types are the $\lambda/4$ plate, which transforms linear to circular polarization and vice versa; and the $\lambda/2$ plate, which rotates linear polarization by 90°. Retarders are made from birefringent materials, such as mica or quartz, and typically take the form of flat optical plates, 2–4 mm thick.

Conceptually related are optical isolators, which use polarization rotation to allow transmission of an optical beam in only one direction through an optical system. The so-called Faraday isolator, illustrated in Fig. 6.20, uses a material which exhibits the Faraday effect, such as a crystal of YIG (yttrium iron garnet): under application of a magnetic field, the crystal becomes non-reciprocal, meaning it rotates the polarization in the same direction regardless of propagation direction. Thus in the example of the figure, the YIG rotates the polarization to the right for beams coming from the right or the left.

Using input and output polarizers, the Faraday effect may be used for optical isolation. As shown in Fig. 6.20, an input polarizer oriented at 0° allows vertically linearly polarized light, from, for example, a laser diode into the YIG crystal, which rotates the field by 45° to the right (clockwise when looking in at the input, from the right). An output polarizer, oriented at 45°, then lets this field through. Reflections from structures at the output, such as a fiber facet or another planar surface, is then transmitted by the 45° polarizer back into the crystal, whereupon the optical field is subject to another rotation by 45° *in the same direction* (i.e., to the right or clockwise, seen from the input). The backwards propagating field is thus linearly polarized at 90° and is therefore blocked by the input polarizer, which is at 0°; thus light is only transmitted forward through the isolator.

Optical isolators are widely used in applications where reflections back into the laser diode may have detrimental effects: we saw its use in high-resolution linewidth measurements (Sect. 5.3.4) and many laser diode telecommunications modules require an isolator to assure stable device operation, as shown schematically in Fig. 6.1. Due to the need for a magnetic field, large-aperture bulk optical isolators tend to be large and weigh several kilograms; however, fiber-based or waveguide-based isolators are available or in development for use in optical microsystems and these components are in the millimeter size range.

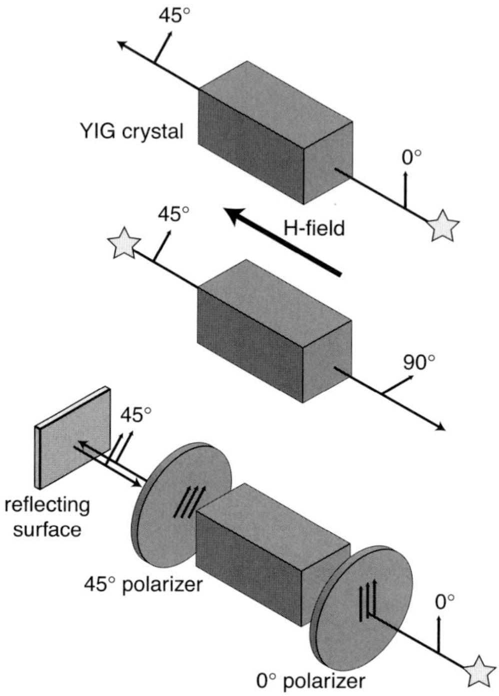

Fig. 6.20. An optical isolator based on the Faraday effect. The YIG crystal rotates the optical field 45° in the same direction, regardless if the field propagates to the left *top* or to the right *center*; light incident from the right and reflected from the left is thus blocked by the input polarizer *bottom*. The star indicates an optical source

Filters

Filters perform a variety of optical functions and typically take the form of a specially-prepared glass or plastic plate; the optical behavior is usually defined by surface coatings or a modification of the filter material itself, such that miniaturized versions may be custom manufactured.

One of the most common filter functions includes wavelength filtering, generally using a dielectric layer stack; the resulting interference filters may be designed to perform a high-pass, low-pass or band-pass function. Advanced filter technologies allow the specification of very abrupt spectral characteristics. A further application, albeit one primarily of interest for laboratory laser characterization, is attenuation uniformly across the spectrum using a neutral density filter. This optical function is often required, for example, to avoid saturation or damage effects when a laser is incident on a CCD imager, high-sensitivity photodetector or other sensitive component. Neutral density, or grey, filters are typically characterized by a density parameter D, such that transmission through the filter, T, is given by

$$T = 10^{-D};\qquad(6.16)$$

generally available neutral density filters have values for D ranging from 0.1 to 4.0.

6.1.6 Laser Diode Optics Applications

We have examined a number of refractive optical structures which, often in their macroscopic implementation, have a myriad of applications in the very broad field of optics. Laser diode microsystems generally employ only a small subset of these functions; we will consider the two most important ones, namely collimation and focusing. Both of these are considered schematically in Fig. 6.1: the emission from the semiconductor laser is collimated, directed through a number of optical components and subsequently focused onto the end facet of an optical fiber. The illustration shows both of these functions as performed by ball lenses, which represents a cheap if not necessarily optimal approach.

Collimation

We have considered collimation implicitly and in the form of a few examples in the previous discussion. In its simplest form, the collimation of a light beam involves converting the spherical wavefronts from a point source, in which the optical rays diverge, into planar wavefronts, in which the rays run parallel. We may think of a collimated beam, in its most rudimentary form, as having a constant beam diameter regardless of propagation distance, such that it may usefully be transmitted through an optical system. In reality, a laser emits a Gaussian beam (Saleh and Teich 1991, Chap. 3; Smith and King 2000, Chap. 15.8) such that it will always diverge to some extent. Nevertheless, the collimation of a laser beam is often the primary function of the first optical elements placed in front of a laser.

The laser diode is not particularly cooperative in this endeavor. As we saw in Sect. 5.6, the emission from a laser diode has a roughly Gaussian intensity distribution, but can be astigmatic and, due to diffraction effects at the aperture, can have a strongly elliptical emission profile. In addition, the laser aperture is not a point source but is rather distributed in space (recall Fig. 5.22) and multi-mode lasers can have complicated near-field intensity distributions. Insofar as the emission profile is concerned, edge-emitting lasers are generally worse than VCLs.

A collimator is subject to several performance requirements. Due to the highly divergent beam a high numerical aperture is required, typically 0.5 – 0.7. In addition, the working distance (the spacing between the last feature of the lens system and the laser facet) should not be too small to allow practical positioning of the laser diode, particularly if mounted in a package. Collimator design may also take into account the optical effects of a cover glass on the laser package; the thickness of the planar lens surface defining the end of the collimator facing the laser diode may be reduced to compensate for glass thickness.The result of these requirements is that the lens system typically has a large aperture (8 – 10 mm); the resultant collimated beam then also has roughly this diameter. If a beam with a small diameter (say 1 mm) is needed, it is more practical to collimate into a large-diameter beam and to de-expand this to the desired width.

Verifying the quality of collimation takes some practice. As we briefly mentioned, Gaussian beam propagation implies that an optical field always diverges. The extent of collimation may be checked by moving a screen along the beam and verifying that the spot size does not appreciably change with the dimensions of the system. A more reliable approach is by use of auto-collimation: the beam is directed back toward the source using a mirror, with a slight lateral offset. By minimizing the resultant spot size, optimal collimation may be assured.

Focusing

Some applications require that the emission from a laser diode is focused to a point. Lasers used for machining, marking or surgical uses fall into this category, as does the seemingly simple task of coupling the emission into an optical fiber. We could use a single bi-convex lens and position the laser source at $2f$ from the lens; as we saw in Example 6.1, and referring to Eq. 6.5, the light is then focused to a point at a distance $2f$ from the other side of the lens. Such a $4f$ arrangement, or the equivalent using ball lenses as shown in Fig. 6.16, is simple but not optimal. More practical is the use of a collimator to generate the best approximation to a spatially uniform plane wave and then a focusing lens to focus to a point. As we saw in Eq. 6.11, the smallest focus point size of a diffraction-limited lens is a function of the wavelength and numerical aperture of the lens.

Complete Coupling Optics

A complete optical system to, for example, couple light from a laser diode into a fiber, can range from simple and cheap (such as the example of Fig. 6.16) to almost arbitrarily complicated and expensive. Microsystem applications will tend to the former, if not for cost then certainly for size reasons; the performance compromises are often balanced against complete system considerations.

If performance considerations are paramount, then such an optical system will need to undertake a number of functions. Starting at the laser diode, the astigmatism will be corrected before collimation, typically using a plano-concave cylinder lens to bring the lateral and transverse foci into agreement. Collimation will then take place as described above, using a plano-convex, meniscus and cemented doublet lenses. The collimator is then followed by an anamorphic prism pair to correct the ellipticity of the beam; the resulting circular beam may then be expanded, using a bi-concave lens and subsequently focused to a point focus using a large-aperture lens arrangement to assure diffraction-limited performance. Clearly such a system is not generally microsystem-compatible, but the experimenter should be aware of the demands placed on laser diode optics.

6.2 Diffractive Optics

Diffractive optics, sometimes known as physical optics, relies on diffraction effects to perform a desired optical function. Physical structures with either a certain periodicity or accurately designed features with size scales on the order of the wavelength give rise to interference which is employed to yield the desired optical behavior.

Diffractive optics encompasses a wide variety of structures based on diffraction effects (Turunen and Wyrowski 1997). These range from the relative simplicity of periodic gratings to diffractive lenses and holograms. Although the physics of these devices and their design can become relatively involved, we will present an overview of what types of diffractive optical elements have applications in laser diode microsystems.

6.2.1 Diffraction

Diffraction is based on the interference of electromagnetic waves as we discussed in Sect. 2.2.6. It may perhaps be most easily understood by considering Huygen's principle, named after Christiaan Huygens (see Sect. 2.2), which states that each point on the wavefront of a propagating wave may be considered a point source. By adding the contributions from each of these point sources further along in the propagation direction, the new wavefront may be calculated. At any position on the new wavefront, then, the total field is the sum of the contributions of the fields which meet and interfere at that point, such that consideration of the phase of the waves is essential for determining propagation.

Optical diffraction effects come into play in a number of ways. Given an obstruction to an optical wave, the edge of this obstruction will act as a point source for the field giving rise to interference effects which manifest themselves as periodic intensity variations at the edge of the optical field. The finite aperture of a lens provides an example: the Airy function intensity distribution which results when focusing light to a point using a lens with a non-infinite diameter, as we saw in Sect. 6.1.1, is the result of diffraction from the edges of the aperture. This diffraction effect is unavoidable and, as we mentioned above, an optical element or system in which all other aberrations are negligible in comparison is termed diffraction limited.

Periodic structures also exhibit strong diffraction effects. By adding the phases of the fields which are "emitted" from each point of such a periodic structure, the resultant interference gives rise to wavefronts with strongly varying intensity distributions. Since this effect is highly wavelength dependent, periodic structures are often used to separate wavelength components in an optical field. In general, diffraction effects come into play whenever optical structures contain wavelength-size features; as we will see below, the resulting interference may be used to generate almost arbitrary wavefronts, which is the principle behind, for example, holograms.

6.2.2 Gratings

A diffraction grating is a periodically modulated corrugation with feature size and period size scale ranging from sub-wavelength to several hundreds of wavelengths (Hutley 1982). We have already seen gratings as the resonator mirror structures for distributed feedback lasers in Sect. 4.4; in that example, the grating was used as a wavelength-dependent waveguide reflector. Gratings and their fabrication have a rich history dating back to Joseph von Fraunhofer (1787-1826) and Albert Michelson (1852-1931) and today have a wide variety of uses in optical microsystems including spectral analysis, coupling of optical fields into and out of waveguides and as wavelength-selective reflectors.

Grating Behavior

For a light beam incident normally on a transmission grating, as shown schematically in Fig. 6.21, diffraction results in a deflection of part of the beam at an angle θ from the normal. This effect leads to a spatially-variant, wavelength-dependent intensity distribution which is also a function of the grating period, Λ. Angles which result in an intensity maximum define so-called diffraction orders, identified by an integer m.

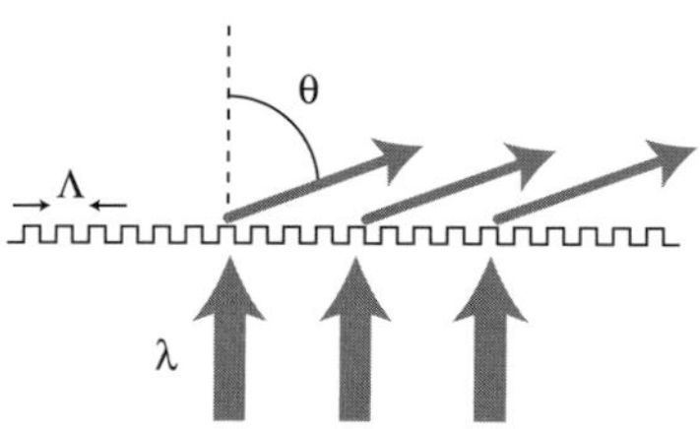

Fig. 6.21. Light incident normally on a transmission grating; a portion of the incident light is diffracted at an angle θ

The relationship between grating period, wavelength and transmission angle of a a diffracted order m is given by the grating equation, namely

$$\Lambda \sin\theta = m\lambda . \tag{6.17}$$

For an incident beam at an arbitrary angle, as shown in Fig. 6.22 for the case of a reflection grating, the grating equation is slightly modified to read

$$\Lambda(\sin\theta_o - \sin\theta_i) = m\lambda ; \tag{6.18}$$

in both of these expressions, m is an index representing the diffraction order and can take integer values $m = \pm 1, \pm 2, \pm 3, \ldots$.

We see from the grating equations Eq. 6.17 and Eq. 6.18 that λ and Λ must be of the same order of magnitude to generate usable diffraction angles; in addition, we

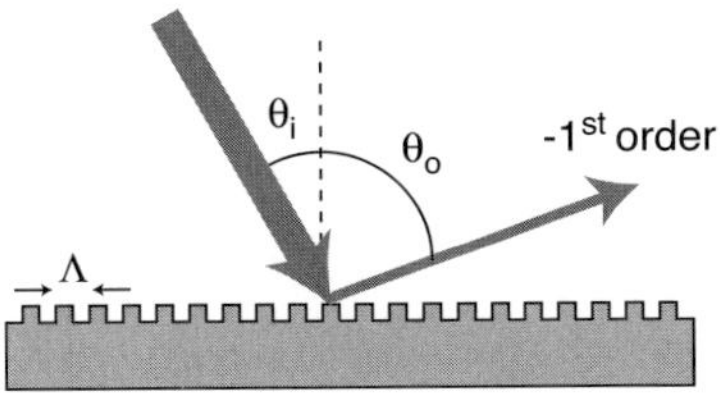

Fig. 6.22. A beam incident at an arbitrary angle θ_i onto a reflection grating giving rise to a diffracted order at θ_o

see that, insofar as θ remains less than 90°, more than one diffraction order may be present, with diffraction angles on both sides of the normal to the interface. This effect is seen in the schematic representation of Fig. 6.23; a portion of the illumination beam, incident at an arbitrary angle, is directly transmitted and portions of the field are diffracted into the various orders on either side of the 0th order, or undiffracted, beam.

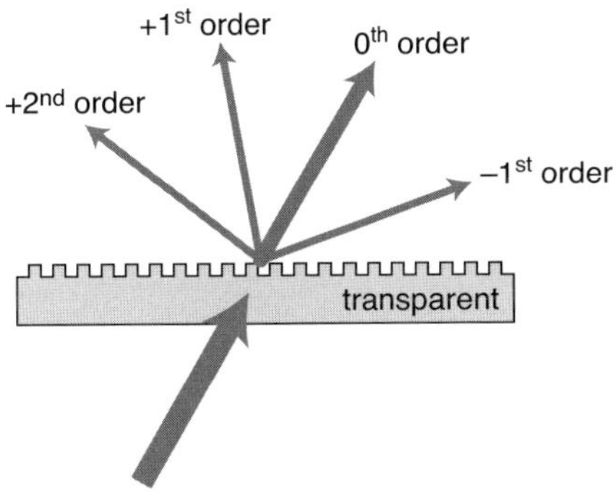

Fig. 6.23. The propagation of various diffraction orders by illumination of a transmission grating with a monochromatic light source

Our consideration of gratings to this point has tacitly assumed that illumination is monochromatic at a wavelength λ. For sources with a distribution of wavelengths, either the narrow distribution of a laser with finite linewidth or the broad spectrum of an optical field which is to be analyzed, each diffraction order contains an angular spread of wavelengths. This angular dispersion, the change in diffraction angle with wavelength, has the form (Hutley 1997)

$$\frac{\partial\theta_o}{\partial\lambda} = \frac{m}{\Lambda\cos\theta_o}; \tag{6.19}$$

we see thus that higher diffraction orders have a higher angular dispersion.

Corrugation and Grating Shape

The shape of the grating corrugation plays a significant role in determining the grating efficiency; efficiency is a measure for how much light is diffracted into a given order. Whereas the figures above have shown a rectangular tooth shape, this form is sometimes difficult to fabricate in gratings with periods below 1 μm. Sinusoidal and triangular corrugations are quite common; the former often results when holographic techniques are used to define the grating structure.

A specialized corrugation form is that of the blazed grating. In this case, the "tooth" takes the form of a prism with the top surface defined at an angle α to the plane of the grating; see Fig. 6.24 in Example 6.5 for a typical structure. The blazed grating combines specular reflection with diffraction to direct the majority of beam energy into one of the diffracted orders, rather than remaining in the undiffracted 0th order. As a result, diffraction efficiency is strongly increased for incidence at the blaze angle.

The grating structures we have considered have been assumed to be uniform: the period does not vary with position either along the corrugation direction or normal to it. A curved grating, however, for which the corrugations bend in the plane of the grating, may be used for focusing or collimating the diffracted beams. Alternatively, a chirped grating, one in which the period slowly varies with position, gives rise to a position-dependent Λ and thus a spatially varying diffraction pattern or a broadened spectral response. Such variations in the standard grating structure have been shown to be very useful for a variety of optical functions (Nishihara, Haruna and Suhara 1989).

Example 6.5

Assignment: *Use a grating to make a tunable laser.*

Solution: *We have seen that the diffraction angle of a grating varies with wavelength; for a spectrally broad input, the various wavelength components are then diffracted at different angles. We also recall that the gain spectrum of a semiconductor laser diode is relatively broad; looking back to Fig. 3.17, for example, we see that with sufficient pumping, a width of 20–25 nm is reasonable. In a normal laser diode, the cavity formed by the two mirrors selects the wavelength with the highest gain leading to the typical narrow-band emission.*

If we could use a "wavelength selector" to tune through the wide gain peak, we would have a tunable laser. A grating may be used for exactly this function: by choosing a mounting angle so that the energy from the first order mode exactly overlaps the incident field, the grating becomes a wavelength-selective mirror. By physically rotating the grating, the reflected wavelength may be shifted.

Such an arrangement is shown in Fig. 6.24. We start with a laser diode which has one facet coated with an anti-reflection layer, thereby destroying the laser cavity; as a result, the diode chip emits a broad LED-like spectrum with a shape and width roughly equal to that of the gain spectrum. This emission is collimated with a microscope objective and directed through a plate beam splitter onto a blazed grating. The grating is mounted at the blaze angle α *from the vertical, as seen in the detail of the figure; this configuration is known as a Littrow mount. In this case, the first order diffraction peak is directed back in the same direction as the incident beam and is thus reflected back into the laser cavity. The wavelength-dependent properties of the grating imply that only a single wavelength (with a finite linewidth) given by*

$$\lambda = 2\Lambda \sin\alpha$$

is reflected from the grating at the correct angle to return to the laser cavity such that this external cavity arrangement then lases at that wavelength. By rotating the grating, the reflected wavelength and thus the laser wavelength shifts. The beam splitter couples a portion of the beam out of the extended cavity corresponding to the useful laser emission.

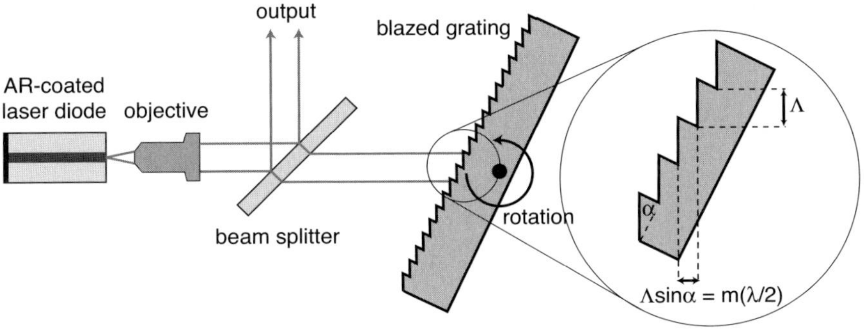

Fig. 6.24. An example configuration for an external cavity tunable laser employing a blazed grating. The laser diode has an anti-reflection coating on the right side; the grating is positioned at an angle α, corresponding to the Littrow mounting configuration, and is rotated about its center to tune the lasing wavelength

This arrangement represents an external cavity tunable laser and can either (relatively) easily be built in the laboratory or purchased as a commercial product. Tuning ranges of 30–40 nm with SMSR better than 40 dB are easily achieved. Clever techniques for maintaining a stable beam direction during tuning (Hawthorn, Weber and Scholten 2001) or for high speed tuning (Boggs et al. 1998) have been demonstrated.

If this arrangement seems rather macroscopic, it most certainly is. However, micro-mechanical implementations of external cavity tunable lasers using advanced MEMS technologies have been demonstrated (Syms and Lohmann 2002). The grating, a rotating or bending mount and actuators have been fabricated in silicon with total device sizes on the order of 1 mm. Using a hybridly mounted, anti-reflection coated laser diode, a ball lens for collimation and a fiber output, the entire microsystem was tunable over a range exceeding 140 nm in the 1.55 μm wavelength regime.

Grating Fabrication

Gratings have been fabricated since the 19th century using ruling engines; these are mechanical devices which use precise positioning of a milling tool to carve the grating into the surface of a substrate, generally glass. Whereas these techniques are still used to manufacture macroscopic gratings with relatively large periods, microfabrication techniques are dominant for the production of gratings with feature sizes at or below visible wavelengths.

Optical lithography techniques, using exposure through a mask, are only useful for gratings with $\Lambda \gg \lambda$; electron beam systems, on the other hand, have the nanometer resolution to allow directly writing grating structures in photoresist with dimensions down to the 100 nm range. The high resolution of an e-beam system contrasts with the long writing times and high costs of such a system such that this approach is only viable for certain specialized, highly demanding applications, such as definition of waveguide gratings for distributed feedback laser mirrors.

A popular alternative is grating definition using holography (Hutley 1982; IIzuka 1987). In this approach, the light from a relatively high-power laser, typically a HeCd or Ar gas laser, is split into two beams which recombine at a precisely specified angle at a defined point in space. Interference between these two beams gives rise to a stable standing wave pattern which is a function of the wavelength and beam angle. By placing a photoresist-coated substrate at that position, the interference pattern results in the exposure of a grating with well-defined period. Holography has the advantage that, unlike electron-beam exposure, it is a parallel technique and true masters of this art can expose plates up to a square meter in size with sub-micron gratings which have nanometer accuracy. However the achievable resolution is fundamentally limited by the exposure wavelength such that structures with periods of several 100 nm or below are difficult if not impossible to define using this approach.

6.2.3 Diffractive Optical Components

The use of wavelength-sized physical features to adjust the phase of a component of an optical field may be employed to make fully diffractive lenses. There is a considerable diversity in the types of structures that have been developed and the nomenclature is correspondingly varied.

Holograms and Kinoforms

Holograms are well-known to most readers and consist of two- or three-dimensional phase information which may be used to construct an arbitrary image when illuminated with a (generally) coherent light source; the images typically give the impression of being "three-dimensional" as the image changes with viewing angle. Holography was one of the first applications for lasers in the 1960s and is being used now for complex image generation as well as for optical memory.

A special case of the hologram is often called a kinoform, a planar structure which, upon illumination, generates an arbitrary two-dimensional phase profile and thus an arbitrary intensity distribution in the image. Kinoforms are usually referred to by their application: wavefront modulators may be used to correct a the wavefront emitted from a laser diode or other light source whereas pattern generators are employed to split a field into several components for optical fanout or to generate an arbitrary two-dimensional intensity distribution (Aagedal, Wyrowski and Schmid 1997; Leger 1998). The kinoform is generally a complex, computer-generated structure, an example of which is shown in Fig. 6.25; this structure ideally only changes the phase of the optical field which traverses it and absorption effects should be minimized. In this way, an arbitrary phase function may be added to an optical field, thus yielding a desired intensity distribution due to interference effects at the image plane.

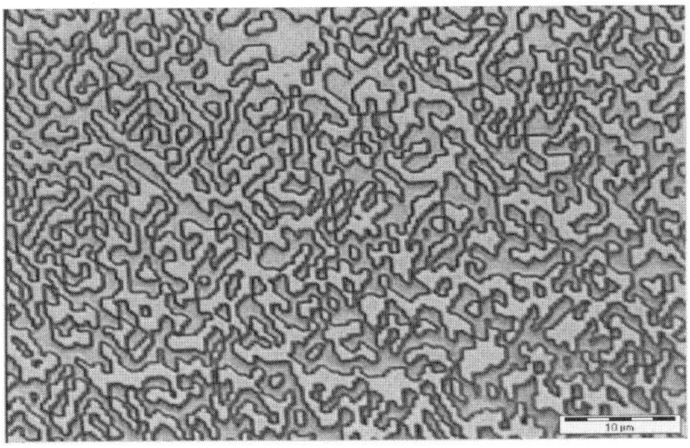

Fig. 6.25. A two-dimensional diffractive optical pattern generator using a computer-generated two-dimensional phase distribution for performing a desired optical function. The structure is a UV replica of an original fabricated in fused silica. Scale bar is 10 μm. Photo courtesy of Heptagon Oy

Fresnel Zone Plate

Already demonstrated by Lord Rayleigh (1877 - 1919), the Fresnel zone plate is a diffractive element which uses concentric rings of a (typically absorbing) material on transparent substrate. The sizes of the rings and spaces between them are calculated to yield constructive interference of the transmitted optical field at points on the optical axis, thereby focusing the beam onto the axis. The advantage of the Fresnel zone plate is that very simple structuring may be used in its fabrication: patterned surface metallization, photoresist or etched recesses will lead to the desired optical function, if with varying efficiencies.

Diffractive Lenses

Lenses using diffraction effects are important components for optical microsystems. Due to their structure, which seems reminiscent of the Fresnel lenses discussed in Sect. 6.1.3, these are often somewhat erroneously referred to as "Fresnel lenses" or "Fresnel-like lenses". We recall that the true Fresnel lens is purely refractive; in this section we consider diffractive structures, which we thus refer to as "diffractive optical lenses".

In direct analogy to the real Fresnel lens, a diffractive optical lens may be designed by "folding back" the surface of a refractive lens; a spherical or arbitrary aspheric lens surface may be used as a model. Consider the diagram of Fig. 6.26: a calculated refractive lens profile, shown by the dotted line, is folded back to the surface whenever the thickness is such that an incident wave undergoes a phase shift yielding a multiple of 2π. Since the phase of the wave is given by the product of the propagation constant k and distance z,

$$kz = \frac{2\pi n}{\lambda_0} z \tag{6.20}$$

for lens material index n and vacuum wavelength λ_0, a new segment is folded back whenever the z position in the original lens reaches the value

$$z_{segment} = i\frac{\lambda_0}{n} \tag{6.21}$$

for index $i = 1, 2, 3, \ldots$. The resulting segments have a thickness of an integer multiple of the wavelength and become narrower farther from the lens center such that the entire lens has a thickness equal to one wavelenghth or an integer multiple thereof.

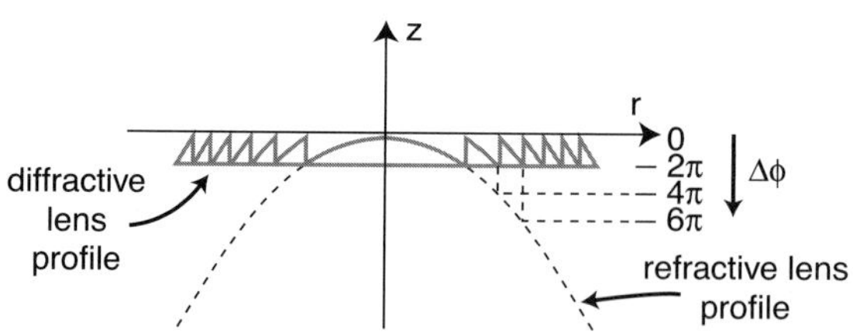

Fig. 6.26. The diffractive optical lens as generated from segments of a refractive lens folded back onto itself every 2π of phase shift

The original profile of the refractive lens is maintained, albeit split into segments, but the diffractive optical lens relies on both diffraction and refraction effects to operate. Each segment retains the curvature of the original model refractive lens and thus focusses incoming light by refraction; however, the rays from the various segments need to be in phase at the focal point. This latter requirement is satisfied due to the exact thickness of the segments (an integer multiple of the wavelength) by which constructive interference is then assured for all components of the beam meeting at the optical axis. As a result, the diffractive optical lens is very wavelength-sensitive such that chromatic aberration, as with most diffractive elements, is very strong (Morris and McIntyre 1997).

Fabrication

As with gratings, the feature sizes of diffractive optical elements can be in the range of the wavelength and the resulting requirements of fabrication tolerance are only met by advanced microfabrication techniques. The diffractive optical lens is particularly demanding, since the lens surface profile must be accurately reproduced while the segments reach micrometer widths at the edges of the lens. Such a profile may be written directly into photoresist using continuous-relief exposure techniques (Gale 1998), typically involving photoresist exposure by a laser with precisely modulated intensity. A measure for the quality of a diffractive lens is the efficiency, namely how much of the incident field is brought to the focus, and continuous relief micro-lenses with aspheric zone profiles can achieve values close to 100%.

An approximation to the continuous relief of a diffractive optical lens as shown in Fig. 6.26 may be fabricated by using a binary pattern, which results in a Fresnel zone plate as described above. The segments of the lens are then replaced by concentric rings of varying width as shown in the bottom of Fig. 6.27, with the advantage that only two exposure levels (i.e., exposed and unexposed) are required. Such a lens will indeed perform the required optical function, but with an efficiency typically in the range of 40%.

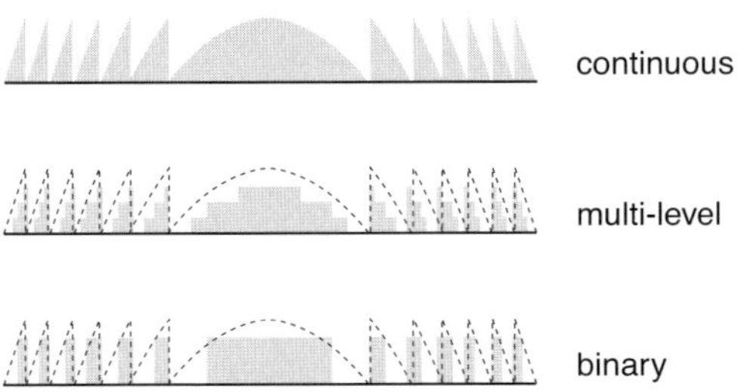

Fig. 6.27. A continuous relief diffractive optical lens *top* and approximations to it: multi-level exposure *center*; and binary exposure *bottom*; the ideal profile is shown by the dotted line

Such "binary" structures may be further refined by using more than two exposure levels to approximate the continuous shape of the lens. As we see in the center diagram of Fig. 6.27, the continuous contour of the ideal lens may be approximated by a step-like profile (Stern 1998). Such a profile may be generated by repeated exposure of a photoresist layer using masks with differing zone widths. The number of levels scales with mask number N as 2^N, such that a good approximation to the continuous structure may be achieved with a nominal number of masks. This scheme is limited by the need for very precise alignment between layers. Lenses of this type have been shown to have efficiencies on the order of 80% for a 4 layer structure.

Applications

Diffractive optical lenses may be employed in place of refractive micro-lenses in many cases; their chromatic aberration is less of an issue for use in laser diode-based optical systems since the lens may be designed for the relevant laser wavelength. For collimation, diffractive lenses have been designed with high numerical aperture ($NA > 0.75$) and short focal length, in the range 10 μm to 2 mm. Of particular value for laser diode optics is the ability to perform beam shaping and correction using diffractive structures such as kinoforms (Tschudi and Wang 1997); these may also be applied for collimation and reduction of beam astigmatism. Finally, diffractive optical fanout elements are of great utility in optical systems where the emission from a laser diode needs to be split into several distinct beams.

Diffractive structures may be combined with refractive components to increase their performance or correct aberrations (Morris and McIntyre 1997). By patterning diffractive elements onto the surface of a refractive lens, single lenses may be fabricated which have the optical performance of doublets or triplets in which only refractive effects are used. In addition, diffractive elements may be used for compensation of temperature and wavelength drifts, yielding compact achromatic or athermatic lenses.

6.3 Optical Fibers

It is sometimes of advantage to undertake the transmission of an optical field into, out of or occasionally within an optical system by means of optical fibers. Transmission of optical signals over longer distances is almost always by means of fiber and even within smaller optical microsystems, coupling by means of fiber rather than waveguide is often practical (Hecht 1999; Rogers 2001). In addition, there is a variety of fiber-based optical components available, including splitters, combiners, isolators, amplifiers and modulators in addition to fiber-coupled lasers and detectors, allowing numerous useful optical functions on a signal to be performed without the need to couple out of the fiber.

6.3.1 Fiber Structures

Optical fibers are waveguide structures fabricated using highly-developed glass manufacturing technology. The fiber itself consists of a glass cylinder with a standardized diameter of 125 μm; the material is essentially SiO_2 of truly astonishing purity. At the center of this cylinder is the core, a region typically doped with Ge, B or P, which has a slightly higher refractive index than the rest of the glass fiber. The core region is that which guides light and the physics of total internal reflection and the transmission by discrete optical modes is the same as for the semiconductor waveguides discussed in Sect. 3.4.

Core Configurations

As with the waveguides we discussed previously, the dimensions of the fiber core critically affect the propagation conditions in the fiber. The diagrams of Fig. 6.28 show the refractive index profiles of three basic fiber types. The multi-mode fiber has a relatively thick core, typically 50 μm or 62.5μm in diameter with some even larger diameters available. The refractive index step between core and cladding, $\Delta n = 0.02$ in the figure, is typical; the fact that the refractive index changes abruptly often leads to the designation "step-index" fiber. True to its name, the multi-mode fiber allows a multitude of lateral modes to propagate.

The graded-index fiber has a parabolic index profile quite similar to that found in GRIN lenses (see Eq. 6.15 in Sect. 6.1.4) and the propagation conditions are identical. For a graded-index fiber, the refractive index profile leads to equal propagation velocities for all the modes the fiber supports, such that multi-path dispersion (the differing longitudinal velocity of modes which undergo different numbers of reflections as they propagate down the fiber) is strongly reduced.

The third fiber type we consider in Fig. 6.28 is the single-mode fiber which has a step-index core region but with a relatively small diameter, between 7 and 10 μm. This narrow core region allows only a single mode to be supported such that only that mode propagates in the fiber. The exact dimensions are a function of the wavelength regime used: most single-mode fibers are designed for the 1.55 μm telecommunications wavelength regime and thus have larger core diameters than single-mode fibers for, say, the 800 nm wavelength band. Single-mode fibers are essential for all long-range optical communications systems and these often employ more complex fiber structures which use intricately designed refractive index profiles to yield dispersion-shifted, dispersion-flattened or polarization-maintaining fibers (Gowar 1993).

A growing fiber technology is that which uses plastic rather than glass for the fiber medium (Weinert 1999). Using materials such as PMMA or fluorinated plastics, these optical fibers are potentially considerably cheaper than glass versions, benefit from improved robustness and generally have very large fiber cores (85 – 3 000 μm) resulting in improved ease in coupling. Limitations of the use of plastic optical fiber include the relatively high losses and the restricted temperature

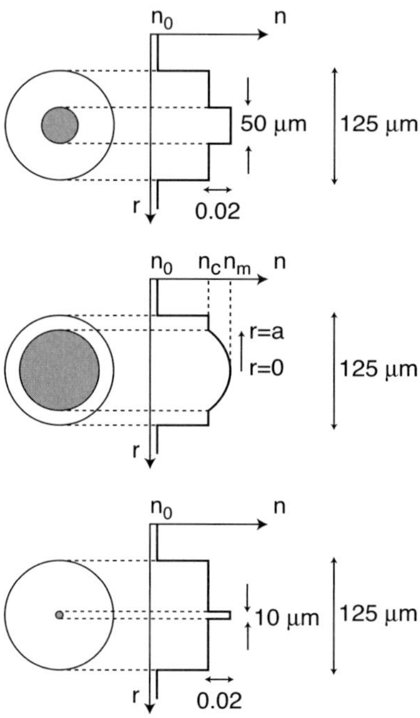

Fig. 6.28. The refractive index profiles of three basic fiber types seen in cross-section: a multi-mode fiber *top*; a graded-index fiber *center*; and a single-mode fiber *bottom*. The *n*- and *r*-axes are not to scale

range typically requiring operation below 85°C, although recent materials advances have extended the useful temperature and humidity ranges (Ishigure et al. 2002).

Cables

A bare fiber consists of the glass cylinder with 125 µm diameter and is typically covered with a few tens of microns of a polymer coating designed primarily to prevent moisture diffusion into the glass. Fiber in this form is relatively fragile and unwieldy when handled macroscopically but is used directly for installation in enclosed systems and for coupling into microsystems.

For laboratory or external use, the fiber is typically used as a cable: the glass fiber is embedded in several layers of plastic and surrounded by a polyurethane jacket (Mahlke and Gössing 2001). The resulting structure is relatively large (with a diameter of 3 mm) but quite robust and is typically terminated by one of a series of standardized connectors. An alternative for systems with multiple optical inputs and/or outputs is a fiber ribbon, a two-dimensional fiber array with a standardized spacing of 250 µm. The fibers in a ribbon have an additional surface insulation and may be terminated in a two-dimensional connector.

6.3.2 Characteristics

Several issues should be considered before designing fiber into an optical microsystem: losses, dispersion and mode profile are the most important. Losses, as for the waveguide, define the maximum transmission distance before the intensity of the optical signal is reduced to unacceptable levels; dispersion defines the extent to which the various frequency components of pulses move at different velocities; and the mode profile allows determination of the coupling efficiency. Optical fiber characterization is of particular importance for long-distance applications but also required for its use in short-range microsystems (Derickson 1998).

Losses

The losses in optical fibers result in a reduction in optical signal strength. Losses have a number of origins of which the most significant is material absorption. Material absorption in SiO_2 is wavelength-dependent, with a minimum around 1.5 μm and a strong increase toward longer wavelengths due to lattice vibrations and the resonant absorption effects of the oxide bonds. The most significant impurity in glass is water since the hydroxyl radicals (OH^-) absorb strongly at a number of resonance peaks down into the near-infrared wavelength regime, most prominently at 1390 nm. For this reason, the water concentration in the fiber, established during the manufacturing process, must be kept below 10^{-8}.

As with waveguides, other loss mechanisms can become significant under certain circumstances. Material inhomogeniety gives rise to Rayleigh scattering which increases rapidly with decreasing wavelength. This effect dominates the UV absorption and leads to the fiber "glow" that is easily seen in bare fiber transmitting visible wavelengths.

Optimally-fabricated optical fiber can have remarkably low loss levels in particular wavelength ranges. The natural minimum of SiO_2 absorption around 1.5 μm has resulted in the bulk of the efforts being expended on optimization of loss in that wavelength regime; the use of this wavelength band for long-range optical telecommunications is due to the minimized fiber losses, which limit the system performance more than, for example, dispersion. High-end fiber available has loss levels below 0.2 dB/km at $\lambda = 1.55$ μm. Fibers in the 850 nm range have at least an order of magnitude higher loss levels.

The optical losses of plastic fiber are considerably higher than in glass: PMMA fibers, suitable only for use in the 650 nm visible range, exhibit losses of about 150 dB/km and plastic optical fibers for the 1.3 μm wavelength range have values exceeding 50 dB/km. Primarily due to the optical losses, plastic fibers are presently limited to uses where the maximum transmission distance is less than 100 m.

Dispersion

Dispersion, the variation of propagation velocity with optical signal wavelength, is a significant issue for long-range fiber applications but is generally of secondary importance in small-scale systems. Non-zero dispersion results from a number of sources. Material dispersion is due to the fact that the refractive index of SiO_2 varies with wavelength, such that material propagation velocity is wavelength dependent. As we saw in Eq. 3.72, the longitudinal propagation constant k_z in a waveguide, which has a direct analogue in the fiber waveguide, varies with wavelength, which leads to waveguide dispersion. Finally, for multi-mode fibers, the various path lengths taken by the various modes down the fiber leads to multi-path dispersion. Dispersion is not an issue for continuous transmission of single-frequency light; however, since pulsed signals contain, by virtue of their Fourier transform, a spectrum of frequencies, these frequency components will move with differing velocities in a dispersive fiber resulting in a spreading of the pulse.

The total fiber dispersion, D_T [ps/km nm], is the sum of these contributions. The units of dispersion refer to the induced time delay (ps) per fiber length (km) per spectral width (nm), hence the somewhat unusual designation. Whereas material dispersion is a function of material type and doping and is always positive, waveguide dispersion can be positive or negative. As a result, the sum of the dispersion components can be made zero for one particular wavelength, particularly in single-mode fibers for which multi-path dispersion is inexistent. This dispersion minimum typically occurs in the 1.3 μm wavelength band; fibers whose core structure shifts this toward 1.55 μm are known as dispersion-shifted fibers.

Typical dispersion values in normal fiber are in the range $D_T \leq 3.5$ ps/km nm for $1285 \leq \lambda \leq 1330$ nm. Thus for microsystem applications, save those in which long fiber paths or very high speed pulse modulation are used, dispersion is generally not an issue.

Mode Profiles

The optical fiber has cylindrical symmetry and thus requires solution of Maxwell's equations in cylindrical coordinates to determine the electromagnetic field distribution and the form of the mode. Using essentially the same approach as taken in Sect. 3.4.2, the modal solution for the fiber takes the form of Bessel functions in the core region (an oscillatory solution) and Hankel functions in the cladding (exponentially decaying solutions).

The circularly symmetric core results in nicely circularly symmetric mode profile as shown in Fig. 6.29. If we compare the fiber mode with that emitted by semiconductor light emitters, we see that coupling the emission into a fiber can be more and less problematic. The near-field profile of an LED, such as that shown in Fig. 5.28, has qualitatively the same shape as that of the fiber, implying that mode overlap between the LED and fiber should be good. The same is usually true for a VCL, whose far-field profile is generally Gaussian and circularly symmetric; the

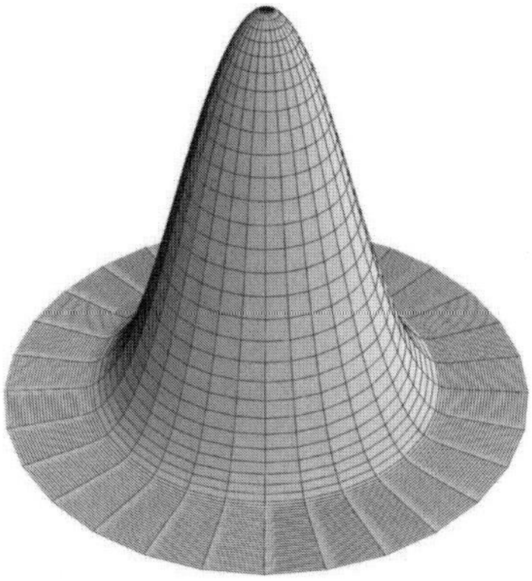

Fig. 6.29. Calculated mode profile for the lowest-order mode in a single-mode fiber

plot of Fig. 5.37 was taken for the case where higher-order lateral modes in the VCL adversely affect the far-field distribution and for this case, fiber-coupling could be more difficult.

The real problem arises when considering the emission from edge-emitting lasers with a typical far-field distribution as shown in Fig. 5.33. As we saw in Sect. 5.6 and again above in the discussion of laser diode optics in Sect. 6.1.6, edge-emitting lasers have highly elliptical, astigmatic output. Thus coupling from one of these sources into an optical fiber is a challenging problem explaining why such effort has gone into the design of high-performance laser diode collimation systems.

6.3.3 Coupling Techniques

Assuming that the coupling optics has generated a suitably formed beam from the laser diode emission, coupling this into a fiber is then typically by the end-fire technique: the incident beam is focused onto the cleaved fiber facet. A fiber facet, generated by scribing and cleaving the glass using a specialized apparatus, is very planar and uniform. Lensed fiber ends are also available, in which the fiber tip is rounded to achieve a lensing function and thus improved coupling efficiency. One step further in complexity is a fiber collimator, essentially a refractive lens arrangement permanently coupled to a fiber end.

An important consideration is the numerical aperture of the fiber which, in analogy to a lens, defines the maximum angle at which light is accepted to the fiber core. Coupling light into a fiber at the facet must be done within a cone of angles which assures that total internal reflection takes place at the core/cladding boundaries. Referring to Fig. 6.30, θ_i must be small enough to assure that θ'_i is greater than the critical angle. For the typical fiber structure shown in the figure, where the refractive index step between core and cladding is 0.004, the critical angle is quite large such that the range of θ_i is quite small.

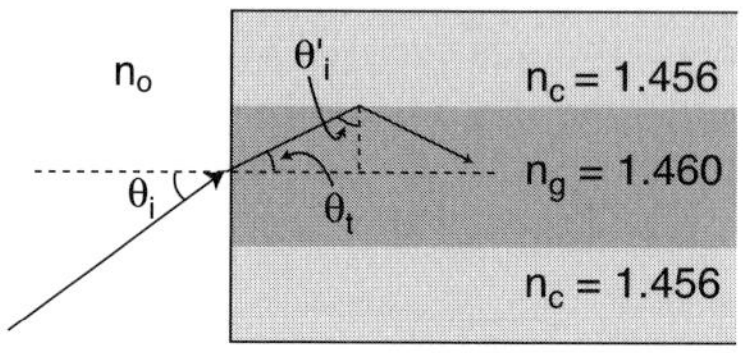

Fig. 6.30. Numerical aperture of a fiber defined as the sine of the maximum input angle for which total the internal reflection condition is satisfied

From basic trigonometry and the definition of the critical angle (Eq. 3.70), the construction of Fig. 6.30 allows us to derive

$$NA = n_0 \sin\theta_{imax} = \sqrt{n_g^2 - n_c^2} . \tag{6.22}$$

The small typical *NA* values for optical fibers imply that normal incidence must be closely approximated and that light is ideally coupled into a fiber using a lens or lens system with a long focal length. Good positioning accuracy of the optics and fiber is thus required. The arrangement shown in Fig. 6.1 is a typical fiber coupling setup using ball lenses to achieve end-fire coupling directly into the fiber facet.

Example 6.6

Assignment: *Determine the numerical aperture and maximum input angle of the fiber in Fig. 6.30.*

Solution: *Since the refractive indices of the core and cladding are given, we may simply apply Eq. 6.22. Given $n_g = 1.460$ and $n_c = 1.456$, we find that $NA = 0.108$ from which, for incidence in air, the maximum half angle is found to be about 6°. This value indicates a very limited cone of input rays.*

6.4 Photodetectors

Despite the fact that most optical microsystems use light, the input and output to the system are often electrical. Thus for many optical systems, the optical signal needs to be converted to an electrical one, a function generally undertaken by a photodetector. These devices often form the last stage of an optical system and have a great variety of structures and levels of performance (Kingston 1978; Derniak and Boreman 1996).

6.4.1 Photon Detection

Most photodetectors of interest for microsystems are semiconductor-based devices based on absorption of photons in the material. We looked at optical absorption in semiconductors in Sect. 3.2.2 and it is this physical mechanism on which semiconductor photodetectors are based: photons create electron-hole pairs and these form a detector current which is the electrical output signal.

Materials

Absorption varies strongly between material types and is highly wavelength-dependent, as seen in Fig. 6.31 in which the absorption spectra of a number of the more popular semiconductors for photodetectors are shown. Due to its relatively low cost and advanced fabrication technology, silicon is attractive as a photodetector material for the visible wavelength ranges, for $\lambda \leq 1.1$ μm (Kaneda 1985). The III-V semiconductors, including GaAs and InGaAs, cover wide ranges of wavelengths into the mid infra-red; the direct bandgap leads to strong absorption at the expense of higher material cost (Pearsall and Pollack 1985). Germanium is popular for wavelengths below 2 μm and the II-VI semiconductors, not shown in the figure, have been developed for detection at long wavelengths, well into the mid-IR; these types of detectors need to be operated cooled (typically 77 K) and are relatively expensive. The materials selected in Fig. 6.31 represent only a small subset of those available.

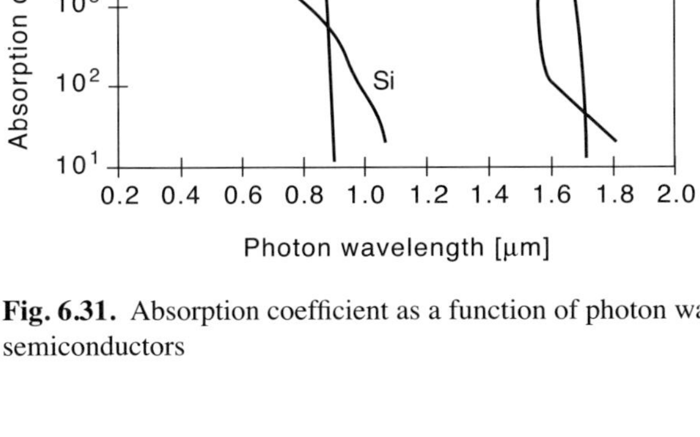

Fig. 6.31. Absorption coefficient as a function of photon wavelength for several popular semiconductors

Photodiodes

The electron-hole pairs generated when photons are absorbed in a semiconductor need to removed from the material to generate a detector current. To this effect, electrons and holes should be separated to avoid recombination and thus carrier and subsequent signal loss in the material. For this reason, most semiconductor-based photodetectors are based on the pn diode, as we saw in Sect. 3.1.6, and share many similarities with the laser diode structure; indeed, a reverse-biased laser may be operated as a photodetector (Moss et al. 1992). For carriers generated at or near the pn junction, the built-in or externally applied electric field sweeps the electrons to the n-side of the junction and holes to the p-side, from which the carriers can diffuse or drift to the contacts and contribute to the detector current.

The resultant photodiode structure is the basis for a large fraction of semiconductor photodetectors. The diode is typically operated in reverse bias, such that a wide depletion region gives rise to a large absorption volume for photons and the electric field in the junction efficiently sweeps the carriers to separate sides of the junction. The IV characteristic of the photodiode, as shown in Fig. 6.32, shows the expected exponential increase in forward bias and the saturation current in reverse bias. Under illumination, the reverse bias current increases in proportion to the number of carriers generated and thus ultimately the number of photons absorbed. The reverse bias current when no illumination is present, due to thermal generation and the thus the temperature-dependent value of n_i^2 in Eq. 3.36, is known as the dark current and represents a noise source in photodiodes which we discuss further in Sect. 6.4.3 below.

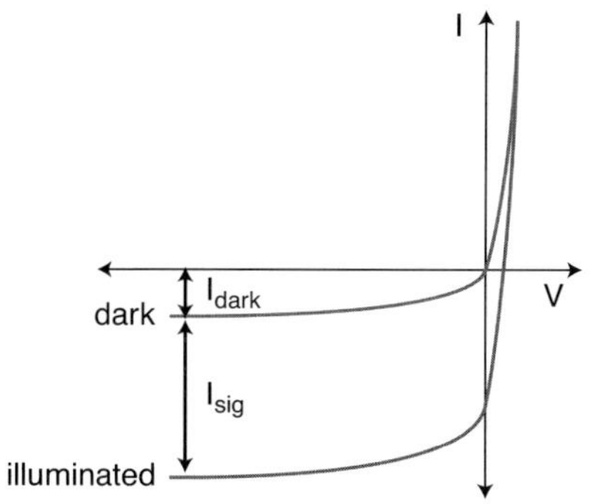

Fig. 6.32. The IV plot of a photodiode with and without illumination. The reverse bias current represents the detector signal

6.4.2 Photodiode Structures

The most rudimentary detector is then a pn junction fabricated in a material which absorbs photons at the wavelength of interest. There is a large spectrum of photodiode structures which have been developed to optimize detector performance. Most practical for many applications is a large-area pn junction such as that shown in Fig. 6.33. The figure shows a diffused p-i-n junction, namely one with a thick intrinsic (nominally undoped) region between the p- and n-regions. The undoped region serves to increase the width of the depletion region, in which photons are absorbed and carriers generated, and increase the maximum electric field (and thus bias voltage) which may be applied (Miyamoto, Hagimoto and Kato 1994). Photon illumination is from thc surface and the contacts are structured to allow maximum semiconductor area to be illuminated. Typical dimensions range from sub-millimeter to centimeters laterally with thickness around 300–400 μm.

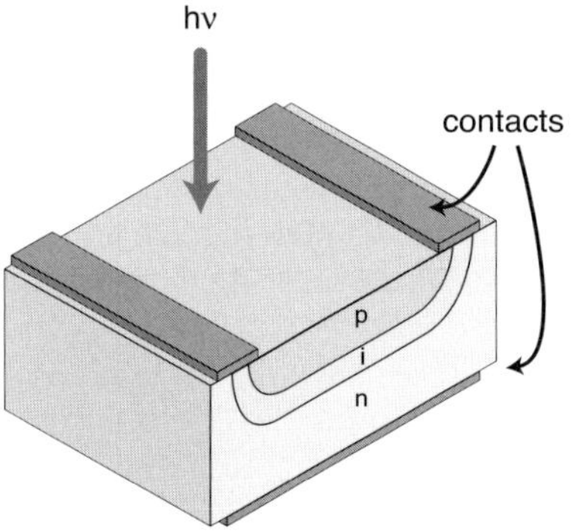

Fig. 6.33. A planar photodiode using a diffused p-i-n junction for absorption of photons. Illumination is from the surface where the contacts are structured to allow exposure of a maximum semiconductor area

Variations on this structure may be achieved when using III-V semiconductors for photodiode fabrication. A single or double heterostructure pn diode may be employed in a layer configuration similar to that of the semiconductor laser. The improved carrier confinement, large absorption region and the ability to grow absorbing layers epitaxially on a transparent substrate, and thus allow backside illumination, can lead to performance improvements at the expense of fabrication cost.

There are further photodiode types which may be of interest to the microsystem engineer. Avalanche photodiodes typically use a p-i-n diode structure with a large intrinsic region, very high quality material and careful surface passivation; application of a high reverse bias results in a large electric field in the junction which leads to avalanche multiplication of the generated electron-hole pairs through impact ionization. The avalanche photodiode then has internal carrier multiplication with a gain of up to 10^2 making it suitable for high-sensitivity detection in low-light conditions (Owens et al. 1994). Advances in fabrication technology have led to the availability of large-area avalanche photodiodes with diameters exceeding 15 mm and devices exhibiting very low noise (Wang et al. 2002).

The metal-semiconductor-metal, or MSM, photodiode uses an arrangement of Schottky diodes on the surface of the material for rectification, rather than a bulk pn junction (Rhoderick and Williams 1988). The Schottky diodes, fabricated by the deposition of metals with the appropriate work function directly onto the surface of a lightly-doped semiconductor (for example AuTi or the transparent indium tin oxide – ITO – on silicon), are typically arranged in a finger pattern to allow absorption of light while maintaining an efficient collection of carriers. MSM detectors have very low capacitance and are thus suitable for detection of high-speed optical signals (Soole and Schumacher 1991).

Further photodiode variations use more complex structures for imaging applications. Charge coupled devices (CCDs) are fabricated as two-dimensional arrays of light-sensitive capacitors, typically based on CMOS technology; these are standard technology for imaging spatially extended optical fields and generating two-dimensional intensity profiles. Of further interest for optical microsystems is the position-sensitive detector (PSD), essentially a long extended detector structure on which the varying position of a focused light beam leads to shifts in an array of output signals, such the light peak position may be determined to high accuracy. PSDs are useful in arrangements in which the optical output signal takes the form of the physical movement of an intensity maximum.

6.4.3 Characteristics

As we saw in the IV plot of Fig. 6.32, the output characteristic of a photodiode is the exponential diode characteristic. The diode is biased into reverse bias and the reverse-bias current, which increases due to generation of carriers by absorbed photons, is the detector signal. Several aspects of the resultant device characteristic are relevant for the user.

Efficiency

In an ideal photodetector, every photon incident on it would lead to the generation of at least one electron in current. Photodiode efficiency, η [ ], is a measure of how much current is actually generated for a given incident photon flux or simply how many electrons are generated per photon. Incident optical power P [W] represents $P/(h\nu)$ photons, for photon energy $h\nu$ [J] given by

$$E = h\nu = \frac{hc}{\lambda} \tag{6.23}$$

and detector current I_d [A] corresponds to I_d/q electrons, given electronic charge q [1.6×10^{-19} C]. The efficiency may then be defined as

$$\eta = \frac{\text{electrons}}{\text{photons}} = \frac{I_d/q}{P/h\nu} = \frac{h\nu}{q}\left(\frac{I_d}{P}\right) . \tag{6.24}$$

We note that, contrary to what we might think Eq. 6.24 is telling us, η is not a function of photon energy or wavelength. The reason is that the generated current is related to the *number* of photons absorbed, not their energy, but optical input power P is a function of both photon number and their energy. If $E_{photon} > E_g$, a higher-energy photon generates the same number of electrons as a lower-energy photon, namely one in both cases, such that the photon flux (photons/time) and not their energy is the relevant parameter for determining detector current.

A number of factors can result in a decrease of η. Starting with light incident on the semiconductor, some photons may be reflected by Fresnel reflection at the surface in the absence of anti-reflection coatings; further photons may be absorbed directly at the surface, where surface recombination directly annihilates the carriers; some photons may be absorbed deeper in the material, far from the pn junction; and some carriers may recombine before reaching the contacts. Each of these processes results in a reduction in detector efficiency and photodiode designers have striven to diminish their effect.

Efficiency of a photodetector may be determined by measurement of the detector current as a function of calibrated input power. The result should be a linear characteristic with a slope R [A/W], a parameter known as responsivity; unlike η, R is a function of wavelength. From knowledge of the optical input wavelength and thus the energy per photon, efficiency may then easily be determined from

$$\eta = \frac{h\nu}{q}R = \frac{h\nu}{q}\left(\frac{I_d}{P}\right) . \tag{6.25}$$

Noise

One of the primary limitations on detector performance is noise, the physics of which is similar to that considered for laser diodes in Sect. 5.4.2. Particularly for applications in which low optical signal levels are to be measured, detector noise will provide a lower limit on sensitivity. A number of noise types and sources are active in photodetectors, the relative importance of which depends on the structure, the materials and the operating conditions; we discuss two of the more important ones here.

Shot noise is inherent to optical signals and derives from the fact that photons are discrete entities and arrive at the detector with a stochastic distribution in time. Thus even a "constant" optical signal fluctuates about a mean value, the width of this fluctuation giving a measure of the magnitude of shot noise. Recalling from Eq. 5.36 that the mean square value of a signal ζ is given by

$$\langle \Delta\zeta^2 \rangle = \overline{(\zeta - \bar{\zeta})^2} , \tag{6.26}$$

namely the average of the square of the difference between the signal and its average, we can write a value for the mean square value of the noise current, $\langle I_N^2 \rangle$ [A^2] as

$$\langle I_N^2 \rangle = \frac{2q^2\eta}{h\nu} BW \cdot P \,, \tag{6.27}$$

where BW [Hz] is the bandwidth of the detector and P the optical input power. The bandwidth, which provides a measure for the highest temporal frequency components (shortest pulse widths) to which the detector can respond directly affects the magnitude of noise, since the ability to detect a larger portion of the frequency spectrum implies that the noise components of those spectral portions also contribute to the total noise figure.

The shot noise level may be characterized by the signal to noise ratio, S/N, defined as

$$S/N = \frac{\eta}{2h\nu}\left(\frac{P}{BW}\right) \,, \tag{6.28}$$

which relates the average of the optical signal level to the noise level. A high value for S/N implies that the signal level is significantly above that of the noise such that reliable detection is expected. We note that S/N is inversely related to bandwidth such that for a given optical input signal, a high-speed detector will have poorer noise performance.

A second noise source of relevance for photodiodes is due to the detector itself. As we saw in equations which define the IV characteristic of a diode (Eq. 3.35 and Eq. 3.36), the reverse bias current depends on the concentration of electron-hole pairs in the depletion region; in the absence of optical generation, carriers are created by thermal generation. Thermal generation of carriers leads to a detector current even in the absence of an optical signal, a parameter known as dark current. For small optical input intensities, a large dark current can dominate the detector output thus limiting the photodiode sensitivity.

The dark current is most critically a function of material, detector area and temperature. Narrow-bandgap semiconductor materials, such as those used for detection at infra-red wavelengths, have a relatively high thermal carrier generation rate at a given temperature since the thermal energy is a significant fraction of the bandgap energy. The dark current is a strong function of the intrinsic carrier concentration, n_i (see Eq. 3.20), which in turn increases rapidly with temperature. As a result, decreasing the operating temperature of the photodiode is the most effective means to reduce the dark current, especially for narrow-bandgap materials. For this reason, infrared detectors based on Ge or II-VI compounds such as HgCdTe are almost always operated cooled at 77 K. Finally, dark current also scales directly with the junction area such that smaller-area detectors have an advantage in this regard, albeit of course at the expense of reduced detector aperture.

6.5 Modulators

Within an optical system it is frequently necessary to change parameters of the optical signal dynamically in a controlled manner: examples include changes of intensity, phase or direction. Modulators are a family of opto-electronic or opto-mechanical components which perform one of these functions. A very wide variety of optical effects, materials and device structures have been developed for these purposes, often very specialized for a particular application (Wakita 1997).

The three most important modulation parameters are probably those we just mentioned. Intensity modulation may be as simple as turning a laser beam on and off, albeit often at GHz modulation rates; alternatively, continuous modulation of intensity with a high resolution of intensity levels also has a range of applications. Modulation of the phase of a light beam is frequently required in interferometric systems such that a controlled and dynamically variable phase shift may be introduced in one portion of the optical field with respect to an unmodulated reference field. Finally, changes in optical beam propagation direction using dynamic directional modulators is essential for beam scanning as well as optical switching applications.

6.5.1 Propagation Modulation

Modulators change the propagation conditions of a light beam; as a result, we may consider the effect of modulation on an optical field by looking at the effect of a modulator on the propagation constant, k; we introduced this parameter in Eq. 2.21 for a wave propagating according to the propagation equation Eq. 2.27. Writing the propagation equation in a simplified form for an optical field with intensity S [W/m^2], we have

$$S = S_0 e^{j2kz} \tag{6.29}$$

for a field propagating in the z direction. In the same manner as for the calculation of mode propagation in a laser (Eq. 2.131 in Sect. 2.6.1), the propagation constant is in general complex and may be written as

$$k = \beta + j\frac{\alpha}{2} . \tag{6.30}$$

The imaginary part of k, α [m^{-1}], defines the absorption and thus intensity of the wave whereas the real part, $\beta = (2\pi n)/\lambda_0$ [m^{-1}] for refractive index n, defines the propagation velocity and thus phase of the wave.

The function of an intensity modulator is to change α in a controlled manner leading to

$$\alpha \rightarrow \alpha_0 + \Delta\alpha ; \tag{6.31}$$

in a phase modulator, the real refractive index is modulated, such that

$$\beta \rightarrow \beta_0 + \Delta\beta\,. \tag{6.32}$$

In the last two expressions, $\Delta\alpha$ and $\Delta\beta$ are the changes in propagation constant induced by the modulator.

Given a modulator with an interaction length L [m], as shown in a schematic form in Fig. 6.34, into which an optical field with intensity S is incident, the modulated field S' is then given by

$$S' = S_0 \exp\left[j\frac{4\pi}{\lambda_0}(n_0 + \Delta n)L\right] \cdot \exp[-(\alpha_0 + \Delta\alpha)L]\,. \tag{6.33}$$

where ideally either $\Delta\alpha$ or Δn but not both are non-zero. The extent of phase modulation is then typically given by the induced phase shift

$$\Delta\phi = \frac{4\pi}{\lambda_0} L \Delta n \tag{6.34}$$

and the extent of intensity modulation by the ratio of the input and output intensities

$$\frac{S'}{S} = e^{-\Delta\alpha L}\,. \tag{6.35}$$

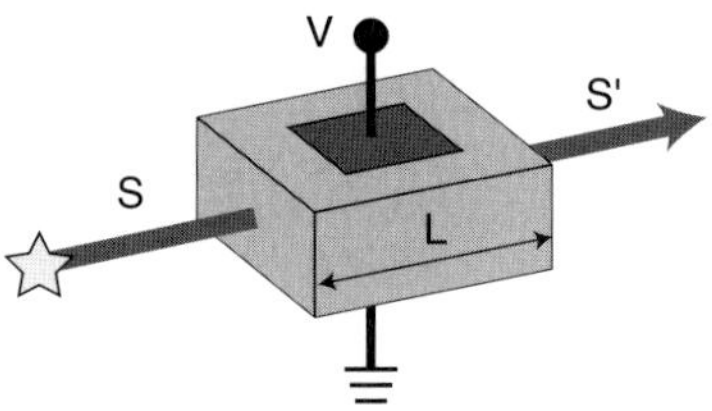

Fig. 6.34. Schematic representation of a modulator of length L; an applied voltage V changes the propagation conditions of the wave with incidence intensity S to S'

Chirp

As briefly mentioned above, an ideal phase or intensity modulator modulates only phase or intensity, respectively, such that $\Delta\alpha = 0$ for an ideal phase modulator and $\Delta n = 0$ for an ideal intensity modulator. In reality, many electro-optic-based modulators can be optimized for phase or intensity modulation but *both* Δn and $\Delta\alpha$ are in general non-zero. The extent of "parasitic" modulation is given by the chirp parameter, defined as

$$\text{chirp} = \frac{\Delta n}{\Delta k} = \frac{\Delta n}{\Delta\alpha}\left(\frac{4\pi}{\lambda_0}\right)\,. \tag{6.36}$$

We encountered chirp previously, in the discussion of laser linewidth in Sect. 5.3.4, and the physical origin of the effect in laser diodes and semiconductor-based modulators is identical. Intensity modulators should have low chirp (i.e., small parasitic phase shift) and phase modulators are designed for high chirp (i.e., small parasitic intensity change). The physical relationship between real and imaginary components of the refractive index of any modulator material limits the values of chirp attainable.

6.5.2 Physical Effects for Modulation

The induced change in refractive index or absorption required for the functionality of a modulator may be achieved through a number of physical effects; refractive index is a function of numerous parameters, including temperature, electric field, strain and free carrier concentration. The function of a modulator is to allow controlled variation of one of these parameters to yield a refractive index with the desired value.

By inducing a known Δn or $\Delta\alpha$ in a piece of material through which an optical field has a well-defined propagation length, the phase or intensity of the field may be modulated. The goal in modulator design is to achieve this effect with a minimum of required input power, in the smallest device possible and often with the ability to vary the modulation at high speed.

Electro-Optics

One of the most established and widely applied approaches for modulation is through the use of electro-optical effects (Narasimhamurty 1981; Walker 1987). Electro-optics employs optically non-linear materials such that the complex refractive index varies with applied electric field and the optical propagation constant may be electrically modulated. In non-linear materials, the refractive index n is a function of the applied electric field, such that we may approximate this parameter with a series expansion as

$$n = \sqrt{1 + \chi_1 + \chi_2 \boldsymbol{E} + \chi_3 \boldsymbol{E}^2 + \ldots} \qquad (6.37)$$

in which the χ terms represent different components of the susceptibility. Comparing with the definition of Eq. 2.60, we see that non-linear materials have higher-order terms in susceptibility which, if they are non-zero, imply that n is a function of the electric field $\boldsymbol{E}$.

As a result, an externally applied electric field can give rise to a change in the refractive index and the last two terms of Eq. 6.37 are of primary interest. In materials where $\chi_2 \neq 0$, the linear electro-optical (or Pockels) effect, named after Freidrich Pockels (1865–1913), gives rise to a change in refractive index of the form

$$\Delta n = \frac{1}{2} n_0^3 r \cdot \boldsymbol{E} \tag{6.38}$$

in which $\boldsymbol{E}$ represents the field in the crystal due to, for example, an externally-applied voltage and n_0 is the refractive index in the absence of such an applied electric field. The constant r [m/V] represents the linear electro-optical coefficient which may be greater or less than zero and is, in general, crystal-line-direction-dependent as well as slightly wavelength-dependent (Eimerl 1987). A typical value for r, for example in GaAs, is $r = -1.44\times10^{-12}$ m/V; r is zero in centro-symmetric crystals such as Si.

The quadratic electro-optic effect or Kerr effect, named for John Kerr (1824–1907), is present in materials for which $\chi_3 \neq 0$. In this case,

$$\Delta n = \frac{1}{2} n_0^3 s \cdot \boldsymbol{E}^2 \tag{6.39}$$

in which s [m^2/V^2] is the quadratic electro-optical coefficient, strongly wavelength-dependent in most materials. This effect is relatively small in semiconductor materials such as Si or the III-Vs (GaAs at $\lambda = 1.09$ μm has $s = -5.4\times10^{-20}$ m^2/V^2) but is on the order of $s = 2\times10^{-17}$ m^2/V^2 in specialized modulator materials such as PLZT ($(Pb_{1-x}La_x)(Zr_yTi_{1-y})_{1-x/4}O_3$).

Plasma Effects

In semiconductor materials, both the refractive index and the absorption may be modulated by changing the carrier concentration. Using injection of charge carriers from a pn junction to increase the electron concentration n_e [cm^{-3}], the resulting Δn and $\Delta\alpha$ are then given by (Soref and Lorenzo 1986)

$$\Delta n = -\frac{q^2\lambda_0^2}{8\pi^2 c^2 n_0 \varepsilon_0}\left(\frac{\Delta n_e}{m_e^*}\right) \tag{6.40}$$

and

$$\Delta\alpha = \frac{q^2\lambda_0^2}{4\pi^2 c^3 n_0 \varepsilon_0}\left(\frac{\Delta n_e}{m_e^{*2}\mu_n}\right), \tag{6.41}$$

where n_0 is the background (unmodulated) refractive index, m_e^* [kg] the electron effective mass and μ_n [m^2/Vs] the electron mobility.

We see that an increase in carrier concentration results in an increase in optical absorption but a decrease in refractive index. Evaluating the expressions above also shows that the effect is relatively small; using the material parameters for silicon, we find that $\Delta n_e = 10^{18}$ cm^{-3} gives rise to a refractive index shift of about $\Delta n = 10^{-3}$. Plasma-effect modulators based on silicon may be used for phase or intensity modulation (Cutolo et al. 1997).

Thermal Effects

Refractive index is temperature-dependent such that varying the temperature of most optical media results in a modulation of Δn. As with plasma modulation, the effect is relatively small: for example in silicon, the change in refractive index with temperature is about of $\Delta n/\Delta T \sim 10^{-5}\ \mathrm{K}^{-1}$ and for SiO_2 $\Delta n/\Delta T \sim 2\times10^{-4}\ \mathrm{K}^{-1}$ (Takato et al. 1988)

Since modulation using thermal effects requires high temperature differences (tens to hundreds of degrees), relatively high electrical input powers are often required to yield usable modulation levels. In addition, the thermal mass of the material leads to limitations on the modulation speed, as heating and cooling tend to be processes with time constants in the millisecond to second range, depending on the mass and the material used. Micro-fabricated thermal modulators, an example of which we will see in Sect. 6.5.3, have reduced the effect of these limitations such that usable thermal modulators are now available.

Acousto-Optics

Inducing strain in a crystalline material also results in a change of real refractive index in the strained regions. This effect is often used indirectly, not necessarily for phase modulation but for beam deflection. If a time-varying voltage is locally applied to a piezoelectric material, alternating regions of compressive and tensile strain may be induced in the material leading to the generation of traveling surface acoustic waves (SAW) (Tsai 1990). Such an acoustic wave may be represented by a harmonic function with a spatial period Λ.

The SAW thus takes the form of a region of periodically varying refractive index, since the variation in strain leads to a variation in n. An effective grating structure with period Λ results and this may be used to deflect an optical field using the same principles as discussed in Sect. 6.2.2. Whereas the period of the SAW "grating" is fixed by the physical structure of the transducer which applies the voltage to the crystal, the strength of the grating (i.e., the maximum modulation in Δn) may be varied by changing the applied voltage. As a result, the SAW may be continuously varied between non-existent and some maximum value thereby allowing continuous variation in the magnitude of the deflected optical field.

SAW modulators are popular for continuous intensity modulation, where the optical intensity is not reduced by absorption but through deflection of all or a portion of the beam. Acousto-optic modulators are standard laboratory components and installed in micro-systems in which continuous intensity modulation is required. SAW modulators may also be used for scanning applications in which the deflected beam may be re-directed over a given angular range.

6.5.3 Modulator Structures

A modulator consists of a length of material through which the optical field propagates and a means to vary the propagation conditions; the latter may include application of an electric field or a voltage, a change in temperature or injection of current. There is a wide variety of configurations employed to accomplish this and we look at a few of the more generally applicable structures.

Bulk Structures

The simplest type of structure which may be used, for example, for electro-optic modulation, is that of the type shown in Fig. 6.34: a block of material with defined length and thickness with contacts allowing the application of a voltage to generate the required electric field. Given the modulator length L, the phase shift induced in an optical signal due to a modulation in the refractive index Δn is given by

$$\Delta\phi = \frac{2\pi}{\lambda_0} \cdot L \cdot \Delta n = \pi \frac{V}{V_\pi} \tag{6.42}$$

where V represents the applied voltage. The parameter V_π [V] is the half-wave voltage, namely the voltage which needs to be applied to yield a phase shift of π. Power requirements usually stipulate that V_π should be small.

Waveguide and Interferometer Structures

An optical waveguide fabricated using a material with externally variable refractive index may directly be used as a modulator; III-V-based waveguides are particularly popular for this application. The configuration is conceptually similar to the bulk waveguide, save that the optical field is guided and the electric field needs only to be established in the waveguide region. Since waveguide cross-sections generally have sizes in the micrometer range, relatively small voltages can lead to appreciable electric fields, thereby reducing V_π. Thus a waveguide-based modulator can generally achieve a given phase shift with a shorter device length and/or a smaller applied voltage than a bulk modulator; a typical value for V_π for a 1 mm long modulator using a III-V semiconductor is about 5 V. The attendant disadvantage is that the optical field needs to be coupled into the waveguide in the first place, such that waveguide modulators are primarily of interest for integrated-optical based microsystems. Waveguide-based electro-absorption modulators are, however, frequently used for intensity modulation of laser output, such as, for example, in a monolithically integrated DBR laser configuration (Lammert et al. 1997).

Waveguide-based structures may also be configured as interferometers to allow the use of phase-modulation to yield highly-efficient intensity modulation (Soref, McDaniel and Bennett 1988). In the example shown in Fig. 6.35, a waveguide-based Mach-Zehnder interferometer is monolithically integrated with a DFB laser on a single InP substrate. One arm of the interferometer has a phase modulator which takes advantage of the electro-optical behavior of InP and the ability to achieve high electric fields with low voltages in the thin waveguide. If the phase modulator is unbiased, the optical fields in the two arms of the interferometer are in phase at the output and constructive interference results in an intensity maximum. If the phase modulator shifts the phase in one arm by π, then destructive interference results and the output is minimized and close to zero for a well-balanced system.

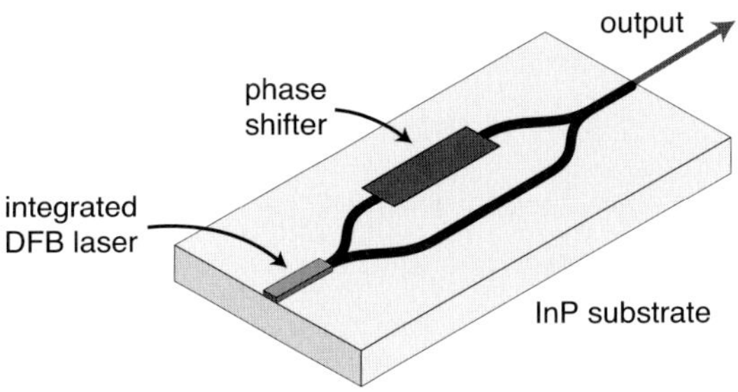

Fig. 6.35. A waveguide-based Mach-Zehnder interferometer used for high-bandwidth intensity modulation of the emission from an integrated DFB laser monolithically integrated in an InP substrate

In this manner, phase modulation, which can be accomplished with good efficiency and at large bandwidth, may be used for high-speed modulation of the laser output. With maximum bias voltages less than 5 V and modulation bandwidths on the order of 10 GHz, an integrated modulator of this type is often used in monolithically integrated, high-speed laser modules for telecommunications.

Example 6.7

Assignment: *Compare the half-wave voltages for a bulk and a waveguide-based modulator which both use the linear electro-optical effect at a wavelength of 850 μm. Both devices are 1 mm long; the bulk device consists of a 100 μm thick slab and the waveguide is 2 μm thick. Assume* $r = -5\times10^{-12}$ *m/V and* $n_0 = 3.5$ *in both cases.*

Solution: *If we are using the linear electro-optic effect for phase modulation, then Eq. 6.38 will give us the relevant change in refractive index with applied field. We may relate this to the applied voltage simply as*

$$\Delta n = \frac{1}{2}n_0^3 r \cdot \boldsymbol{E} = \frac{1}{2}n_0^3 r \cdot \frac{V}{t}$$

for applied voltage V and thickness t. We can assume that the applied voltage is dropped across the entire substrate thickness of the bulk modulator and over the waveguide thickness for the second modulator.

Given the modulator of length L, a phase shift of π occurs when

$$\Delta\phi = \pi = \Delta k \cdot L = \frac{2\pi}{\lambda_0}\Delta n \cdot L = \frac{2\pi}{\lambda_0}\left(\frac{1}{2}n_0^3 r \cdot \frac{V}{t}\right) \cdot L.$$

The applied voltage V for which the previous expression is satisfied is then the half wave voltage, V_π, such that we solve this to yield

$$V_\pi = \frac{t}{L}\left(\frac{\lambda_0}{n_0^3 r}\right).$$

Using the numerical values given above, we see that for the bulk device with $t = 100\ \mu m$

$$V_\pi\big|_{bulk} = 397\ V$$

and for the waveguide device with $t = 2\ \mu m$

$$V_\pi\big|_{waveguide} = 7.9\ V.$$

It is clear that the reduced dimensions of the waveguide modulator yield a significant reduction in the half-wave voltage, assuming that all the other device and material parameters stay the same.

Micro-Machined Structures

Advances in micro-machining technology have allowed the fabrication of modulators which employ effects that, used in a macroscopic device, would require prohibitive power input levels and have poor operating characteristics. One example is the family of modulators based on thermal effects: the ability to fabricate structures of small dimension and thus low thermal mass implies that the required large temperature range can be achieved with low electrical input power.

An example of such a structure is given in Fig. 6.36 which shows a thermally-tunable filter based on silicon technology (Hohlfeld, Epmeier and Zappe 2003). The filter is based on a 1.8 μm thick membrane consisting of Si and SiO_2 layers which form a pair of Bragg mirrors and an optical cavity such that the filter transmits only the wavelength of the cavity resonance. Due to change in the refractive indices of the cavity and Bragg mirrors with temperature, the transmission wavelength may be tuned thermally. Temperature changes are induced using resistive heaters, visible as the coil-like structure on the membrane surface.

Since the membrane containing the filter layers has a very small volume ($1.8 \times 250 \times 250\ \mu m^3$), temperature changes of up to 500°C may be achieved using electrical inputs below 10 mW and these temperature variations may be switched on millisecond time scales. The filter was shown to have a tuning efficiency of 0.071 nm/K for wavelengths in the 1.55 μm wavelength range. Optimized designs which use, for example, cantilevers to mount the membrane to the

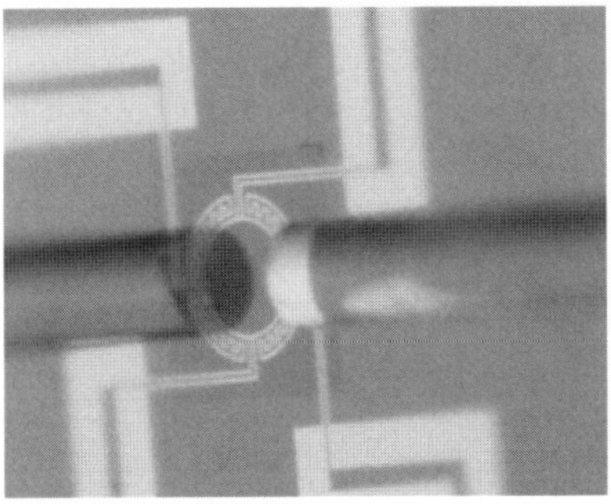

Fig. 6.36. Photograph of a silicon-based, thermally-modulated filter which relies on change in refractive index with temperature. Light is input from the optical fiber at the right; its reflection is visible at the left. The diameter of the inner ring is 125 μm and the extent of the membrane is shown by the darker surface color

substrate, require even lower input powers, at the expense of slightly longer cooling times.

6.5.4 Scanners

Deflection of an optical field through the use of movable mirrors is an increasingly popular means for modulation of beam propagation direction. Whereas the macroscopic equivalents are well-established for laser beam deflection (for example in bar-code scanners), the availability of micro-mechanical mirrors, particularly as individually-adressable arrays, has made even complex beam deflection processes feasible as part of an optical microsystem (Su, Toshiyoshi and Wu 2001).

Research and development activities on the development of new types of MEMS-based micro-mirrors and means for their actuation are at present quite extensive. Commercial products are nonetheless available so that the microsystem engineer has a number of options for beam deflection and scanning. A popular example is the Digital Mirror Device (DMD) developed by Texas Instruments whose structure is shown schematically in Fig. 6.37. Arrays of 848×600 mirrors, each with an area of $17 \times 24\ \mu m^2$, are manufactured using micro-fabrication techniques on a $1.6 \times 1.6\ mm^2$ chip. Each mirror is individually adressable and electrostatically actuated to achieve a maximum tilt angle of 10° along a single axis.

The most popular application for the DMD at present is in digital display projectors in which each mirror is used to define one pixel of a two-dimensional image. The fast mirror response time (20 μs) allows definition of up to 256 grey levels for each of the three fundamental colors in an image and thus an extensive color palette. Illumination is by a broad-band thermal light source but the DMD concept may also be applied to laser beam scanning or positioning.

Alternative mirror technologies have been developed for optical switching, a function with considerable importance in optical telecommunications networks.

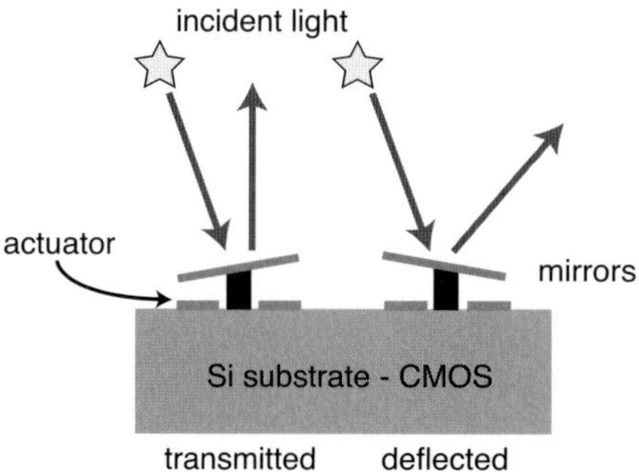

Fig. 6.37. Schematic diagram of two micro-mechanical mirrors used for beam deflection

Gimbal-mounted mirrors which allow movement on two axes, and thus full three-dimensional scanning of an input optical beam, are in use for dynamically reconfigurable coupling of output from one two-dimensional fiber array into another. These and related beam deflection and scanning mechanisms are of great utility for many further laser-based microsystems applications.

7 Integration and Assembly

Our goal in studying semiconductor lasers in this textbook has been to use them as light sources in optical microsystems. Key to achieving this is the availability of techniques for integrating laser diodes with the remaining optical and non-optical components of a more complex optical configuration. In this final chapter, we examine some of the technologies and processes available for integration and assembly of such microsystems.

We may distinguish between two primary approaches for fabricating a complete laser diode-based microsystem: hybrid assembly and monolithic integration. Hybrid techniques assemble disparate optical components, each fabricated by an individually optimized technology, into a complete system using a variety of means for assuring optical alignment through accurate device positioning and mounting. Monolithic integration, on the other hand, fabricates the complete system through a single process on a single piece of semiconductor; post-process alignment and positioning are thus no longer required. Whereas the hybrid approach offers greater flexibility and the ability to choose optimized individual components at the expense of the considerable effort required to align and assemble these, monolithic technologies yield complete, robust and compact systems on a chip albeit at considerable technological expense. Which approach is more suitable for a particular microsystem implementation is generally a function of which technologies are available and, often predominantly, what level of costs are allowable.

In the following sections, we will outline some of the relevant issues in system assembly and discuss some of the demands placed on laser diode packaging. Under the heading of hybrid assembly techniques we include discussion of V-grooves, machined passive alignment structures as well as silicon-optical-bench and MOEMS technologies. We conclude with an overview of what is achievable with monolithic approaches and some of the technological challenges involved.

7.1 Laser Diode Integration Issues

When deciding on the optimal approach for including a semiconductor laser in a microsystem, a number of issues need to be considered. The requirements on ultimate system size, laser diode power, coupling efficiency and overall system robustness need to be balanced with the technologies available and restrictions on system price. The final technology decision is usually unique for each application such that each microsystem project must consider these questions individually.

7.1.1 Mounting Requirements

The packaging, mounting or integration technologies for laser diodes or LEDs must fulfill a number of requirements. Most basically, the mount must physically position the laser relative to the optical system to assure maximal coupling of the light into the required optics. Once optimized, this position should be stable with respect to system vibration, temperature changes, operating conditions of the laser and any external influences such as humidity. The relative placement of laser and optics should remain stable over the lifetime of the device which in some cases may be years to tens of years.

In addition, the mounting technology has to provide electrical and thermal contact to the laser diode. Electrical contact is necessary to provide the laser drive current; thermal contact is essential to assure that the heat generated during laser operation may be removed from the chip. The latter contributes to a maintenance of a stable operating temperature, which can generally only be assured if the package itself is mounted on a heat-sink or is otherwise temperature stabilized.

Specialized applications may place greater demands on the laser package. High-speed laser modulation requires specialized wiring and contact technology to allow the application of microwave electrical signals to the laser. For high power lasers, the high currents required stipulate the use of considerably thicker wiring and the significant thermal generation implies that advanced heat sinking technologies need to be applied to the laser package. Finally, the packaging of laser arrays comes with its own set of problems, including extra demands on positioning accuracy with respect to a lens or fiber array, high wiring density and also the thermal management of the laser chip.

In summary, the laser package needs to provide stable positioning as well as electrical and thermal contact in a setup which provides maximum performance in the minimum of volume at lowest possible cost. Which of these performance demands is paramount usually depends on the application.

7.1.2 Package or Chip?

When using a laser diode as a component in an optical microsystem, the user must decide in which form she should obtain the laser. Packaged semiconductor lasers are mounted in a macroscopic housing whereas laser chips are exactly that, namely tiny bits of patterned semiconductor.

Packaged Laser Diodes

Most users of laser diodes encounter the devices already packaged by the manufacturer. A variety of standard package options are industrially available; the most popular are probably the TO packages that we saw in Sect. 5.1.1 and an example of which is shown schematically in Fig. 5.1. This type of package comes in a range of sizes and typically features the laser mounted on metallic submount hermetically sealed in an inert atmosphere and with either a planar glass or lensed aperture. Industrial products also typically include a photodetector inside the package allowing direct monitoring of the laser emission power. Both edge-emitting and VCL lasers are available in a TO package.

Other types of laser diode packages include open sub-mounts, generally used for edge emitting lasers; these are compact metallic blocks providing electrical and thermal contact for the laser but without a hermetic seal. The surface mount device (SMD) package may be used for LEDs or possibly VCLs; this package is suitable for mounting directly on a printed circuit board and is very compact. LEDs, finally, are available in lensed plastic packages as well as SMDs.

Laser diodes or LEDs in one of these industrially-manufactured packages are macroscopic components which may be incorporated into an optical or optoelectronic system using standard assembly equipment. Advantages of using the laser in this form include stable and optimized electrical and thermal contact, established with well-characterized and reproducible industrial processes. The hermetic seal and inert atmosphere of the TO package generally improves laser lifetime due to the absence of an oxidizing atmosphere which can result in a degradation of the laser facets. Packaged lasers purchased from a reputable manufacturer are tested and often burnt-in thereby increasing manufacturing yield of the sub-systems in which they are employed. Finally, in an many cases not to be underestimated, packaged laser diodes are macroscopic components with sizes in millimeter range; handling and mounting are typically unproblematic.

The limitations encountered when using packaged laser diodes are typically size: the macroscopic dimensions limit the size to which the resultant system may be reduced. In some cases cost may also be factor: packaging and testing are expensive and the total laser costs reflect these contributions. However packaged lasers are not always more expensive than chip versions: the high volumes sold of some types of packaged laser diodes may in some cases make the "simpler" chips a more expensive alternative for the manufacturer.

Laser Diode Chips

An alternative to the packaged laser is the use of laser diode chips directly. As we saw in Chapter 4, the active region of a semiconductor laser (the laser cavity) is typically quite small but for reasons of manufacturability and handling practicality the typical laser chip has dimensions on the order of $300 \times 300 \times 100\ \mu m^3$; a typical edge emitting laser chip was shown in Fig. 5.3. The small size of the unpackaged laser diode chip implies that it may in many cases be ideal for direct incorporation into a microsystem.

Use of a semiconductor chip in a highly-compact laser-based microsystem has advantages in overall achievable system size as well as added flexibility in the types of substrates used and in positioning the laser and other optical components. The latter aspect may be of particular relevance if laser or LED arrays are used, since high-density 1D or 2D arrays may achieved directly on-chip or through the close placement of adjacent chips.

The limitations or problems encountered in the use of semiconductor laser chips generally revolve about handling, positioning and mounting. Students ordering a selection of laser diode chips are frequently astonished at the miniscule motes on blue tape which they receive; manipulation of these laser dust specks requires specialized micro-manipulators and vacuum pencils in the laboratory and pick-and-place equipment in an industrial setting. The relevant mounting techniques are the subject of Sect. 7.2. As the resultant microsystems are typically not packaged in a hermetically-sealed inert atmosphere, concerns about robustness and reliability of chip-based microsystems may also be relevant.

7.1.3 Thermal and Electrical Contact

As we outlined briefly in Sect. 7.1.1 above, two of the primary physical functions of a laser diode mounting technology are to provide the best possible thermal and electrical contact to the laser. These functions allow the transfer of power into the laser and removal of excess heat from it. We saw in Sect. 5.5 that changes in laser diode temperature acutely affect its operating characteristics: output power level, efficiency and wavelength are among the important parameters which change with varying laser substrate temperature. Maintaining a stable and, in high-end applications, known temperature is thus essential for reliable laser operation and the heat-sinking properties of the laser mount are essential in guaranteeing reproducible device behavior.

Lasers in industry-standard packages are typically mounted in materials with good thermal conductivity (σ_{th}[Wm^{-1}K^{-1})]), such as copper, diamond, copper-tungsten alloys or advanced composite materials; silicon or SiC are also suitable and are of particular interest as microsystem substrates. In addition to the requirements on thermal conductivity, the mount materials must have a thermal coefficient of expansion, $\Delta L/L\Delta T$ [K^{-1}], closely matched to that of the laser substrate (typically a III-V semiconductor) and must have a low electrical resistivity. Advanced packaging technologies will also frequently use a sub-mount, between the laser and the package itself, which is designed to act as a heat spreader for the laser chip: CVD diamond, AlN and composites are often employed, where the former is of particular utility due to the very high thermal conductivity of diamond. The thermal performance of some popular mount and sub-mount materials is given in Table 7.1.

Table 7.1. Thermal performance of several popular laser diode mounting materials; data from (Merritt et al. 1997)

Material	σ_{th} [$Wm^{-1}K^{-1}$]	$\Delta L/L\Delta T$ [K^{-1}]
diamond	1 000 – 1 600	2×10^{-6}
GaAs	44	6×10^{-6}
Cu	398	1.7×10^{-5}

The laser chip is generally attached to the sub-mount or mount using flux-less solder, typically SnPb or In. The laser contact typically consists of a multi-layer alloy, such as GeAuNiAu or TiPtAu, but has a thick Au outer layer to which solder contact is made. AuSn is also occasionally deposited on the heat spreader to facilitate contact. The contact metals and solders are chosen to provide good adhesion, wettabililty and thermal stability and should have minimal interdiffusion and long-term degradation.

The laser chip is thus mounted using a selection of these materials with either the top- or back-side attached to the submount. As electrical contact is in most, but not all, cases also required to the opposite surface, this latter contact is typically made by standard wire bonding techniques. Since GaAs and InP substrates are physically more fragile than Si, thermosonic ball bonding is the generally preferred technique.

7.1.4 Optical Coupling

In addition to getting electrical power in and thermal power out of the laser, the laser assembly technology must also assure that the useful optical power emitted by the laser is coupled with greatest possible efficiency into the following optical components. In an optical microsystem, these components may for example be a microlens, an optical fiber, a waveguide or a detector. Whereas the required positioning tolerances for lasers in macroscopic systems have correspondingly macroscopic values, the small physical dimensions of optical components, particularly fiber and waveguide cores, in microsystems implies that the positioning accuracy of the laser diode and these components must generally be in the micrometer regime to ensure useful levels of optical coupling.

Coupling Efficiency

The efficiency of coupling between a laser diode and a subsequent component is typically characterized by the losses generated in the coupling process. Given power emitted by the laser P_0 [W], the insertion loss [dB] relates this to the fraction transmitted into the optical system, P_t [W], and is defined as

$$\text{insertion loss} = -10\log\left(\frac{P_t}{P_0}\right). \qquad (7.1)$$

Alternatively, the return loss [dB] is a measure of the power reflected, P_r [W], due to coupling, namely

$$\text{return loss} = -10\log\left(\frac{P_r}{P_0}\right). \qquad (7.2)$$

We see that these parameters are typically characterized logarithmically and that both insertion and return loss have positive values. Since transmitted power is given by incident power minus the reflected power, the two loss parameters are of course related. A small value (ideally 0) for insertion loss is the goal of any optical coupling process.

Low values for insertion loss may be achieved by optimal mode matching, which we will discuss immediately below, and through a reduction in losses due to reflections at the various surfaces encountered by the optical beam. The latter are most often due to simple Fresnel reflection at the surfaces of lenses, fibers or waveguide facets and may be reduced by the use of anti-reflection coatings; cleaved fiber ends and semiconductor chip facets are particularly problematic due to the high optical quality of the reflecting surface. In some cases, an external Fabry–Perot resonator may be established between laser diode and the external element, the ensuing resonance phenomenon then being highly position- and temperature-dependent. The resulting optical feedback into the laser cavity can have significant detrimental effects on laser operating stability, such that the use of an optical isolator (which we saw in Sect. 5.3.4) may be required for critical applications.

Mode Matching

Optimal coupling of the laser emission into another optical component is also critically defined by the overlap of the optical modes in the two devices. Whereas this consideration is of limited relevance for the coupling between a laser diode and a lens, which is typically placed in the far-field of the laser emission, mode matching is of primary importance for coupling laser emission into an optical device placed in the near-field, such as encountered when butt-coupling into a fiber or waveguide.

In Sect. 5.6.1 we learned that the near-field intensity distribution, or modal profile, of a semiconductor laser may be nicely circularly symmetric (in a VCL, for example) or unpleasantly asymmetric (for most edge-emitting lasers). For the former case, a simple analysis of coupling may be undertaken by assuming that the emitted optical power distribution is Gaussian in shape, namely

$$P(r) = P_0 e^{-2(r/r_0)^2} \qquad (7.3)$$

for power magnitude P_0 [W] and circular mode field diameter (MFD) at the laser facet $2r_0$ [m]; r is the radial variable normal to the propagation direction z. Since the electric field distribution, $\boldsymbol{E}(r)$, is of the form

$$\boldsymbol{E}(r) = \boldsymbol{E}_0 e^{-(r/r_0)^2}, \tag{7.4}$$

we see that r_0 defines the $1/e$ half-width of the optical field and the $1/e^2$ half-width of the optical power[1].

As a Gaussian beam propagates (Saleh and Teich 1991, Chap. 3; Smith and King 2000, Chap. 15.8), the MFD varies with propagation distance z as

$$2r_0(z) = 2\sqrt{r_0^2 + \frac{\lambda^2}{\pi^2}\left(\frac{z}{r_0}\right)^2} \tag{7.5}$$

for laser emission wavelength λ; the Gaussian field thus increases in width but maintains its shape as the field moves from the laser diode facet. In the case where this light is to be coupled into an optical fiber, the intensity distribution of the mode in the fiber core also has a circularly-symmetric Gaussian form, albeit with a MFD which is a function of the core structure. Optimal coupling between the laser diode and the fiber is then achieved when these mode sizes match, in other words for a laser diode-fiber spacing z' for which

$$2r_0(z') = 2r_0\big|_{fiber} . \tag{7.6}$$

This simple analysis gives an approximate indication of the relative longitudinal position of a laser and a fiber for best possible mode matching; the required positioning accuracy is typically several micrometers. In this calculation, the transverse and lateral positions (normal to the propagation direction) are assumed to be such that the optical axes of the laser diode and facet align. In fact, transverse and lateral offsets are much more critical than variations in longitudinal position in determining laser/fiber coupling efficiency; the requirements for transverse and lateral positioning are typically a fraction of the laser waveguide and fiber core diameters and are thus typically below 1 µm.

Such a rudimentary calculation becomes of limited value for systems for which the modes (of either the laser diode or the subsequent optical component) are no longer circularly symmetric or if multi-mode devices are considered. For a rectangular waveguide with dimensions $t \times h$, as an example, we may approximate the effective MFD by $r_0^2 \approx th$. In cases such as this, admittedly the more common, an overlap integral of the modal profiles or a more involved calculation (Lierstuen and Sudbø 1995) should generally be done to optimize and characterize the coupling efficiency.

---

[1] Some users and laser diode manufacturers use the FWHM value rather than the $1/e^2$ width to define the extent of the optical beam.

Coupling Approaches

The coupling of optical power from a laser diode into an optical component may be accomplished through two basic techniques: butt-coupling and end-fire coupling. The simplest conceptually and mechanically, butt-coupling places the two components in close proximity to each other such that each is in the near-field of the other (Hunsperger, Yariv and Lee 1977). Equation 5.51 in Sect. 5.6 defined the spacing required to remain in the near field regime of an optical emitter and values for typical laser diode apertures are in the range of tens of microns. Without any interceding optics, butt-coupling relies on ideal mode-overlap to ensure good optical coupling such that positioning, particularly lateral and transverse, is of prime importance; despite its simplicity, the approach can yield coupling efficiencies of over 90% (Heinrich, Zeeb and Ebeling 1997). It is just this simplicity that makes butt-coupling attractive for laser diodes employed in optical microsystems, since the component sizes and positioning accuracy achievable allow the realization of compact and manufacturable assemblies.

Improvement in butt-coupling efficiency may be achieved through the use of optical structures integrated into the laser or optical component. The fiber tip, for example, may be lensed so as to improve collection efficiency of the coupled light (Kuwahara, Sasaki and Tokoyo 1980) and reduce the alignment tolerances. For the waveguide of the edge-emitting semiconductor laser, the use of lateral (Zengerle et al. 1992) or vertical (Koch et al. 1990) tapers or spot-size converters (Uchida. et al. 1996) allows the expansion of the mode size to accommodate that of the following optical component.

End-fire coupling is a more involved approach and uses lenses to facilitate the coupling process; the ball lenses used in the structure of Fig. 6.1 represent an example. The lenses are used to focus or collimate the light emitted from the laser diode, often allowing for considerable improvement in coupling efficiency when compared to butt-coupling approaches. The system and alignment complexity, however, also increase and end-fire approaches are significantly more expensive than equivalent butt-coupling configurations. A number of end-fire coupling schemes have been demonstrated for microsystem assembly, including the placement of GRIN lenses between the laser diode and a fiber (Nakagawa et al. 1996) or the use of a cylinder lens to allow efficient coupling of a laser diode array with a fiber array (Crow et al. 1978).

7.2 Hybrid Technologies

Having seen some of the demands placed on a mounting and packaging technology for laser diodes in general, we may now look at some of the techniques which have been developed for achieving this. We consider in this section those technologies which may be used to assemble a laser diode into a hybrid system; separate fabrication processes are used for the fabrication of each optical component and for the construction of the complete system. The following techniques are often used in combination with one another to yield a complete system technology and we consider some of the more important ways in which optical components may be combined, aligned and mounted.

Assembly techniques may be broadly divided into two categories: active and passive. Active alignment requires that the laser diode be operated and the emitted light coupled into the optical system, from which a maximization of output power allows determination of optimal component positions; this process is considered in Sect. 7.2.2. In passive alignment, mechanical features such as grooves, stops and spacers are used to define the physical position of the components which may be "dropped" into place on an appropriate substrate; the relevant technologies are discussed in Sect. 7.2.3 – Sect. 7.2.5.

7.2.1 Substrates

Laser diodes and the optical components of an optical microsystem are generally combined using a mechanical substrate material. This substrate should fulfill a number of requirements which we saw in part above: good physical and mechanical stability, high thermal conductivity, good electrical conductivity and compatibility with attachment technologies, such as soldering. A further consideration is machinability: for use in microsystem assembly, the material must be structurable at the micrometer level. As we will see below, the ability to accurately and reproducibly fabricate mechanical features at these size scales is essential for achieving precise system alignment.

A final consideration is cost: the substrate material itself and, as importantly, its structuring technology should be affordable; since system assembly is often the most significant cost factor in a microsystem product, this aspect is not to be underestimated.

Metals and Ceramics

We saw above that the standard material technology for a commercial laser diode or LED package is usually based on metals. Due to the limited requirements on positioning accuracy in a standard laser package, achievable with standard pick-and-place equipment, the substrates do not require precision structuring with submicrometer accuracy. A number of specialized metallic and ceramic materials for microsystem substrates are available, with tailored mechanical, thermal and electrical properties. Whereas ultra-precision machining techniques are available, the attainable precision is limited and is not at the level of that achievable with silicon; these processes also generally require individual fabrication of each mount and are thus costly when produced in large volumes.

Silicon

As a semiconductor with a highly-developed fabrication technology and a number of attractive physical characteristics, silicon is in many cases well suited for use as an optical microsystem substrate material. Mechanically very stable and with a relatively high thermal conductivity, silicon may also be doped to yield defined regions of good electrical conductivity or combined with metallic wiring for more optimal electrical behavior. In addition, the material is inert or may easily be passivated such that long-term stability concerns are minimal.

As important as the intrinsic physical properties of silicon is its fabrication technology. It is unlikely that any other material on the planet is subject to such an extensive and highly-precise manufacturing technology: definition and fabrication of electrical and physical structures with nanometer accuracy across a 300 mm silicon substrate is virtually routine as is the fabrication of a wide variety of static and dynamic mechanical structures with micrometer size scales. Equally relevant is the highly-parallel manner in which this is done: batch processes allow mass-fabrication of these components and features such that economies of scale figure strongly in any cost calculations.

Advanced silicon technology is available for the fabrication of mechanical, electrical and optical features on the substrate. We will examine several of the more important mechanical structures in Sect. 7.2.3 and Sect. 7.2.5 below: these are generally used to accurately position a laser diode and any optical components relative to each other. Electrical components range from relatively simple metallic contacts and wiring, with which electrical contact to a laser or other active optical component may be made, to complete driver circuitry. In the latter case, the electronics required, for example, to bias and modulate a laser or generate and amplify a detector signal may be incorporated directly into the silicon substrate. Optical devices, finally, may include waveguides, typically fabricated using silicon dielectrics (SiO_2, Si_3N_4, SiON and others) deposited on the substrate, and complex integrated optical circuits (Mino et al. 1995). Mirrors, lenses, gratings and movable opto-mechanical structures may also be included in the substrate.

These capabilities come at a cost: combining mechanical, electrical and optical functions on a single silicon substrate on which a laser diode and other optical devices are subsequently mounted can become as technologically complex as the monolithic technologies which we will discuss in Sect. 7.3. In addition, the significant initial capital investment required to establish a silicon process line, even a relatively simple one, can be prohibitive. The alternative to investing in the establishment of a silicon process line is the use of foundry processes in which many of these capabilities are made available to users at a fraction of the cost.

Polymers

For the hybrid assembly of very low cost laser-diode based microsystems, many users are considering the use of polymer-based substrates. Using one of a wide variety of plastic materials, the mechanical features for component alignment may be fabricated using low-cost industrial injection molding or embossing technologies; these offer physical feature resolutions in the micrometer range while requiring the manufacture of only one high-quality master structure from which a large quantity of replicated substrates may be manufactured. The master may be fabricated using precision machining in metals, silicon micro-machining (Klein and Neyer 1994) or through speciality processes such as LIGA (Brenner et al. 1993). An alternative technology is the use of photo-structurable polymers which employ photolithographic techniques to define high-aspect-ratio structures on a silicon or glass substrate which may subsequently be hardened by a post-bake process to yield robust three-dimensional features (Lorenz et al. 1997).

The advantage of low cost is somewhat offset by the performance limitations of most polymers when used in this capability. Plastics generally have very poor thermal and electrical characteristics, such that particularly thermal contact to a laser diode mounted on a polymer substrate is difficult to accomplish. Electrical and optical functionality may be attained through modification of the polymer or, more popularly, through deposition and patterning of surface layers, albeit at additional process cost (Kaino 2000).

7.2.2 Active Alignment

The relative positioning of laser diodes and the optics into which the light is coupled is often the most critical process step in the hybrid assembly of an optical microsystem. As we saw in Sect. 7.1.4, positioning accuracy better than 1 μm is often required in fiber- or waveguide-based systems, a demand which sometimes difficult to achieve reproducibly with the passive alignment schemes we discuss in subsequent sections. As an alternative, active alignment couples the light from the laser into the fiber or waveguide, optimizes the transmitted intensity and then fixes the position of the components. Since active alignment typically positions the element with the larger MFD (i.e., the fiber rather than the laser diode, for example), required positioning accuracy is lower, by up to a factor 10, than for passively aligned systems, which typically position the component with the smaller MFD.

The active alignment process requires that the laser diode and the optics are roughly correctly positioned on a submount with respect to one another, typically with an accuracy of 5 – 10 μm; this step may be realized by positioning under a microscope or with industrial pick-and-place equipment. The laser is then biased and the emitted light coupled into the optical system; the transmitted intensity is monitored at the output of the optics, for example at the other end of the fiber. By varying the position of the optical component, using micro-manipulators which allow movement in the required six degrees of freedom ($x, y, z, \theta_x, \theta_y, \theta_z$), the transmitted light intensity is maximized whereupon the positions of the components are fixed, typically using UV-curable epoxies or laser welding. Six positional variables present a daunting prospect for optimization, but automated systems employ clever search routines to find an optimal position.

The advantage of this process is that a very high degree of coupling efficiency may be realized; for high-value, complex laser diode transmitter systems, such as that presented schematically in Fig. 6.1, this approach is often used industrially for packaging into a metal-based butterfly package. Active alignment is, however, complicated and slow: each laser diode module must essentially be assembled individually with a process characterized by very limited throughput. When this consideration is coupled with the relatively large size of the required submounts, necessary to allow access by the micro-manipulators, what results is not really an optical microsystem. For most microsystem applications, as a result, active alignment is not really an option.

7.2.3 V-Grooves

Passive alignment technologies use mechanical features to position the optical components in a microsystem; the laser diodes, waveguides, lenses and fibers need appropriately dimensioned structures to facilitate this process. One of the most established of these is the V-groove for positioning optical fibers (Crow et al. 1978). A fiber can be somewhat tricky to manipulate but has the advantage that its cross-sectional dimension (fiber diameter = 125 μm) is both well-standardized and highly reproducibly manufactured.

The V-groove, as shown in Fig. 7.1, is etched into the surface of (100) silicon using wet etches which essentially stop etching on the (111) crystal plane; these processes are very well established (Bean 1978). KOH in water, for example, has an etch ratio of about 200:1 between the (100) and (111) planes and EDP (ethylene diamine pyrochatechol) in water has an etch ratio of about 40:1. Starting from the (100) surface, the etchant then virtually stops on the (111) planes resulting in an etched surface which makes an angle of 54.74° ($\tan^{-1}\sqrt{2}$) with the horizontal. If a (110) surface is used, these etchants will yield vertical sidewalls, normal to the surface, and thus a U-groove (Bassous 1978); however, rectangular patterns, as seen from the surface, are not possible with this crystal orientation which, when combined with the rarity and thus cost of (110) substrates, limits the use of this substrate type.

Since the angle of the etched surface is defined by the crystal planes, elementary geometry allows the calculation of the required etch mask width to yield a V-groove of a certain size. The core of a fiber with known diameter placed in the groove thus has a known height above the substrate surface; a long (into the page) groove allows placing the fiber into a rigid mechanical mount thus providing an accurate, stable position. Core height may be positioned to an accuracy of ±1 μm and micro-mechanical techniques have been developed to assist mounting and holding the fibers in the V-grooves (Strandman and Bäcklund 1997).

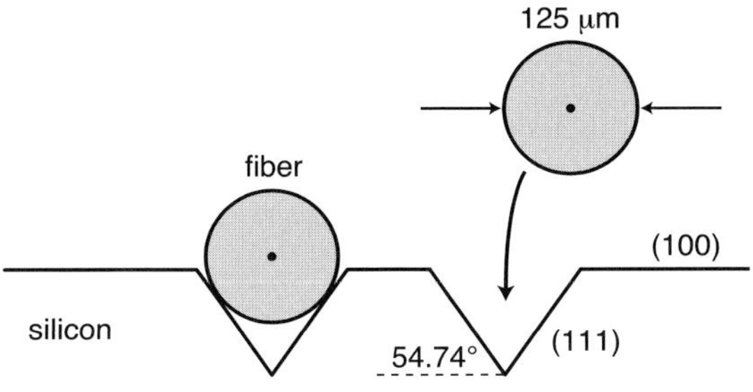

Fig. 7.1. V-grooves etched into (100) silicon used for positioning standard optical fiber

V-grooves may also be used for alignment of one- or two-dimensional arrays of fibers; a variety of these is available commercially. An industry-standard pitch of 250 μm between adjacent fibers is used and such 1D arrays are useful for coupling laser arrays into fiber arrays. Industrially fabricated fiber ribbons (see Sect. 6.3.1) may be terminated in a V-groove array, allowing direct coupling to a laser array with the same standard spacing. Two-dimensional arrays are also becoming more common; these are fabricated by thinning the Si V-grooved substrates to a standard value, stacking the resultant 1D fiber arrays and bonding these to each other. Although a limited number of commercial 2D products is available, the technology is at present still under development.

7.2.4 Flip-Chip

Positioning the optical components of a laser-diode based microsystem is only one part of the hybrid assembly problem: fixing these positions and mounting the devices permanently onto the substrate is another. One common alternative is through the use of UV-curable epoxies (Yamada et al. 1992) which allow free positioning of the components until the adhesive is solidified by exposure to ultraviolet light. Unfortunately, uneven and uncontrollable contraction of the glue during curing will often result in a shift of the component from the desired position.

As an alternative, flip-chip technology combines accurate positioning with electrical contact between the substrate and an optical chip, such as a laser diode or semiconductor photodetector. Developed over 30 years ago for packaging electronics (Miller 1969a), the technique relies on surface tension of molten solder on accurately defined contact pads to pull two components into alignment and these are subsequently fixed in position and electrically connected upon solder solidification. Flip-chip is now a standard mounting technology for laser diode chips and other optoelectronic components (Edge et al. 1991).

The flip-chip process is shown schematically in Fig. 7.2. Both the substrate (often silicon but occasionally glass or another semiconductor) and the optical component (assumed to be a laser diode chip in this example) have metal contact pads photolithographically defined and patterned on the surface. A solder ball is deposited on the contact pad of the substrate and the optical chip is "flipped" and brought upside-down into rough alignment and contact with the substrate. The assembly is heated to above the melting temperature of the solder: when molten, surface tension pulls the laser chip into position above the substrate, the relative positions defined by the contact pads. Upon cooling, the two components are mechanically aligned as well as electrically and thermally contacted.

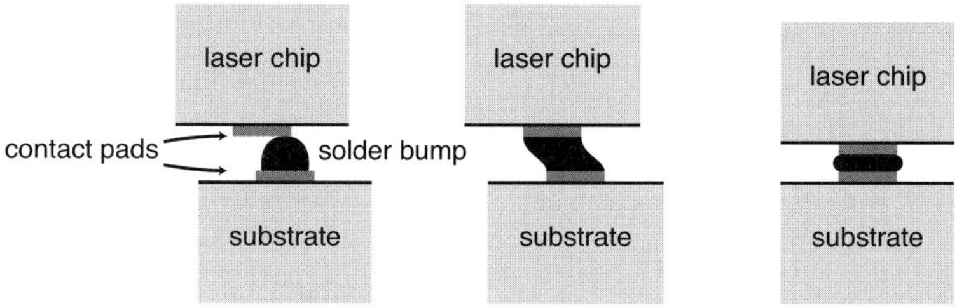

Fig. 7.2. Schematic view of the flip-chip alignment process. The laser chip is "flipped" and brought into rough contact with the silicon substrate *left*; the solder is heated above the melting point such that surface tension pulls the two components into alignment *center* and these are fixed in position upon cooling *right*. Not to scale

With gold as the outer metallization material on the contact pads, the solder used for the flip-chip process is typically either a Au-Sn eutectic or In. A flux-less atmosphere is essential as flux vapors are known to result in contamination of the laser facet with a concomitant reduction in laser lifetime. Solderless flip-chip approaches are likewise in development; these use thermocompression or thermosonic processes.

The great advantage of flip-chip technology is that the components need only be very roughly aligned to one another; the accuracy of pick-and-place equipment is not even required. Final alignment is accomplished by surface tension and the alignment tolerances are ultimately a function of the accuracy of the contact pad placement and dimensions. Positioning tolerances below 1 μm are routinely achievable using this approach (Wale and Edge 1990) and the availability of flip-chip bonders which facilitate the positioning and rough pre-alignment has made this process an industrial standard. Flip-chip technology is expandable to include the assembly of laser diode, fiber and optical component arrays (Basavanhally and Nordin 1997; de Labachelerie et al. 2001). Having a technique to simultaneously optically align and electrically contact an array of optical components to one another is of great value in the assembly of highly-parallel optical microsystems; such aggregates are frequently impractical if not impossible to assemble by standard means.

7.2.5 Silicon Optical Bench

One of the basic tools of the optical engineer, even one working on optical microsystems, is the optical table. Used as a surface to allow stable positioning of optical components with respect to one another, the optical table is typically a complex composite structure floating on a pneumatic suspension. But with dimensions measured in meters and mass measured in tons, it is a tad difficult to imagine integrating such a component into a microsystem.

An alternative using silicon as an "optical table" has become established in the repertoire of optical microsystem assembly technologies. Known as the silicon optical bench (and frequently abbreviated SiOB), this "technique" is actually a general agglomeration of technologies (including V-grooves, flip chip, eutectic-metal bonding (Venkatasubramanian and Timmons 1994) or polyimide-based bonding (Matsuo et al. 1997)) which employ silicon to mechanically align, electrically and thermally contact and physically support several optical components (Hashimoto et al. 1998) or use patterned alignment marks on the silicon surface to align components onto it (Cohen et al. 1992). As we discussed in Sect. 7.2.1, the advanced state of silicon fabrication technology allows the definition and structuring of mechanical, electrical and optical components with sub-micrometer accuracy. By using micro-machining to fabricate mechanical alignment posts and spacers (Armeinto et al. 1991; Jones et al. 1994), the silicon substrate may be used for passive alignment of laser diodes, photodetectors, lenses, prisms and other optical devices to one another and subsequently acts as a micro-optical bench with 100 micrometer to millimeter dimensions (Park et al. 2002).

A schematic example of how the silicon optical bench concept may be employed is shown in Fig. 7.3. The goal of this assembly is to align an array of four edge-emitting lasers to four corresponding optical fibers. V-grooves are etched into the silicon substrate with the standard spacing of 250 μm. A mechanical alignment stop is also fabricated using micro-machining processes, most practically by dry etching. Finally, a set of contact pads and metallization stripes is defined on the silicon surface.

The laser chip is patterned correspondingly: the spacing between adjacent laser stripes is also set to 250 μm and the surface metallization chosen to be compatible to that on the silicon substrate; materials which allow good solder reflow are used. Solder bumps are then deposited on the silicon contact pads and the laser chip flipped onto the substrate. During the solder reflow step, surface tension then moves the laser chip into position in front of the optical fibers; the chip is pulled against the alignment stop which serves to define the lateral position with greater accuracy than achievable with the flip-chip process alone. The system is then completed with the fibers placed in the V-grooves and fixed using a UV-curable epoxy. The entire SiOB assembly is then typically packaged; wire bonding is used to contact the laser using the contact pads and a bond is frequently required to the laser chip back-side as well.

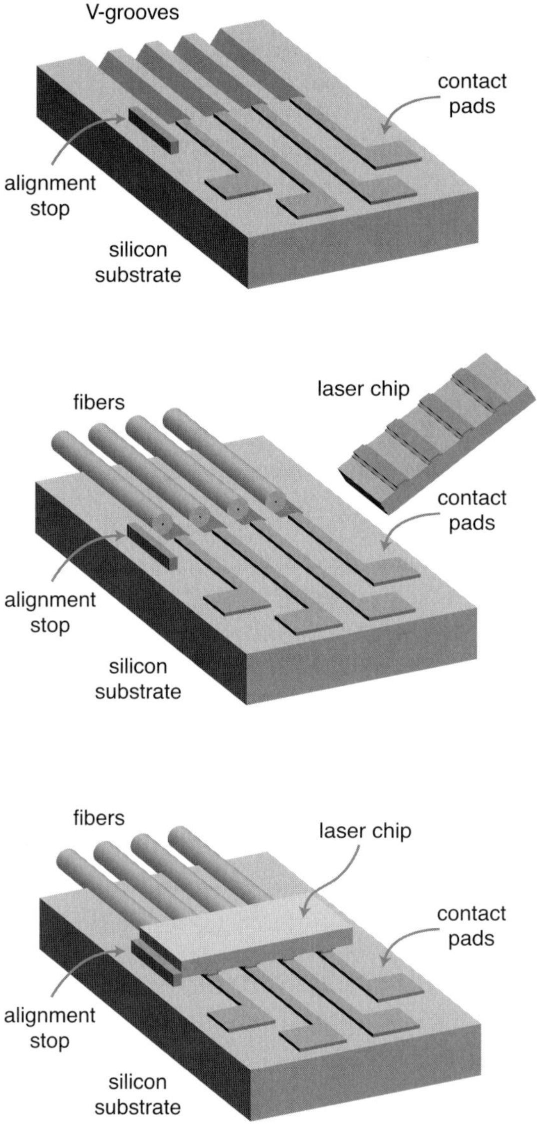

Fig. 7.3. Schematic view of the alignment of a laser chip and optical fibers onto a silicon optical bench. The silicon substrate is patterned with V-grooves, contact pads and alignment stops *top* with corresponding contact pads on the laser chip *center* which is mounted by flip-chip techniques onto the substrate *bottom*. Not to scale

The SiOB concept may be extended to include even greater functionality on the silicon substrate than shown in this example; the microsystem applications considered in Sect. 1.2 all used some form of the silicon optical bench as a basis for the optical microsystem. The silicon substrate, for example, may be combined with a polymer-based structure such that mirrors may be placed into a free-space optical path (Wallrabe et al. 2002) or glass substrates may be included to allow optical transmission at visible wavelengths. Other microsystem developers have incorporated liquid crystals into the hybrid system for variable optical attenuation (Hirabayashi, Wada and Amano 2001). It is expected that, as technology develops, the inclusion of silicon electronics on the same substrate, which has been demonstrated in the laboratory, will also become commercially feasible.

Of utility for many applications is the inclusion of waveguides on the silicon surface for coupling, splitting or combining the optical fields; the resulting substrate structure is frequently referred to as a planar lightwave circuit (PLC) (Himeno, Kato and Miya 1998). Processes using polyimide (Higurashi, Sawada and Ito 2002) or flame hydrolysis-deposited silica (Blume et al. 2000) waveguides have been demonstrated. A PLC using a terraced silicon substrate with silica waveguides for coupling a laser diode, two photodiodes and input/output fibers together using a wavelength-selective waveguide coupler has been developed for low-cost optical communications systems (Okano et al. 1997; Okano et al. 1998). The $14.5 \times 1.4 \times 1.0\ mm^3$ silicon PLC chip also includes a vertically-mounted dielectric multi-layer filter which is positioned in a 200 μm deep, 20 μm wide slot cut into the Si substrate using a diamond saw. Mounting the components is through a flip-chip approach which also uses imaged fiducial marks to improve the rough alignment before bonding. Spot-size converters are frequently used in the laser diodes integrated onto a PLC to optimize laser/waveguide coupling (Uchida et al. 1996).

A number of commercial vendors offer SiOB processes on a custom basis; the substrate and its mechanical alignment features, metallization and other structures are fabricated based on the demands of a microsystem developer. The costs are a function of complexity and volume. Combined with the availability of complementary foundry processes as well as assembly and packaging services, even the fabless user may thus design and conceive a laser diode-based optical microsystem with some promise (if not necessarily assurance) of industrial viability.

7.2.6 MOEMS

For the microsystem engineer, silicon is not only of great value for its highly-developed electronic functions but also for its capabilities in micro-mechanics. MEMS technology has demonstrated that Si may be used for the fabrication of highly functional and robust micro-mechanical components using a broad palette of etch and deposition technologies (Menz, Mohr and Paul 2001). When applied to optical functions, silicon has also proven of value for the manufacture of ultra-miniaturized opto-mechanical components giving rise to the concept of micro-opto-electromechanical systems (MOEMS), sometimes maternally shortened to micro-optomechanical systems (MOMS).

The optical systems based on MOEMS are generally free-space optical configurations which use ultra-miniaturized opto-mechanical components such as mirrors, lenses and beam scanners to yield highly-compact systems. These components are manufactured using specialized surface- or bulk-micro-machining technologies (Wu 1997; Tabib-Azar and Beheim 1997) and generally include electrostatic actuators to allow controlled movement of the silicon-based micro-optical elements; clever micro-fabricated hinges, latches and other mechanical accessories permit lateral positioning and rotation. Insofar as the light source for the MOEMS system is integrated onto the Si substrate, this generally takes the form of a hybridly mounted laser diode or LED, usually mounted using the SiOB techniques discussed above and bonded using flip-chip technology. MEMS-fabricated latches are sometimes used to fix the laser chip into position.

MOEMS technology is generally complex and highly specialized for a particular application, yet examining a few case studies is instructive. In one example, active alignment between a laser diode and an optical fiber may be accomplished using a MOEMS "chip" which includes mounts for the laser and fiber in addition to an electrostatically-actuated micro-mechanical movable mirror (Solgaard et al. 1995). The mirror, which is positioned normal to the silicon substrate surface, may be moved on two axes and is used to couple the light from the laser into the fiber. By monitoring the output intensity, as we saw in Sect. 7.2.2, optimal coupling may be achieved, in this case with an opto-mechanical system which is part of the microsystem substrate.

A complete optical pickup system for reading optical storage media (such as a CD) has been demonstrated using a complex MOEMS technology (Lin et al. 1996). The light from a laser diode, mounted on the silicon substrate using hinged latches, is coupled through a rotatable beam splitter, Fresnel lenses and angled mirrors to result in a collimated beam emitted perpendicularly to the plane of the substrate. Reflected light is coupled backwards through the system to yield an optical interference signal at a buried photodetector. Most of the optical components "pop up" from the substrate; this assembly must at present be done manually, possibly limiting the industrial viability of the technique.

As a final example, a miniaturized MOEMS-based external cavity tunable laser (using the same approach as macroscopically employed in Example 6.5 of Sect. 6.2.2) has been fabricated using an electrostatically-actuated Ni mirror fabricated on a silicon substrate (Uenishi, Honma and Nagaoka 1996). An anti-reflection coated laser diode is hybridly mounted on the substrate and a tuning range of $\Delta\lambda = 20$ nm was demonstrated with a physical system volume considerably below that of an external cavity laser using macroscopic optics. Similar approaches have been used to fabricate integrated tunable etalons for wavelength tuning of ultra-compact laser diode-based optical systems (Sawada, Higurashi and Itoh 1999) and an external-cavity DBR laser using gratings written into a silicon waveguide (Tanaka et al. 1996).

7.3 Monolithic Integration

With a sufficiently advanced technological base, a number of optical and optoelectronic functions may be integrated onto the same substrate as that in which the laser diode is fabricated. The resulting monolithic structure results from the fabrication of several components concomitantly with the semiconductor laser, such that subsequent alignment, positioning and hybrid assembly steps are no longer required. The result is, usually, an even more compact laser diode-based optical microsystem with advantages in robustness, stability and size over hybrid systems at, however, considerable technological (and thus pecuniary) cost.

Since semiconductor lasers must be fabricated using optically active materials, monolithically integrated optical systems are generally made using III-V semiconductors such as GaAs or InP. The monolithic combination of laser diode light sources with photodetectors, modulators and other active optical components, all linked by waveguides, has led to the concept of photonic integrated circuits (PICs) (Koch and Koren 1991) which, in analogy to electrical integrated circuits, use photons rather than electrons as information carriers in the "circuitry". Related to these are opto-electronic integrated circuits (OEICs) (Wakao 1992) which combine photonic components with driver or amplifier electronics.

We will briefly look at two general types of monolithically integrated components. Waveguide-based or integrated optical approaches use optical waveguides as photonic circuitry to perform a variety of optical functions; free-space approaches use laser diodes with monolithically-integrated components to facilitate the assembly of unguided optical microsystems.

7.3.1 Integrated Optical Approaches

Integrated optics refers to the well-established use of waveguides for guiding light on a suitable substrate thus allowing the conception of purely optical circuits (Miller 1969b). Using semiconductor-based integrated optics technology (Zappe 1995), lasers may be monolithically attached to a photonic integrated circuit, obviating any need for post-process alignment and assembly. We anticipated a popular example of an application for this technology in Sect. 6.5.3 (Fig. 6.35) which we present again in Fig. 7.4: a laser diode is followed by a waveguide-based Mach-Zehnder interferometer with phase modulators in each arm and the modulated light coupled into an optical fiber. This configuration may be used for high-speed intensity modulation of the laser emission in a form which is considerably more compact and robust than any bulk-optical implementation, even if such were practical.

Such a monolithic laser/modulator chip is frequently used in high-speed InGaAsP on InP-based laser systems, modulated in the GHz regime for optical communications applications (Zucker et al. 1992). Although not shown in the figure, the photonic laser chip and the optical fiber would in practice be mounted on, for example, a structured silicon substrate using the flip-chip and V-groove tech-

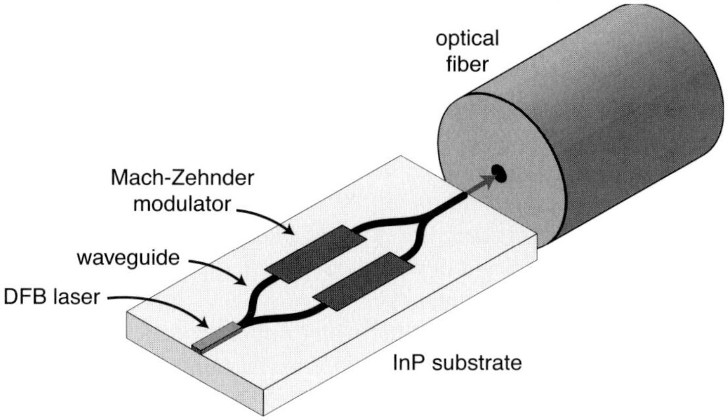

Fig. 7.4. A DFB laser monolithically-integrated with a Mach-Zehnder interferometer-based intensity modulator for high-speed laser modulation; the emitted light is coupled into an optical fiber. Not to scale

niques and possibly other SiOB technologies discussed in Sect. 7.2. Similar PIC technologies have been developed for the fabrication of monolithic chips with 16 distinct laser wavelengths for wavelength division multiplexing (WDM) in optical communication systems (Young et al. 1993), integration of lasers with photodiodes for continuous on-chip monitoring of laser output (Liou 1990) and complete transceiver chips for low-cost optical networks (Matz et al. 1994).

Monolithic Lasers

A number of technology issues arise when we consider the example monolithic structure of Fig. 7.4: the nature of the laser is one central consideration. Since the waveguide interferometer circuit is monolithically attached to the waveguide forming the laser cavity, it is not possible to use to use a cleaved-facet edge-emitting Fabry–Perot laser, since cleaving would result in at least two chips. For monolithic systems, DFB or DBR lasers are thus the structures of choice, since the grating mirrors may be formed as part of the monolithic waveguide structure. Although technologies for dry-etching conventional mirror facets have been developed (Vettiger et al. 1991), grating-based laser diodes have proven to lead to superior system performance in monolithic optical communications chips.

Transparency

The waveguide circuit monolithically attached to the DFB laser in a PIC is fabricated concurrently with the laser itself. As we recall from the discussion of basic laser physics in Sect. 2.4.1 and as we considered for the fabrication of the grating section of a DBR laser in Sect. 4.4.3, an unpumped section of material from which a laser diode is fabricated is strongly absorbing at the emission wavelength of the laser. As a result, the waveguide sections attached to the laser in Fig. 7.4 need to be made transparent at the wavelength of interest.

Transparency may be achieved a number of ways: the use of dielectric waveguides (Bouadma et al. 1994) or quantum-well intermixing techniques (Marsh 1993; Zappe, Hofstetter and Arnot 1994), as described in Sect. 4.4.3, has been widely applied to the manufacture of PICs. Advanced epitaxial techniques which allow the simultaneous growth of waveguide regions with varying absorption wavelength have also been demonstrated (Lammert et al. 1994) as has the use of vertically-separated waveguides of different absorption wavelengths between which light may be transferred using evanescent coupling (Remiens et al. 1990).

The transparency issue is a central one in the development and applicability of PICs using integrated lasers. The ultimate length of a waveguide section which may be monolithically attached to a laser diode, and thus the photonic circuit size and functionality, is a function of the optical losses such that technologies which provide high transparency are essential for the fabrication of more complex photonic integrated microsystems.

Electronics

The III-V semiconductors used for fabrication of laser diodes and PICs have also been extensively developed for electronics; the combination of optical and electronic components on to a single chip is thus a further technological extension, leading to the fabrication of OEICs. Most practical from the point of view of the user of laser diodes in more advanced optical microsystems is the inclusion of laser drive and stabilization circuitry directly on the laser chip; not only is compactness an issue, but the ability to closely monitor chip temperature and regulate laser output accordingly is a decided advantage. For high-speed modulation applications, the electrical connection between driver circuitry and the laser may be made quite short and designed with a layout which optimizes high-bandwidth signal transmission.

Such integrated photonic/electronic systems have been demonstrated (Nichols et al. 1994) predominantly for high bandwidth operation. The technology has been extended to the integration of photodetectors with amplifier circuitry (Hurm et al. 1993) as well as the fabrication of complete transceivers which are then assembled with input and output fibers using flip-chip technology and integrated waveguides for optical coupling (Jackson et al. 1994).

The required process complexity to achieve high levels of integration, both photonic and electronic, is often at the core of the technological cost of a monolithic system and has proven to be a limitation for many optical applications. Whereas monolithic laser diode-based chips have been employed in high-end systems (predominantly in optical communications), economic considerations have resulted in a leaning toward hybridly-assembled optical microsystems for most other applications.

7.3.2 Free-Space Approaches

A "free-space" *monolithic* optical system seems to be a misnomer; what we mean to address in this final section is the use of monolithically integrated optical components to facilitate the fabrication of optical microsystems based on free-space (as opposed to waveguide) optics.

The vertical cavity laser, for example, has numerous characteristics which make it quite suitable for inclusion in an optical microsystem; these include low power consumption, high efficiency, a symmetric beam profile and the ease with which two-dimensional arrays may be fabricated. The vertical emission of the VCL, however, makes coupling into a waveguide considerably more difficult than for, say, a monolithically integratable DFB laser. As a result, VCLs are used with advantage in free-space micro-optical systems.

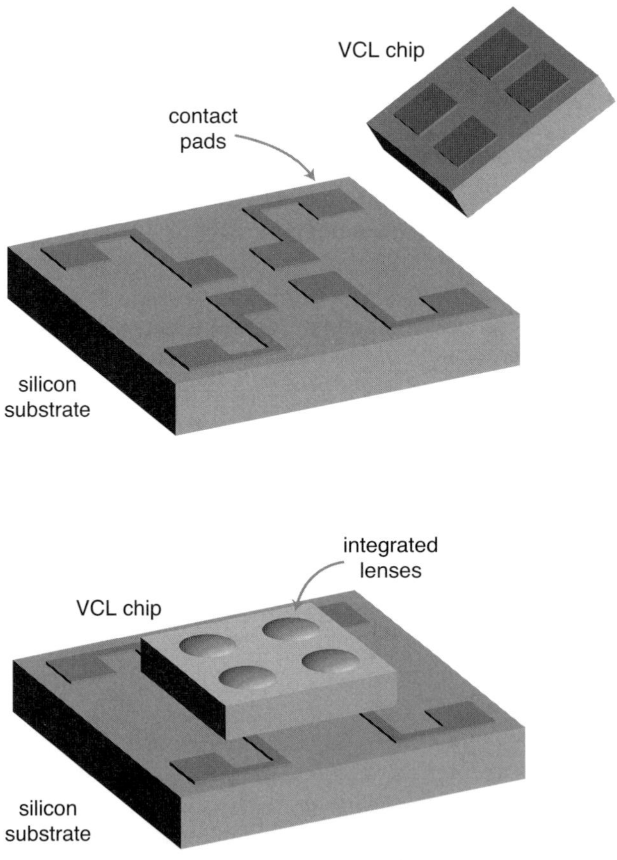

Fig. 7.5. A 2×2 VCL array with monolithically-integrated collimation lenses fabricated on the back-side of the substrate mounted onto a silicon optical bench using flip-chip techniques

The assembly of a VCL-based optical microsystem may be eased by monolithically integrating collimation lenses onto the VCL substrate (Blum et al. 1995; Strzelecka et al. 1995). The example of Fig. 7.5 shows such a chip being mounted onto a silicon optical bench using a flip-chip approach (Louderbach et al. 1999). In this case the VCLs are designed to emit at a wavelength for which the substrate material is transparent; the upper mirrors are thus highly reflecting and may even be covered with contact metal, as shown in this configuration. Light emission through the back-side of the substrate is then collimated by the lens structures etched directly into the substrate material. The light from the lasers may thus be coupled into fibers or other optical elements without the need to align the collimation lenses, thus simplifying the complete system assembly. Related schemes for accomplishing this form of integration include the use of dielectric layers deposited on the surface or back-side of the VCL chip for fabrication of the lenses or hybrid assemblies using diffractive optical components for VCL fanout and coupling (Prather et al. 2001).

There is a variety of monolithically integrated components which facilitate the assembly of an optical microsystem in this manner. Grating couplers are very convenient for inducing surface emission of light from an edge-emitting laser diode waveguide; the form of the coupler may be designed to collimate or focus the beam (Shiono and Ogawa 1994; Ura and Sheard 1998). Properly designed waveguide tapers, as discussed in Sect. 7.1.4 (Zengerle et al. 1992; Koch et al. 1990), when combined with a technology for inducing transparency over their length, are likewise useful monolithic components which ease construction of a more complex optical system.

At the risk of becoming tedious on the last page of the text, we aver again that the optimal assembly or integration technology, of which we have seen a variety, is ultimately a function of required optical system performance and affordable system costs. We close on this point to provide the reader with encouragement to apply her or his ingenuity to the development of novel and clever laser diode-based optical microsystems using all the tools at ones disposal.

References

Aagedal, H., F. Wyrowski and M. Schmid. 1997. Paraxial beam splitting and shaping. In *Diffractive Optics*, edited by J. Turunen and F. Wyrowski. Berlin: Akademie Verlag.

Adachi, S. 1985. GaAs, AlAs and $Al_xGa_{1-x}As$: Material parameters for use in research and device applications. *Journal of Applied Physics*. 58:R1–R29.

Adachi, S. 1992. *Physical Properties of III-V Semiconductor Compounds*. New York: John Wiley and Sons.

Agrawal, G.P., and N.K. Dutta. 1986. *Long-Wavelength Semiconductor Lasers*. New York: Van Nostrand Reinhold.

Aizenberg, J., A. Tkachenko, S. Weiner, L. Addadi and G. Hendler. 2001. Calcitic microlenses as part of the photoreceptor system in brittlestars. *Nature*. 412:819–822.

Alferov, Z.I., V.M. Andreev, D.Z. Garbuzov, Y.V. Zhilyaev, E.P. Morozov, E.L. Portnoi and V.G. Trofin. 1971. Investigation of the influence of the AlAs-GaAs heterostructure parameters on the laser threshold current and the realization of continuous emission at room temperature. *Soviet Physics–Semiconductors*. 4:1573–1575.

Alferov, Z.I., V.M. Andreev, E.L. Portnoi and M.K. Trukan. 1970. AlAs-GaAs heterojunction injection lasers with a low room-temperature threshold. *Soviet Physics–Semiconductors*. 3:1107–1110.

Alferov, Z.I., V.M. Andreev, V.I. Korol'kov, D.N. Tret'yakov and V.M. Tuchkevich. 1968. High voltage p-n junctions in $Ga_xAl_{1-x}As$ crystals. *Soviet Physics–Semiconductors*. 1:1313–1314.

Armiento, C.A., M. Tabasky, C. Jagannath, T.W. Fitzgerald, C.L. Shieh, V. Barry, M. Rothman, A. Negri, P.O. Haugsjaa and H.F. Lockwood. 1991. Passive coupling of InGaAsP/InP laser array and singlemode fibres using silicon waferboard. *Electronics Letters*. 27:1109–1111.

Arnaud, J. 1996. Classical theory of laser linewidth. *Optical and Quantum Electronics*. 28:1589–1615.

Ashcroft, N.W. and N.D.Mermin. 1976. *Solid State Physics*. Washington D.C.: International Thomson Publishing.

Astles, M.G. 1990. *Liquid-Phase Epitaxial Growth of III-V Compound Semiconductor Materials and Their Device Applications*. Bristol: Adam Hilger.

Babic, D.I., J.J. Dudley, K. Streubel, R.P. Mirin, J.E. Bowers and E.L. Hu. 1995. Double-fused 1.52 μm vertical cavity lasers. *Applied Physics Letters*. 66:1030-1032.

Baney, D.M. and W.V. Sorin. 1990. Linewidth and power spectral measurements of single-frequency lasers. *Hewlett-Packard Journal*. February:92–96.

Basavanhally, N.R. and R.A. Nordin, 1997. Array device packaging. In *Optoelectronic Packaging*, edited by A.R. Mickelson, N.R. Basavanhally and Y.-C. Lee. new York: John Wiley and Sons.

Basrur, J. P., F. S. Choa, P.-L. Liu, J. Sipior, G. Rao, G. M. Carter and Y. J. Chen. 1997. The process and efficiency of ultraviolet generation from gallium nitride blue light emitting diodes. *Applied Physics Letters*. 71:1385-1387.

Bassous, E. 1978. Fabrication of novel three-dimensional microstructures by the anisotropic etching of (100) and (110) silicon. *IEEE Transactions on Electron Devices*. ED-25:1178–1184.

Bastard, G. 1988. *Wave mechanics applied to semiconductor heterostructures*. Les Ulis: Les Editions de Physique.

Baur, J., P. Schlotter and J. Schneider. 1998. White light emitting diodes. *Advances in Solid State Physics*. 37:67–78.

Bean, K.E. 1978. Anisotropic etching of silicon. *IEEE Transactions on Electron Devices*. ED-25:1185–1192.

Betschon, S. 2002. 20 Jahre 20 Zentimeter. *Neue Zürcher Zeitung*. 4 October. 69.

Bigo, S., Y. Frignac, G. Charlet, W. Idler, S. Borne, H. Gross, R. Dischler, W. Poehlmann, P. Tran, C. Simonneau, D. Bayart, G. Veith, A. Jourdan, J.-P. Hamaide. 2001. 10.2 Tbit/s (256x42.7 Gbit/s PDM/WDM) transmission over 100 km TeraLight™ fiber with 1.28 bit/s/Hz spectral efficiency. *Proceedings of the Optical Fiber Conference 2001*. PD25-1–PD25-3.

Blakemore, J.S. 1982. Semiconducting and other major properties of gallium arsenide. *Journal of Applied Physics*. 53:R123-R181.

Blatt, F.J. 1968. *Physics of Electronic Conduction in Solids*. New York: McGraw Hill.

Blum, O., S.P. Kilcoyne, M.E. Warren, T.C. Du, K.L. Lear, R.P. Schneider, R.F. Carson, G. Robinson and F.H. Peters. 1995. Vertical-cavity surface-emitting lasers with integrated refractive microlenses. *Electronics Letters*. 31:44–45.

Blume, O., A. Ambrosy, D. Ferling, K. Matthies, K. Rueß, M. Schilling, S. Schneider, G. Robert, J.L. Nicque, F. Boubal, S. Rabaron, D. Tregoat and C. Artigue. 2000. High Performance Hybrid Integrated Transmit Receive Device. *Proceedings of the 2000 European Conference on Optical Communications (ECOC 2000)*. 87-88.

Boggs, B., C. Greiner, T. Wang, H. Lin and T.W. Mossberg. 1998. Simple high-coherence rapidly tunable external-cavity diode laser. *Optics Letters*. 23:1906–1908.

Born, M. and E. Wolf. 1980. *Principles of Optics*. 6th ed. Oxford: Pergamon Press.

Bouadma, N., J. Liang, R. Levenson, S. Grosmaire, P. Boulet and S. Sainson. 1994. Integration of a laser diode with a polymer-based waveguide for photonic integrated circuits. *IEEE Photonics Technology Letters*. 6:1188–1190.

Brenner, K.-H., M. Kufner, S. Kufner, J. Moisel, A. Müller, S. Sinzinger, M. Testorf, J. Göttert and J. Mohr. 1993. Application of three-dimensional micro-optical components formed by lithography, electroforming and plastic molding. *Applied Optics*. 32:6464–6469.

Burrus, C.A. and D.A.B. Miller. 1971. Small-area, double-heterostructure aluminum-gallium arsenide electroluminescent diode sources for optical-fiber transmission lines. *Optics Communications*. 4:307–309.

Campbell, S.A. 2001. *The Science and Engineering of Microelectronic Fabrication*. Oxford: Oxford University Press.

Capasso, F., C. Gmachl, R. Paiella, A.Tredicucci, A.L. Hutchinson, D.L. Sivco, J.N. Baillargeon, A.Y. Cho and H.C. Liu. 2000. New frontiers in quantum cascade lasers and applications. *IEEE Journal on Selected Topics in Quantum Electronics*. 6:931–947.

Caracci, S.J., F.A. Kish, N. Holonyak, S.A. Maranowski, S.C. Smith, and R.D. Burnham. 1992. High-performance planar native-oxide buried-mesa index-guided AlGaAs-GaAs quantum well heterostructure lasers. *Applied Physics Letters*. 61:321-323.

Casey, H.C. and M.B. Panish. 1978a. *Heterostructure Lasers*. Part A. New York: Academic Press.

Casey, H.C. and M.B. Panish. 1978b. *Heterostructure Lasers*. Part B. New York: Academic Press.

Chen, C.H., S.J. Chang, Y.K. Su, J.K. Sheu, J.F. Chen, C.H. Kuo and Y.C. Lin. 2002. Nitride-based cascade near white light-emitting diodes. *IEEE Photonics Technology Letters*. 14:908-910.

Choquette, K.D. and H.Q. Hou. 1997. Vertical-cavity surface emitting lasers: moving from research to manufacturing. *Proceedings of the IEEE*. 85:1730–1739.

Choquette, K.D., G. Hasnain, Y.H. Wang, J.D. Wynn, R.S. Freund, A.Y. Cho and R.E. Leibenguth. 1991. GaAs vertical-cavity surface emitting lasers fabricated by reactive ion etching. *IEEE Photonics Technology Letters*. 3:859–862.

Choquette, K.D., M. Hong, R.S. freund, J.P. Mannaerts, R.C. Wetzel and R.E. Leibenguth. 1993. Vertical-cavity surface-emitting laser diodes fabricated by *in situ* dry etching and molecular beam epitaxy regrowth. *IEEE Photonics Technology Letters*. 5:284–286.

Chow, W.W. and H. Amano. 2001. Analysis of lateral-mode behavior in broad-area InGaN quantum-well lasers. *IEEE Journal of Quantum Electronics*. 37:265–273.

Chow, W.W. and S.W. Koch. 1999 *Semiconductor-Laser Fundamentals: Physics of the Gain Materials* Berlin: Springer.

Chow, W.W., K.D. Choquette, M.H. Crawford, K.L. Lear and G.R. Hadley. 1997. Design, fabrication and performance of infrared and visible vertical-cavity surface-emitting lasers. *IEEE Journal of Quantum Electronics*. 33:1810–1823.

Chraplyvy, A.R. and R.W. Tkach. 1998. Terabit/second transmission experiments. *IEEE Journal of Quantum Electronics*. 34:2103–2108.

Cohen, M.S., M.F. Cina, E. Bassous, M.M. Opyrsko, J.L. Speidell, F.J. Canora, M.J. De Franza. 1992. Packaging of high-density fiber/laser modules using passive alignment techniques. *IEEE Transactions on Components, Hybrids and Manufacturing Technology*. 15:944–954.

Coldren, L.A. and S.W. Corzine. 1995. *Diode lasers and photonic integrated circuits*. New York: John Wiley and Sons.

Crow, J.D., L.D. Comerford, J.S. Harper, M.J. Brady and R.A. Laff. 1978. Gallium arsenide laser-array-on-silicon package. *Applied Optics*. 17:479–485.

Crow, J.D., L.D. Comerford., J.S. Harper, M.J. Brady and R.A. Laff. 1978. Gallium-arsenide laser-array-on-silicon package. *Applied Optics*. 17:479–485.

Cutolo, A., M. Iodice, P. Spirito and L. Zeni. 1997. Silicon electro-optic modulator based on a three terminal device integrated in a low-loss single-mode SOI waveguide. *IEEE Journal of Lightwave Technology*. 15:505–518.

Czotscher, K., S. Weisser, A. Leven and J. Rosenzweig. 1999. Intensity modulation and chirp of 1.55-μm multiple-quantum-well laser diodes: modeling and experimental verification. *IEEE Journal of Selected Topics in Quantum Electronics*. 5:606–612.

de Labachelerie, M., N. Kaou, V. Armbruster, J.-C. Jeannot, P. Moillier, H. Porte and N. Devoldère. 2001. A micro-machined connector for the coupling of optical waveguides and ribbon optical fibers. *Sensors and Actuators A*. 89:36–42.

Delorme, F., G. Alibert, P. Boulet, S. Grosmaire, S. Slempkes and A. Ougazzaden. 1997. High reliability of high-power and widely tunable 1.55-μm distributed Bragg reflector lasers for WDM applications. *IEEE Journal of Selected Topics in Quantum Electronics*. 3:607–614.

Dereniak, E.L. and G. Boreman. 1996. *Infrared Detectors and Systems*. New York: John Wiley & Sons

Derrickson, D., ed. 1998. *Fiber Optic Test and Measurement*. Upper Saddle River, New Jersey: Prentice Hall PTR.

Dupuis, R.D. 1987. An introduction to the development of the semiconductor laser. *IEEE Journal of Quantum Electronics*. 23:651–657.

Dutta, N.K., J. Lopata, R. Logan, and T. Tanbun-Ek. 1991 Integrated distributed feedback laser and optical amplifier. *Applied Physics Letters*. 59:1676-1677.

Edge, C., R.M. Ash, C.G. Jones and M.J. Goodwin. 1991. Flip-chip solder bond mounting of laser diodes. *Electronics Letters*. 27:499–501.

Eimerl, D. 1987. Crystal symmetry and the electrooptic effect. *IEEE Journal of Quantum Electronics*. QE-23:2104–2115.

Einstein, A. 1917. Zur Quantentheorie der Strahlung. *Physikalische Zeitschrift* 18: 121–128.

Fabry, C. and A. Perot. 1899. Théorie et Applications d'une Nouvelle Méthode de Spectroscopie Interférentielle. *Annales de chimie et de physique*. 16(7) 115–144.

Faist, J., C. Gmachl, F. Capasso, C. Sirtori,. D.L. Sivco, J.N. Baillargeon and A.Y. Cho. 1997. Distributed feedback quantum cascade lasers. *Applied Physics Letters*. 70:2670–2672.

Faist, J., F. Capasso, C. Sirtori, D.L. Sivco, J.N. Baillargeon, A.L. Hutchinson, S.-N.G. Chu and A.Y. Cho. 1996. High power mid-infrared (λ~5 μm) quantum cascade lasers operating above room temperature. *Applied Physics Letters*. 68:3680–3682.

Faist, J., F. Capasso, D. L. Sivco, C. Sirtori, A. L. Hutchinson, and A. Y. Cho. 1994. Quantum cascade laser. *Science*. 264:553–556.

Favre, F. 1987. Theoretical analysis of external optical feedback on DFB semiconductor lasers. *IEEE Journal of Quantum Electronics*. QE-23:81–88.

Fiedler, F. and A. Schlachetzki. 1987. Optical parameters of InP-based waveguides. *Solid-State Electronics*. 30:73-83.

Fukuda, M. 1991. *Reliability and Degradation of Semiconductor Lasers and LEDs*. Boston: Artech House

Fukuda, M. 1999. *Optical Semiconductor Devices*. New York: John Wiley and Sons.

Gale, M.T. 1998. Direct writing of continuous-relief micro-optics. In *Micro-optics*, edited by H.-P. Herzig. London: Taylor and Francis.

Garmire, E. 1973. Optical waveguides in single layers of $Ga_{1-x}Al_xAs$ grown on GaAs substrates. *Applied Physics Letters*. 23:403-404.

Giacomelli, G., F. Marin, M. Gabrysch, K.H. Gulden and M. Moser. 1998. Polarization competition and noise properties of VCSELs. *Optics Communications*. 146:136–140.

Gmachl C., D.L. Sivco, R. Colombelli, F. Capasso and A.Y. Cho. 2002. Ultra-broadband semiconductor laser. *Nature*. 425:883–887.

Gmachl, C., F. Capasso, D. L. Sivco and A. Y. Cho. 2001a. Recent progress in quantum cascade lasers and applications. *Reports on Progress in Physics*. 64:1533–1601.

Gmachl, C., F. Capasso, R. Colombelli, R. Paiella, D.L. Sivco and A.Y. Cho. 2001b. Quantum cascade lasers shape up for trace-gas sensing. *Laser Focus World*. September:65-70.

Gmachl, C., H.Y. Hwang, R. Paiella, D.L. Sivco, J.L. Baillargeon, F. Capasso and A.Y. Cho. 2001c. Quantum cascade lasers with low-loss chalcogenide lateral waveguides. *IEEE Photonics Technology Letters*. 13:182–184.

Gourley, P.L. 1994. Microstructured semiconductor lasers for high-speed information processing. *Nature*. 371:571–577.

Gowar, J. 1993. *Optical Communication Systems*. New York: Prentice Hall.

Grunwald, R., H. Mischke and W. Rehak. 1999. Microlens formation by thin-film deposition with mesh-shaped masks. *Applied Optics*. 38:4117–4124.

Grunwald, R., S. Woggon, U. Griebner, R. Ehlert and W. Reinecke. 1998. Axial beam shaping with non-spherical microoptics. *Japanese Journal of Applied Physics*. 37:3701–3707.

Hall, R.N., G.E. Fenner, J.D. Kingsley, T.J. Soltys and R.O. Carlson. 1962. Coherent light emission from GaAs junctions. *Physical Review Letters*. 9:366-368.

Hamel, W.A., M.P. van Exter and J.P. Woerdman. 1992. Coherence properties of a semiconductor laser with feedback from a distant reflector: Experiment and theory. *IEEE Journal of Quantum Electronics*. 28:1459–1469.

Hartman, R.L., J.C. Dyment, C.J. Hwang and M. Kuhn. 1973. Continuous operation of GaAs-$Ga_{1-x}Al_xAs$ double heterostructure lasers with 30°C half-lives exceeding 1000 h. *Applied Physics Letters*. 23:181–183.

Hartman, R.L., N.E. Schumaker and R.W. Dixon. 1977. Continuously operated (Al,Ga)As double-heterostructure lasers with 70°C lifetimes as long as two years. *Applied Physics Letters*. 31:756–759.

Haselbeck, S., H. Schreiber, J. Schwider and N. Streibl. 1993. Microlenses fabricated by melting a photoresist on a base layer. *Optical Engineering*. 32:1322–1324.

Hashimoto, T., Y. Nakasuga, Y. Yamada, H. Terui, M. Yanagisawa, Y. Akahori, Y. Tohmori, K. Kato and Y. Suzuki. 1998. Multichip optical hybrid integration technique with planar lightwave circuit platform. *IEEE Journal of Lightwave Technology*. 16:1249–1258.

Hawthorn, C.J., K.P. Weber and R.E. Scholten. 2001. Littrow configuration tunable external cavity diode laser with fixed direction output beam. *Review of Scientific Instruments*. 72:4477–4479.

Hayashi, I., M. Panish and P.W. Foy. 1969. A low-threshold room-temperature injection laser. *IEEE Journal of Quantum Electronics*. 5:211–212.

Hayashi, I., M.B. Panish, P.W. Foy and S. Sumski. 1970. Junction lasers which operate continuously at room temperature. *Applied Physics Letters*. 17:109–111.

Hecht, E. 1987. *Optics*. Reading, MA: Addison-Wesley.

Hecht, J. 1999. *City of Light*. Oxford: Oxford University Press.

Hecht, J. 1999. *Understanding Fiber Optics*. 3d ed. Upper Saddle River, New Jersey: Prentice Hall.

Heinrich, J., E. Zeeb and K.J. Ebeling. 1997. Butt-coupling efficiency of VCSELs into multimode fibers. *IEEE Photonics Technology Letters*. 9:1555–1557.

Henry, C.H. 1982. Theory of the linewidth of semiconductor lasers. *IEEE Journal of Quantum Electronics*. QE-18:259–264.

Herman, M.A. and H. Sitter. 1989. *Molecular Beam Epitaxy*. Berlin: Springer-Verlag.

Herrick, R.W. and P.M. Petroff. 1998. Gradual degradation in 850-nm vertical-cavity surface-emitting lasers. *IEEE Journal of Quantum Electronics*. 34:1963–1969.

Herzig, H.-P., ed. 1997. *Micro-optics*. London: Taylor and Francis.

Higurashi, E., R. Sawada and T. Ito. 2002. Hybrid integrated optical sensor for non-invasive blood flow monitoring. *Proceedings of the 2002 IEEE/LEOS International Conference on Optical MEMS*. 33–34.

Hillmer, H., A. Grabmaier, H.-L. Zhu, S. Hansmann and A. Burkhard. 1995. Continuously chirped DFB gratings by specially bent waveguides for tunable lasers. *IEEE Journal of Lightwave Technology*. 13:1905–1912.

Hillmer, H., A. Grabmeier, H.-L. Zhu, S. Hansmann and H. Burkhard. 1995. Continuously chirped DFB gratings by specially bent waveguides for tunable lasers. *IEEE Journal of Lightwave Technology*. 13:1905–1912.

Himeno, A., K. Kato and T. Miya. 1998. Silica-based planar lightwave circuits. *IEEE Journal of Selected Topics in Quantum Electronics*. 4:913–924.

Hirabayashi, K., M. Wada and C. Amano. 2001. Liquid crystal variable optical attenuators integrated on planar lightwave circuits. *IEEE Photonics Technology Letters*. 13:609–611.

Hofstetter, D. and H. Zappe. 1997. Anomalous longitudinal mode hops in GaAs/AlGaAs distributed Bragg reflector lasers. *Applied Physics Letters*. 71:181–183.

Hofstetter, D., B. Maisenhölder and H. Zappe. 1998. Quantum-well intermixing for fabrication of lasers and photonic integrated circuits. *IEEE Journal of Selected Topics in Quantum Electronics*. 4:794-802.

Hofstetter, D., H. Zappe, J.E. Epler and J. Söchtig. 1994. Single-growth-step GaAs/AlGaAs distributed Bragg reflector lasers with holographically-defined recessed gratings. *Electronics Letters*, 30:1858–1859.

Holonyak, N. and S.F. Bevacqua. 1962. Coherent (visible) light emission from $Ga(As_{1-x}P_x)$ junctions. *Applied Physics Letters*. 1:82–83.

Hong, Y., S. Bandyopadhyay, S. Sivaprakasam, P.S. Spencer and K.A. Shore. 2003. Noise characteristics of a single-mode laser diode subject to strong optical feedback. *IEEE Journal of Lightwave Technology*. 20:1847–1850.

Hsieh, J.J., J.A. Rossi and J.P. Donnelly. 1976. Room-temperature cw operation of GaInAsP/InP double-heterostructure diode lasers emitting at 1.1 μm. *Applied Physics Letters*. 28:709–711.

Huang, K. 1987. *Statistical Mechanics*. New York: John Wiley and Sons.

Hunsperger, R.G. 2002. *Integrated Optics: Theory and Technology*. 5th ed. Berlin: Springer-Verlag.

Hunsperger, R.G., A. Yariv and A. Lee. 1977. Parallel end-butt coupling for optical integrated circuits. *Applied Optics*. 16:1026–1032.

Hurm, V., M. Ludwig, J. Rosenzweig, W. Benz, M. Berroth, R. Bosch, W. Bronner, A. Hülsmann, K. Köhler, B. Raynor and J. Schneider. 1993. 14 GHz bandwidth MSM photodiode AlGaAs/GaAs HEMT monolithic integrated optoelectronic receiver. *Electronics Letters*. 29:9-10.

Hutley, M.C. 1982. *Diffraction gratings*. London: Academic Press.

Hutley, M.C. 1997. Diffraction gratings. In *Diffractive Optics*, edited by J. Turunen and F. Wyrowski. Berlin: Akademie Verlag.

Hutley, M.C. 1998. Refractive lenslet arrays. In *Micro-optics*, edited by H.-P. Herzig. London: Taylor and Francis.

Iga, K., S. Ishikawa, S. Ohkuochi and T. Nishimura. 1984. Room-temperature pulsed oscillation of GaAlAs/GaAs surface-emitting injection laser. *Applied Physics Letters*. 45:348–350.

Iizuka, K. 1987. *Engineering Optics*. 2d ed. Berlin: Springer-Verlag.

Ikegami, T., S. Sudo and Y. Sakai. 1995. *Frequency Stabilization of Semiconductor Laser Diodes*. Boston: Artech House.

INSPEC. 1990. *Properties of Gallium Arsenide*. London: INSPEC.

Ishigure, T., M. Sato, A. Kondo and Y. Koike. 2002. High-bandwidth graded-index polymer optical fiber with high-temperature stability. *IEEE Journal of Lightwave Technology*. 20:1443–1448.

Jackson, J.D. 1975. *Electrodynamics*. 2d ed. New York: John Wiley & Sons.

Jackson, K.P., E.B. Flint, M.F. Cina, D. Lacey, Y. Kwark, J.M. Trewhella, T. Caulfield, P. Buchmann, Ch. Harder and P. Vettiger. 1994. A high-density, four-channel, OEIC transceiver module utilizing planar-processed optical waveguides and flip-chip, solder-bump technology. *IEEE Journal of Lightwave Technology*. 12:1185–1191.

Jewell, J.L., J.P. Harbison, A. Scherer, Y.H. Lee and L.T. Florez. 1991. Vertical-cavity surface-emitting lasers: Design, growth, fabrication and characterization. *IEEE Journal of Quantum Electronics*. 27:1332-1346.

Jones, C.A., K. Cooper, M.W. Field, J.D. Rush, R.G. Waller, J.V. Collins and P.J. Fiddyment. 1994. Hybrid integration of a laser diode with a planar silica waveguide. *Electronics Letters*. 30:215–216.

Kaino, T. 2000. Waveguide fabrication using organic nonlinear optical materials. *Journal of Optics A: Pure and Applied Optics*. 2:R1–R7.

Kallweit, D., A. Greiner, J. Korvink, and H. Zappe. 2002. Strain-induced modal and polarization behavior of vertical-cavity lasers. *Journal of Optics*.

Kaneda, T. 1985. Silicon and germanium photodiodes. In *Semiconductors and Semimetals*. Vol. 22D. edited by W.T. Tsang. San Diego: Academic Press 247-328

Kapon, E. 1999. Quantum wire and quantum dot lasers. In *Semiconductor Lasers I*, edited by E. Kapon. San Diego: Academic Press.

Kapon, E. and R. Bhat. 1987. Low-loss single-mode GaAs/AlGaAs optical waveguides grown by organometallic vapor phase epitaxy. *Applied Physics Letters*. 50:1628-1630.

Kashima, Y., A, Matoba and H. Takano. 1992. Performance and reliability of InGaAsP superluminescent diode. *IEEE Journal of Lightwave Technology*. 10:1644-1649.

Kazarinov, R. F. and R. A. Suris. 1971. Possibility of the amplification of electromagnetic waves in a semiconductor with a superlattice. *Soviet Physics–Semiconductors*. 5:707–709.

Kingston, R.H. 1978. *Detection of Optical and Infrared Radiation*. Berlin: Springer-Verlag.

Kittel, C. 1963 *Quantum Theory of Solids*. New York: John Wiley and Sons.

Kittel, C. 1996. *Introduction to Solid State Physics*. 7th ed. New York: John Wiley and Sons.

Kittel, C. and H. Kroemer. 1980. *Thermal Physics*. 2d. ed. San Francisco: W.H. Freeman.

Klein, R. and A. Neyer. 1994. Silicon micromachining for micro-replication technologies. *Electronics Letters*. 30:1672–1674.

Klingshirn, C.F. 1997. *Semiconductor Optics*. Berlin: Springer-Verlag.

Kobayashi, K. and I. Mito. 1988. Single frequency and tunable laser diodes. *IEEE Journal of Lightwave Technology*. 6:1623–1633.

Koch, T.L. and U. Koren. 1991. Semiconductor photonic integrated circuits. *IEEE Journal of Quantum Electronics*. 27:641–653.

Koch, T.L., U. Koren, G. Eisenstein, M.G. Young, M. Oron, C.R. Giles and B.I. Miller. 1990. Tapered waveguide InGaAs/InGaAsP multiple-quantum-well lasers. *IEEE Photonics Technology Letters*. 2:88–90.

Koeth, J., R. Dietrich and A. Forchel. 1998. GaSb vertical-cavity surface-emitting lasers for the 1.5 μm range. *Applied Physics Letters*. 72:1638–1640.

Kogelnik, H. 1988. Theory of optical waveguides. In *Guided-wave optoelectronics*, edited by T. Tamir. Berlin: Springer-Verlag.

Kogelnik, H. and C.V. Shank. 1972. Coupled-wave theory of distributed feedback lasers. *Journal of Applied Physics*. 43:2327–2335.

Köhler, R., A. Tredicucci, F. Beltram, H.E. Beere, E.H. Linfield, A.G. Davies, D.A. Ritchie, R.C. Iotti and F. Rossi. 2002. Terahertz semiconductor heterostructure laser. *Nature*. 417:156–159.

Koyama, F., S. Kinoshita and K. Iga. 1988. Room temperature CW operation of GaAs vertical cavity surface emitting laser. *Transactions of the Institute of Electronics, Information and Communications Engineers*. E71:1089–1090.

Kraus, J.D., and K.R. Carver. 1973. *Electromagnetics*. 2d ed. New York: McGraw Hill.

Kressel, H. and H. Nelson. 1969. Close confinement gallium arsenide pn junction lasers with reduced optical loss at room temperature. *RCA Review*. 30:106–113.

Kroemer, H. 1963. A proposed class of heterojunction injection lasers. *Proceedings of the IEEE*. 51:1782–1783.

Krüger, U. and K. Krüger. 1995. Simultaneous measurement of the linewidth, linewidth enhancement factor a and FM and AM response of a semiconductor laser. *IEEE Journal of Lightwave Technology*. 13:592–597.

Kuchta, D.M., J. Gamelin, J.D. Walker, J. Lin, K.Y. Lau, J.S. Smith, M. Hong and J.P. Mannaerts. 1993. Relative intensity noise of vertical cavity surface emitting lasers. *Applied Physics Letters*. 62:1194–1196.

Kuksenkov, D., S. Feld, C. Wilmsen, H. Temkin, S. Swirhun and R. Leibenguth. 1995. Linewidth and α-factor in AlGaAs/GaAs vertical cavity surface emitting lasers. *Applied Physics Letters*. 66:277–279.

Kumar, M., V. Gupta, G.N. DeBrabander, P. Chen, J.T. Boyd, A.J. Steckl, A.G. Choo, H.E. Jackson, R.D. Burnham and S.C. Smith. 1993. Optical channel waveguides in AlGaAs multiple-quantum-well structures formed by focused ion-beam-induced compositional mixing. *IEEE Photonics Technology Letters*, 4:435-438.

Kuwahara, H., M. Sasaki and N. Tokoyo. 1980. Efficient coupling from semiconductor lasers into single-mode fibers with tapered hemispherical ends. *Applied Optics*. 19:2578–2583.

Lammert, R.M., S.D. Roh, J.S. Hughes, M.L. Osowski and J.J. Coleman. 1997. MQW DBR lasers with monolithically-integrated external-cavity electroabsorption modulators fabricated without modification of the active region. *IEEE Photonics Technology Letters*. 9:566–568.

Lammert, R.M., T.M. Cockerill, D.V. Forbes and J.J. Coleman. 1994. Dual-channel strained-layer InGaAs-GaAs-AlGaAs WDM source with integrated coupler by selective-area MOCVD. *IEEE Photonics Technology Letters*. 6:1167–1169.

Lang, R.J., A.G. Larsson and J.G. Cody. 1991. Lateral modes of broad area semiconductor lasers: theory and experiment. *IEEE Journal of Quantum Electronics*. 27:312–320.

Lau, K.Y. 1989. Short-pulse and high-frequency signal generation in semiconductor lasers. *IEEE Journal of Lightwave Technology*. 7:400–419.

Lau, K.Y. and A. Yariv. 1985. Ultra-high speed semiconductor lasers. *IEEE Journal of Quantum Electronics*. 21:121–138.

Law, J.Y. and G.P. Agrawal. 1997. Effects of optical feedback on static and dynamic properties of vertical-cavity surface-emitting lasers. *IEEE Journal of Selected Topics in Quantum Electronics*. 3:353–358.

Lear, K.L., R.P. Schneider, K. D. Choquette, S.P. Kilcoyne, J.J. Fogiel and J.C. Zolper. 1994. Vertical cavity surface emitting lasers with 21% efficiency by metalorganic vapor phase epitaxy. *IEEE Photonics Technology Letters*. 6:1053–1055.

Leger, J.R. 1998. Laser beam shaping. In *Micro-optics*, edited by H.-P. Herzig. London: Taylor and Francis.

Lenstra, D., B.H. Verbeek and A.J. den Boef. 1985. Coherence collapse in semiconductor lasers due to optical feedback. *IEEE Journal of Quantum Electronics*. QE-21:674–679.

Li, Z.-M. and T. Bradford. 1995. A comparative study of temperature sensitivity of InGaAsP and AlGaAs MQW lasers using numerical simulations. *IEEE Journal of Quantum Electronics*. 31:1841–1847.

Lierstuen, L.O. and Sv. Sudbø. 1995. Coupling losses between standard single-mode fibers and rectangular waveguides for integrated optics. *Applied Optics*. 34:1024–1028.

Lin, C.F. and C. S. Juang. 1996. Superluminescent diodes with bent wave-guide. *IEEE Photonics Technology Letters*. 8:206–208.

Lin, L.Y., J.L. Shen, S.S. Lee and M.C. Wu. 1996. Realization of novel monolithic free-space optical disk pickup heads by surface micromachining. *Optics Letters*. 21:155–157.

Liou, K.-Y., U. Koren, E.C. Burrows, M. Oron, B.I. Miller, M. Young, G. Raybon and C.A. Burrus. 1990. Operation of integrated InGaAsP-InP optical amplifier-monitoring detector with feedback control circuit. *IEEE Photonics Technology Letters*. 2:878–880.

Lipson, S.G., and H. Lipson, D.S. Tannhauser. 1995. *Optical Physics*. 3d ed. Cambridge: Cambridge University Press.

Lorenz, H., M. Despont, N. Fahrni, N. LaBianca, P. Renaud and P. Vettiger. 1997. SU-8: a low-cost negative resist for MEMS. *Journal of Micromechanics and Microengineering*. 7:121-124.

Louderback, D.A., O. Sjölund, E.R. Hegbloom, J. Ko and L.A. Coldren. 1999. Flip-chip bonded arrays of monolithically integrated, microlensed vertical-cavity lasers and resonant photodetectors. *IEEE Photonics Technology Letters*. 11:304–306.

Lu, H., C. Blaauw and T. Makino. 1996. Single-mode operation over a wide temperature range in 1.3 μm InGaAsP/InP distributed feedback lasers. *IEEE Journal of Lightwave Technology*. 14:851–859.

MacDougal, M.H., P.D. Dapkus, V. Pudikov, H. Zhao and G.M. Yang. 1995. Ultralow threshold current vertical-cavity surface-emitting lasers with AlAs-oxide-GaAs distributed Bragg reflectors. *IEEE Photonics Technology Letters*. 7:229-231.

Madelung, O. 1978. *Introduction to Solid State Theory*. Berlin: Springer-Verlag.

Mahlke, G. and P. Gössing. 2001. *Fiber Optic Cables*. Erlangen: Publicis MCD Verlag.

Maiman, T.H. 1960. Stimulated optical radiation in ruby. *Nature*. 187:493-494.

Majewski, M.L., J. Barton, L.A. Coldren, Y. Akulova and M.C. Larson. 2002. Direct intensity modulation in sampled-grating DBR lasers. *IEEE Photonics Technology Letters*. 14:747–749.

Marcuse, D. 1974. *Theory of Dielectric Optical Waveguides*. San Diego: Academic Press.

Marsh, J.H. 1993. Quantum well intermixing. *Semiconductor Science and Technology*. 8:1136–1155.

Martin-Regalado, J., F. Prati, M. San Miguel and N.B Abraham. 1997. Polarization properties of vertical-cavity surface-emitting lasers. *IEEE Journal of Quantum Electronics*. 33:765–783.

Matsumoto, S., M. Fukuda, K. Sato, Y. Itaya and M. Yamamoto. 1994. Highly reliable 1.55 μm GaInAsP laser diodes buried with semi-insulating iron-doped InP. *Electronics Letters*. 30:1305-1306.

Matsuo, S., K. Tateno, T. Nakahara, H. Tsuda and T. Kurokawa. 1997. Use of polyimide bonding for hybrid integration of a vertical cavity surface emitting laser on a silicon substrate. *Electronics Letters*. 33:1148–1149.

Matz, R., J.G. Bauer, P. Clemens, G. Heise, H.F. Mahlein, W. Metzger, H. Michel and G. Schulte-Roth. 1994. Development of a photonic integrated transceiver chip for WDM transmission. *IEEE Photonics Technology Letters*. 6:1327–1329.

McIlroy, P.W., A. Kurobe, and Y. Uematsu. 1985. Analysis and application of theoretical gain curves to the design of multi-quantum well lasers. *IEEE Journal of Quantum Electronics*. QE-21:1958-1963.

Meier, H.P. and V. Graf. 1996. Research lab converts to diode-laser manufacturer. *Laser Focus World*. October:51.

Mensz, P.M. 1994. Prospects for truly blue $ZnSe/Zn_{1-u}Mg_uS_vSe_{1-v}/Zn_{1-x}Mg_xS_ySe_{1-y}$ semiconductor diode lasers. *Applied Physics Letters*. 65:2627–2629.

Menz, W., J. Mohr and O. Paul. 2001. *Microsystem Technology*. Weinheim: Wiley-VCH.

Merritt, S.A., K. Morbarhan, R. Whaley, S. Fox and M. Dagenais. 1997. Semiconductor laser and optical amplifier packaging. In *Optoelectronic Packaging*, edited by A.R. Mickelson, N.R. Basavanhally and Y.-C. Lee. new York: John Wiley and Sons.

Miller, L.F. 1969a. Controlled collapse reflow chip joining. *IBM Journal of Research and Development* 13:239–250.

Miller, S.E. 1969b. Integrated optics: an introduction. *Bell System Technical Journal*. 48:2059–2069.

Mino, S., Y. Yoshino, Y. Yamada, T. Terui, M. Yasu and K. Moriwaki. 1995. Planar lightwave circuit platform with coplanar waveguide for opto-electronic hybrid integration. *IEEE Journal of Lightwave Technology*. 13:2320–2326.

Mitin, V.V., V.A. Kochelap, and M.A. Stroscio. 1999 *Quantum Heterostructures: Microelectronics and Optoelectronics* Cambridge: Cambridge University Press.

Miyamoto, Y., K. Hagimoto and K. Kato. 1994. 15 Gbit/s optical receiver using waveguide pin photodiode. *Electronics Letters*. 30:1622-1623.

Mizutani, A., N. Hatori, N. Nishiyama, F. Koyama and K. Iga. 1998. A low-threshold polarization-controlled vertical-cavity surface-emitting laser grown on GaAs (311)B substrate. *IEEE Photonics Technology Letters*. 10:633–635.

Monti di Sopra, F., H. Zappe, M. Moser, R. Hövel, H.-P. Gauggel, and K. Gulden. 1999. Near-infrared vertical-cavity surface-emitting lasers with 3 MHz linewidth. *IEEE Photonics Technology Letters*. 11:1533–1535.

Monti di Sopra, F., H.-P. Gauggel, M. Brunner, R. Hövel, M. Moser and H. Zappe. 2001. Long-term spectral stability of single-mode VCSELs. *Electronics Letters*. 37:832–834.

Monti di Sopra, F., M. Brunner and R. Hövel. 2002. Polarization control in strained T-bar VCSELs. *IEEE Photonics Technology Letters*. 14:1034–1036.

Monti di Sopra, F., M. Brunner, H.-P. Gauggel, H. Zappe, M. Moser, R. Hövel and E. Kapon. 2000. Continuous-wave operation of phase-coupled vertical-cavity surface-emitting laser arrays. *Applied Physics Letters*. 77:2283–2285.

Moon, R.L., G.A. Antypas and L.W. James. 1974. Bandgap and lattice constant of GaInAsP as a function of alloy composition. *Journal of Electronics Materials*. 3:635–644.

Morgan, R.A., M.K. Hibbs-Brenner, R.A. Walterson, J.A. Lehman, T.M. Marta, S. Bounnak. E.L. Kalweit, T. Akinwade and J.C. Nohava. 1995. Producible GaAs-based top-surface emitting lasers with record performance. *Electronics Letters*. 31:462–463.

Morris, G.M. and K.J. McIntyre. 1997. Optical system design with diffractive optics. In *Diffractive Optics*, edited by J. Turunen and F. Wyrowski. Berlin: Akademie Verlag.

Moser, A. and E.E. Latta. 1992. Arrhenius parameters for the rate process leading to catastrophic damage of AlGaAs-GaAs laser facets. *Journal of Applied Physics*. 71:4848–4853.

Moss, D., D. Landheer, D. Conn, D. Halliday, S. Charbonneau, G. Aers, R. Barber and F. Chatenoud. 1992. High-speed photodetection in a reverse biased GaAs/AlGaAs GRINSCH SQW laser structure. *IEEE Photonics Technology Letters*. 4:609-611.

Nagarajan, R. and J.E. Bowers. 1999. High-speed lasers. In *Semiconductor Lasers I*, edited by E. Kapon. San Diego: Academic Press.

Nakagawa, G., K. Miura, S. Sasaki and M. Yano. 1996. Lens-coupled laser diode module integrated on silicon platform. *IEEE Journal of Lightwave Technology*. 14:1519–1523.

Nakamura, M., H.W. Yen, A. Yariv, E. Garmire, S. Somekh and H.L. Girvin. 1973. Laser oscillation in epitaxial GaAs waveguides with corrugation feedback. *Applied Physics Letters*. 23:224–226.

Nakamura, S. 2000. *The Blue Laser Diode*. Berlin: Springer-Verlag.

Nakamura, S., and S.F. Chichibu, eds. 2000. *Introduction to Nitride Semiconductor Blue Lasers and Light Emitting Diodes*. London: Taylor and Francis.

Nakamura, S., M. Senoh, N. Iwasa, S. Nagahama, TG. Yamada and T. Mukai. 1995. Superbright green InGaN single-quantum-well-structure light-emitting diodes. *Japanese Journal of Applied Physics*. 34:L1332-L1335.

Nakayama, T., Y. Itoh and A. Kakuta. 1993. Organic photo- and electroluminescent devices with double mirrors. *Applied Physics Letters*. 63:594–595.

Namjou, K., S. Cai, E. A. Whittaker, J. Faist, C. Gmachl, F. Capasso, D. L. Sivco and A. Y. Cho. 1998. Sensitive absorption spectroscopy with a room-temperature distributed-feedback quantum-cascade laser. *Optics Letters*. 23:219-221.

Narasimhamurty, T.S. 1981. *Photoelastic and Electro-optic Properties of Crystals*. New York: Plenum.

Nathan, M.I., W.P. Dumke, G. Burns, F.H. Dill and G. Lasher. 1962. Stimulated emission of radiation from GaAs p-n junctions. *Applied Physics Letters*. 1:62–64.

Nazarathy, M., W.V. Sorin, D.M. Baney and S.A. Newton. 1989. Spectral analysis of optical mixing measurements. *IEEE Journal of Lightwave Technology*. 7:1083–1096.

Nichols, D.T., J. Lopata, W.S. Hobson, N.K. Dutta, P.R. Berger, D.L. Sivco and A.Y. Cho. 1994. Monolithic GaAs/AlGaAs optical transmitter circuit using a single growth step. *Electronics Letters*. 30:490–491.

Nishihara, H., M. Haruna and T. Suhara. 1989. *Optical Integrated Circuits*. New York: McGraw Hill.

Öberg, M., P.-J. Rigole, S. Nilsson, T. Klinga, L. Bäckbom, K. Streubel, J. Wallin and T. Kjellberg. 1995. Complete single mode wavelength coverage over 40 nm with a super structure grating DBR laser. *IEEE Journal of Lightwave Technology*. 13:1892–1897.

Ohiso, Y., H. Okamoto, R. Iga, K. Kishi and C. Amano. 2002. Single transverse mode operation of 1.55–μm buried heterostructure vertical-cavity surface-emitting lasers. *IEEE Photonics Technology Letters*. 14:738–740.

Okamoto, H., M. Wada, Y. Sakai, T. Hirono, Y. Kawaguchi, Y. Kondo, Y. Kadota, K. Kishi and Y.Itaya. 1998. A narrow beam 1.3- μm-super luminescent diode integrated with a spot-size converter and a new type rear absorbing region. *IEEE Journal of Lightwave Technology*. 16:1881-1887.

Okano, H., A. Hiruta, H. Komano, T. Nishio, R. Takahashi, T. Kanaya and S. Tsuji. 1998. Hybrid integrated optical WDM transceiver module for FTTH systems. *Hitachi Cable Review*. 17:23–30.

Okano, H., H. Otsuki, H. Uetsuka, T. Teraoka, T. Shiota, S. Aoki and S. Tsuji. 1997. Passive aligned hybrid integrated WDM transceiver module using planar lightwave circuit platform. *IEICE (Institute of Electronics, Information and Communications Engineers) Transactions*. E80-C:112–116.

Okuda, M. and K. Onaka. 1977. Tunability of distributed Bragg-reflector lasers using the electro-optic effect. *Japanese Journal of Applied Physics*. 16:485–490

Ooi, B.S., K. McIlvaney, M.W. Street, A.S. Helmy, S.G. Ayling, A.C. Bryce, J.H. Marsh and J.S. Roberts. 1997. Selective quantum-well intermixing in GaAs-AlGaAs structures using impurity-free vacancy diffusion. *IEEE Journal of Quantum Electronics*. 33:1784-1793.

Orenstein, M., A. C. Von Lehmen, C. Chang-Hasnain, N. G. Stoffel, J. P. Harbison, L. T. Florez, E. Clausen, and J. E. Jewell. 1990. Vertical-cavity surface-emitting InGaAs/GaAs lasers with planar lateral definition. *Applied Physics Letters*. 56:2384-2386.

Owens, P.C.M., J.G. Rarity, P.R. Tapster. D. Knight and P.D. Townsend. 1994. Photon counting with passively quenched germanium avalanche. *Applied Optics*. 33:6895–6901.

Painter, O., A. Husain, A. Scherer, J.D. O'Brien, I. Kim, P.D. Dapkus. 1999. Room Temperature Photonic Crystal Defect Lasers at Near-Infrared Wavelengths in InGaAsP. *IEEE Journal of Lightwave Technology*. 17:2082 - 2095.

Pan, J.-W. and J.-I. Chyi. 1996. Theoretical study of the temperature dependence of 1.3-μm AlGaInAs-InP multiple-quantum-well lasers. *IEEE Journal of Quantum Electronics*. 32:2133–2138.

Panajotov, K., B. Ryvkin, J. Danckaert, M. Peeters, H. Thienpont and I. Veretennicoff. 1998. Polarization switching in VCSEL's due to thermal lensing. *IEEE Photonics Technology Letters*. 10:6–8.

Panish, M. B. 1975 Heterostructure injection lasers. *IEEE Transactions on Microwave Theory and Techniques*. MTT-23:20–30.

Panish, M.B. 1975. Heterostructure injection lasers. *IEEE Transactions on Microwave Theory and Techniques*. MTT-23:20–30.

Pankove, J. 1976 *Optical Processes in Semiconductors*. 2d. ed. New York; Dover Publications.

Pankove, J. A. I. 1968. Temperature dependence of emission efficiency and lasing threshold in laser diodes. *IEEE Journal of Quantum Electronics*. QE-4:119–112.

Park, S.-H., Y. Park, H. Kim, H. Jeon, S.M. Hwang, J.K. Lee, S.O. Nam, B.C. Koh, J. Y. Sohn and D. S. Kim. 2002. Microlensed vertical-cavity surface-emitting laser for stable single fundamental mode operation. *Applied Physics Letters*. 80:183–185.

Park, S.-J., K.-T. Jeong, S.-H. Park and H.-K. Sung. 2002. A novel method for fabrication of a PLC platform for hybrid integration of an optical module by passive alignment. *IEEE Photonics Technology Letters*. 14:486–488.

Pearsall, T.P. 1982. *GaInAsP Alloy Semiconductors*. Chichester, England: Wiley.

Pearsall, T.P. and M.A. Pollack. 1985. Compound semiconductor photodiodes. In *Semiconductors and Semimetals*. Vol. 22D. edited by W.T. Tsang. San Diego: Academic Press. 173-245

Petermann, K. 1991. *Laser diode modulation and noise*. Dordrecht: Kluwer Academic Publishers.

Petermann, K. 1995. External optical feedback phenomena in semiconductor lasers. *IEEE Journal of Selected Topics in Quantum Electronics*. 1:480–489.

Pettit, G.D. and W.J. Turner. 1965. Refractive index of InP. *Journal of Applied Physics*. 36:2081.

Planck, Max. 1948. *Wissenschaftliche Selbstbiographie*. Leipzig: Johann Ambrosius Barth Verlag.

Ponce, F.A. and D.P. Bour. 1997. Nitride-based semiconductors for blue and green light-emitting devices. *Nature*. 386:351–359.

Prather, D.W., S. Venkataraman, M. Lecompte, F. Kiamilev, J.N. Mait and G.J. Simonis. 2001. Optoelectronic multi-chip module integration for chip level optical interconnects. *IEEE Photonics Technology Letters*. 13:1112–1114.

Prosyk, K., J.G. Simmons and J.D. Evans. 1998. A systematic empirical study of the effect of well number and length and the temperature sensitivity of the threshold current in InGaAsP-InP MQW lasers. *IEEE Journal of Quantum Electronics*. 34:535–539.

Quist, T.M., R.H. Rediker, R.J. Keyes, W.E. Krag, B. Lax, A.L. McWhorther and H.J. Zeiger. 1962. Semiconductor maser of GaAs. *Applied Physics Letters*. 1:91–92.

Rapp, S., F. Salomonsson, J. Bentell, I. Sagnes, H. Moussa, C. Mériadec, R. Raj, K. Streubel and M. Hammar. 1999. Near room-temperature continuous-wave operation of electrically pumped 1.55 μm vertical cavity lasers with InGaAsP/InP bottom mirror. *Electronics Letters*. 35:49–50.

Razeghi, M. 1989. *The MOCVD Challenge*. Bristol, England: IOP.

Reinhart, F.K., R.A. Logan and T.P. Lee. 1974. Transmission properties of rib waveguides formed by anodization of epitaxial GaAs on $Al_xGa_{1-x}As$ layers. *Applied Physics Letters*. 24:270-272.

Remiens, D., B. Rose, M. Carre, and V. Hornung. 1990. GaInAsP/InP integrated ridge laser with a butt-jointed transparent optical waveguide fabricated by single-step metalorganic vapor-phase epitaxy. *Journal of Applied Physics*. 68:2450–2453.

Rhoderick, E.H. and R.H. Williams. 1988. *Metal-semiconductor contacts*. 2d ed. Oxford: Clarendon Press.

Ridley, B.K. 1988 *Quantum Processes in Semiconductors*. 2d ed. Oxford: Oxford University Press.

Roentgen, P., W. Heuberger, G.L. Bona, and P. Unger. 1991. MOVPE of AlGaInP/GaInP heterostructures for visible lasers. *Journal of Crystal Growth*. 107:724–730.

Rogers, A. 2001. *Understanding Optical Fiber Communications*. Boston: Artech House.

Rupprecht, H., J.M. Woodall and G.D. Pettit. 1967. Efficient visible electroluminescence at 300°K from $Ga_{1-x}Al_xAs$ p-n junctions grown by liquid-phase epitaxy. *Applied Physics Letters*. 11:81–83.

Ryu, S. and S. Yamamoto. 1986. Measurement of direct frequency modulation characteristics of DFB-LD by delayed self-homodyne technique. *Electronics Letters*. 22:1052–1054.

Ryvkin, B.S. and A.M. Georgievskii. 1999. Polarization selection in VCSELs due to current carrier heating. *Semiconductors*. 33:813–819.

Sadiku, M.N.O. 2001. *Elements of Electromagnetics*. Oxford: Oxford University Press.

Sakuma, I., H. Yonezu, K. Nishida, K. Kobayashi, F. Saito and Y. Nannichi. 1971. Continuous operation of junction lasers at room temperature. *Japanese Journal of Applied Physics*. 10:282–283.

Saleh, B.E.A. and M.C. Teich. 1991. *Fundamentals of Photonics*. New York: John Wiley and Sons.

Sawada, R., E. Higurashi and T. Itoh. 1999. Etalon combining silicon waveguide and silicon plate. *Proceedings of the 3rd International Conference on Micro Opto Electro Mechanical Systems*. 140–144.

Schawlow, A.L. and C.H. Townes. 1958. Infrared and optical masers. *Physical Review*. 112:1940–1949.

Schmid, W., C. Jung, B. Weigl, G. Reiner, R. Michalzik and K.J. Ebeling. 1996. Delayed self-heterodyne linewidth measurement of VCSEL's. *IEEE Photonics Technology Letters*. 8:1288–1290.

Schockley, W. 1950 *Electrons and Holes in Semiconductors* Princeton: D. Van Nostrand Company.

Schreiner, R., P. Nägele, M. Körbl, A. Gröning, J.L. Gentner and H. Schweizer. 2001. Monolithically integrated tunable laterally coupled distributed-feedback lasers. *IEEE Photonics Technology Letters*. 13:1277–1279

Schubert, E.F., Y.H. Wang, A.Y. Cho, L.W. Tu and G.J. Zydzik. 1992. Resonant cavity light emitting diode. *Applied Physics Letters*. 60:921–923.

Scott, J.W., B.J. Thibeault, D.B. Young, L.A. Coldren and F.H. Peters. 1994. High-efficiency sub-milliamp vertical cavity lasers with intra-cavity contacts. *IEEE Photonics Technology Letters*. 6:678–680.

Seeger, K. 2002. *Semiconductor Physics*. 8th ed. Berlin: Springer-Verlag.

Seki, S., H.Oohashi, H. Sugiura, T. Hirono and K. Yokoyama. 1996. Study on the dominant mechanisms for the temperature sensitivity of threshold current in 1.3-μm InP-based strained-layer quantum-well lasers. *IEEE Journal of Quantum Electronics*. 32:1478–1486.

Seurin, J.-F.P., S.L. Chuang, L.M.F. Chirovsky and K.D. Choquette. 2002. Novel VCSEL designs deliver high single-mode output power. *Laser Focus World*. May:119–122.

Shah, V.S., L. Curtis, R.S. Vodhanel, D.P. Bour and W.C. Young. 1990. Efficient power coupling from a 980-nm, broad-area laser to a single-mode fiber using a wedge-shaped fiber endface. *IEEE Journal of Lightwave Technology*. 8:1313–1318.

Shani, Y., R. Rosman, A. Katzir, P. Norton, M. Tacke, and H.M. Preier. 1988. Distributed Bragg reflector $Pb_{1-x}Sn_xSe/Pb_{1-x-y}Eu_ySn_xSe$ diode lasers with a broad single mode tuning range. *Journal of Applied Physics*. 63:5603–5606.

Shannon, R.R. 1997. *The Art and Science of Optical Design*. Cambridge: Cambridge University Press.

Shau, R., M. Ortsiefer, M. Zigldrum, J. Rosskopf, G. Böhm, F. Köhler and M.-C. Amann. 2000. Low-threshold InGaAlAs/InP vertical cavity surface-emitting laser diodes for 1.8 μm wavelength range. *Electronics Letters*. 36:1286–1287.

Shen, C.C., J.J. Hsieh and T.A. Lind. 1977. 1500-h continuous cw operation of double-heterostructure GaInAsP/InP lasers. *Applied Physics Letters*. 7:353–354.

Shiono, T. and H. Ogawa. 1994. Planar-optic-disk pickup with diffractive micro-optics. *Applied Optics*. 33:7350–7355.

Siegmann, A.E. 1986. *Lasers*. Sausalito, CA: University Science Books.

Silfvast, W.T. 1996. *Laser Fundamentals*. Cambridge: Cambridge University Press.

Silver, M., W.E. Booij, S. Malik, A. Galbraith, S.Uppal, P.F. McBrien, G.M. Berry, P.D. Ryder, S.J. Chandler, D.M. Atkin, S.Chan, R. Harding and R.M. Ash. 2002. Wide temperature (–20°C–95°C) operation of an uncooled 2.5 Gb/s 1300-nm DFB laser. *IEEE Photonics Technology Letters*. 14:741–743.

Sinzinger, S. and J. Jahns. 2002. *Microoptics*.New York: John Wiley and Sons.

Smith, F.G. and T.A. King. 2000. *Optics and Photonics*. Chichester: John Wiley and Sons.

Smith, R.A. 1978. *Semiconductors*. 2d ed. Cambridge: Cambridge University Press.

Smith, W.J. 1990. *Modern Optical Engineering*. 2d ed. New York: McGraw Hill.

Soda, H., K. Iga, C. Kitahara and Y. Suematsu. 1979. GaInAsP/InP surface emitting injection lasers. *Japanese Journal of Applied Physics*. 18:2329–2330.

Solgaard, O., M. Daneman, N.C. Tien, A. Friedberger, R.S. Muller and K.Y. Lau. 1995. Optoelectronic packaging using silicon surface-micromachined alignment mirrors. *IEEE Photonics Technology Letters*. 7:41–43.

Soole, J.B.D and H. Schumacher. 1991. InGaAs metal-semiconductor-metal photodetectors for long wavelength optical communications. *IEEE Journal of Quantum Electronics*. 27:737–752.

Soref, R.A. and J.P. Lorenzo. 1986. All-silicon active and passive guided-wave components for $\lambda = 1.3$ and 1.6 μm. *IEEE Journal of Quantum Electronics*. QE-22:873–879.

Soref, R.A., D.L. McDaniel and B.R. Bennett. 1988. Guided-wave intensity modulators using amplitude-and-phase perturbations. *IEEE Journal of Lightwave Technology*. 6:437–444.

Steele, R.V. 2002. Review and forecast of the laser markets. Part II: diode lasers. *Laser Focus World*. February:61–82.

Stern, M. 1998. Binary optics fabrication. In *Micro-optics*, edited by H.-P. Herzig. London: Taylor and Francis.

Strandman, C. and Y. Bäcklund. 1997. Bulk silicon holding structures for mounting of optical fibers in V-grooves. *IEEE Journal of Microelectromechanical Systems*. 6:35–40.

Strzelecka, E.M., G.D. Robinson, M.G. Peters, F.H. Peters and L.A. Coldren. 1995. Monolithic integration of vertical-cavity laser diodes with refractive GaAs microlenses. *Electronics Letters*. 31:724–725.

Su, G.-D. J., H. Toshiyoshi and M.C. Wu. 2001. Surface-micromachined 2-D optical scanners with high-performance single-crystalline silicon micromirrors. *IEEE Photonics Technology Letters*. 13:606–608.

Sun, H. 1997. Measurement of laser diode astigmatism. *Optical Engineering*. 36:1082–1087.

Syms, R.R.A. and A. Lohmann. 2002. Tuning mechanism for a MEMS external cavity laser. *Proceedings of the 2002 IEEE/LEOS International Conference on Optical MEMS*. 183–184.

Sze, S.M. 1981. *Physics of Semiconductor Devices*. New York: John Wiley and Sons.

Tabib-Azar, M. and G. Beheim. 1997. Modern trends in microstructures and integrated optics for communication, sensing and actuation. *Optical Engineering*. 36:1307–1318.

Takato, N., K. Junguji, M. Yasu, H. Toba and M. Kawachi. 1988. Silica-based single-mode waveguides on silicon and their application to guided-wave optical interferometers. *IEEE Journal of Lightwave Technology*. 6:1003-1010.

Takeshita, T., T. Kagawa, K. Tateno, O. Tadanaga, Y. Tohmori and C. Amano. 2002. Degradation behavior of 850-nm vertical-cavity surface-emitting lasers with an air-post index-guide structure. *IEEE Journal of Lightwave Technology*. 20:722–729.

Takeuchi, H. and K. Oe. 1989. Very low loss GaAs/AlGaAs miniature bending waveguide with curvature radii less than 1 mm. *Applied Physics Letters*. 54:87-89.

Tamir, T. ed. 1988. *Guided Wave Optoelectronics*. Berlin: Springer-Verlag.

Tanaka, T., H. Takahashi, M. Oguma, T. Hashimoto, Y. Hibino, Y. Yamada, Y. Itaya, J. Albert and K.O. Hill. 1996. Integrated external cavity laser composed of spot-size converted LD and UV written grating in silica waveguide on Si. *Electronics Letters*. 32:1202–1203.

Tansu, N. and L.J. Mawst. 2002. Temperature sensitivity of 1300-nm InGaAsN quantum-well lasers. *IEEE Photonics Technology Letters*. 14:1052–1054.

Taylor, G.W. and P.A. Evaldson. 1994. Temperature dependent operation of the vertical cavity surface emitting laser. *IEEE Journal of Quantum Electronics*. 30:2262–2270.

Tell, B., Y. H. Lee, K. F. Brown-Goebeler, J. L. Jewell, R. E. Leibenguth, M. T. Asom, G. Livescu, L. Luther and V. D. Mattera. 1990. High-power cw vertical-cavity top surface-emitting GaAs quantum well lasers. *Applied Physics Letters*. 57:1855-1857.

Thurmond, C.D. 1975. The standard thermodynamic functions for the formation of electrons and holes in Ge, Si, GaAs and InP. *Journal of the Electrochemical Society*. 122:1133–1141.

Townes, C. 2000. The birth of the laser. *Opto-Laser Europe*. Millennium issue:20–30.

Tsai, C.S. 1990. *Guided-Wave Acousto-Optics*. Berlin: Springer-Verlag

Tsang, W.T. 1987. Quantum confined semiconductor lasers in Dingle, R. (ed.), *Semiconductors and Semimetals*. 24:397–458.

Tschudi, T. and L. Wang. 1997. Diffractive optics for semiconductor lasers. In *Diffractive Optics*, edited by J. Turunen and F. Wyrowski. Berlin: Akademie Verlag.

Tucker, R.S. 1985. High-speed modulation of semiconductor lasers. *IEEE Journal of Lightwave Technology*. LT-3:1180–1192.

Turunen, J. and F. Wyrowski. eds. 1997. *Diffractive Optics*. Berlin: Akademie Verlag.

Uchida, N., Y. Hibino, Y. Suzuki, T. Kurosaki, N. Ishihara, M. Nakamura, T. Hashimoto, Y. Akahori, Y. Inoue, K. Morwaki, Y. Yamada, K. Kato, Y. Yohmori, M. Wada and T. Sugie. 1996. Passively aligned hybrid WDM module integrated with spot-size converter integrated laser diode and waveguide photodiode in PLC platform for fibre-to-the-home. *Electronics Letters*. 32:1664–1665.

Uenishi, Y., K. Honma and S. Nagaoka. 1996. Tunable laser diode using a nickel micromachined external mirror. *Electronics Letters*. 32:1207–1208.

Ünlü, M.S. and S. Strite. 1995. Resonant cavity enhanced photonic devices. *Journal of Applied Physics*. 78:607–639.

Ura S. and S.J. Sheard. 1998. A configuration for guided-wave excitation into a disposable integrated-optic head. *Optics Communications*. 146:85–89.

Urbach, F. 1953. The long-wavelength edge of photographic sensitivity and of the electronic absorption of solids. *Physical Review*. 92:1324.

Vakshoori, D., P. Tayebati, C.-C. Lu, M. Azimi, P. Wang, J.-H. Zhou and E. Canoglu. 2mW CW singlemode operation of a tunable 1550 nm vertical cavity surface emitting laser with 50 nm tuning range. *Electronics Letters*. 35:900–901.

van der Ziel, J.P., R. Dingle, R.C. Muller, W. Wiegmann, and W.A. Nordland. 1975. Laser oscillations from quantum states in very thin GaAs-$Al_{0.2}Ga_{0.8}As$ multilayer structures. *Applied Physics Letters*. 26:463–465.

van Exter, M.P., M.B. Willemsen and J.P. Woerdman. 1998. Polarization fluctuations in vertical-cavity semiconductor lasers. *Physical Review A*. 58:4191–4205.

van Roosbroeck, W. and W. Schockley. 1954. Photon radiative recombination of electrons and holes in germanium. *Physical Review*. 94:1558–1560.

Varshni, Y.P. 1967. Temperature dependence of the energy gap in semiconductors. *Physica*. 34:149–154.

Venkatasubramanian, R. and M.L. Timmons. 1994. Optoelectronic properties of eutectic-metal-bonded (EMB) GaAs-AlGaAs structures on Si substrates. *Solid-State Electronics*. 37:1809–1815.

Vettiger, P., M.K. Benedict, G.-L. Bona, P. Buchmann, C. Cahoon, K. Dätwyler, H.-P. Dietrich, A. Moser, H.K. Seitz, O. Voegeli, D.J. Webb and P. Wolf. 1991. Full-wafer technology - a new approach to large-scale laser fabrication and integration. *IEEE Journal of Quantum Electronics*. 27:1319–1331.

Wakao, K. 1992. Optoelectronic and photonic integrated circuits. In *Waveguide Optoelectronics* edited by J.H. Marsh and R.M. De La Rue. Dordrecht: Kluwer Academic Publishers.

Wakita, K. 1997. *Semiconductor Optical Modulators*. Dordrecht: Kluwer Academic Publishers

Wale, M.J. and C. Edge. 1990. Self-aligned flip-chip assembly of photonic devices with electrical and optical connections. *IEEE Transactions on Components, Hybrids and Manufacturing Technology*. 13:780–786.

Walker, R.G. 1987. High-speed electrooptic modulation in GaAs/GaAlAs waveguide devices. *IEEE Journal of Lightwave Technology*. LT-5:1444–1453.

Wallrabe, U., U. Hollenbach, P. Krippner, J. Mohr and T. Oka. 2002. Characterization of a micro optical distance sensor. *Proceedings of the 2002 IEEE/LEOS International Conference on Optical MEMS*. 29–30.

Wang, S., J.B. Hurst, F. Ma, R. Sidhu, X. Sun, X.G. Zheng, A.L. Holmes, A. Huntington, L.A. Coldren and J.C. Campbell. 2002. Low-noise impact-ionization-engineered avalanche photodiodes grown on InP substrates. *IEEE Photonics Technology Letters*. 14:1722–1724.

Weigl, B., M. Grabherr, C. Jung, R. Jäger, G. Reiner, R. Michalzik, D. Sowada and K.J. Ebeling. 1997. High-Performance Oxide-Confined GaAs VCSELs. *IEEE Journal of Selected Topics in Quantum Electronics*. 3:409–415.

Weinert, A. 1999. *Plastic Optical Fibers*. Erlangen: Publicis MCD Verlag.

Whitaker, T. 1999. Commercial VCSELs. *Compound Semiconductor*. 5:16–27.

Wilmsen, C.W., H. Temkin and L.A. Coldren, eds. 1999. *Vertical-Cavity Surface-Emitting Lasers*. Cambridge: Cambridge University Press.

Wirth, R., C. Karnutsch, S. Kugler and K. Streubel. 2001. High-efficiency resonant-cavity LEDs emitting at 650 nm. *IEEE Photonics Technology Letters*. 13:421–423.

Wu, M.C. 1997. Micromachining for optical and optoelectronic systems. *Proceedings of the IEEE*. 85:1833–1856.

Yamada, M. and Y. Suematsu. 1981. Analysis of gain suppression in undoped injection lasers. *Journal of Applied Physics*. 52 2653–2664.

Yamada, Y., F. Hanawa, T. Kitoh, and T. Maruno. 1992. Low-loss and stable fiber-to-waveguide connection using UV curable adhesive. *IEEE Photonics Technology Letters*. 4:906–908.

Yariv, A. 1973. Coupled-mode theory for guided-wave optics. *IEEE Journal of Quantum Electronics*. QE-9:919–933.

Yariv, A. 1991. *Optical Electronics*. 4th ed. Philadelphia: Saunders College Publishing.

Yoshikawa, T., T. Kawakami, H. Saito, H. Kosaka, M. Kajita, K. Kurihara, Y. Sugimoto and K. Kasahara. 1998. Polarization-controlled single-mode VCSEL. *IEEE Journal of Quantum Electronics*. 34:1009–1015.

Yoshikawa, T., Y. Sugimoto, H. Hotta, K. Tada, K. Kabayashi, F. Miyasaka and K. Asakawa. 1995. 650 nm AlGaInP visible light laser diode with dry-etched mesa stripe. *Japanese Journal of Applied Physics*. 34:1273–1278.

Young, M.G., U. Koren, B.I. Miller, M.A. Newkirk, M. Chien, M. Zirngibl, C. Dragone, B. Tell, H.M. Presby and G. Raybon. 1993. A 16 x 1 wavelength division multiplexer with integrated distributed Bragg reflector lasers and electroabsorption modulators. *IEEE Photonics Technology Letters*. 5:908–910.

Zappe, H. 1995. *Introduction to Semiconductor Integrated Optics*. Boston: Artech House.

Zappe, H. 2003. Integrated Micro-Optics. In *MEMS Handbook*, edited by J. Korvink and O. Paul. New York: Noyes Publications.

Zappe, H., D. Hofstetter and H.E.G. Arnot. 1994. Quantum well engineering for semiconductor integrated optical sensors. *SPIE Proceedings on Nanofabrication Technologies and Device Integration*. 2213:205–216.

Zappe, H., M. Hess, M. Moser, R. Hövel, K. Gulden, H.-P. Gauggel and F. Monti di Sopra. 2000. Narrow-linewidth vertical-cavity surface-emitting lasers for oxygen detection. *Applied Optics*. 39:2475–2479.

Zeng, L., B. X. Yang, A. Cavus, W. Lin, Y. Y. Luo, M. C. Tamargo, Y. Guo and Y. C. Chen. 1998. Red–green–blue photopumped lasing from ZnCdMgSe/ZnCdSe quantum well laser structures grown on InP. *Applied Physics Letters*. 72:3136–3138.

Zengerle, R., H. Brückner, H. Olzhausen and A. Kohl. 1992. Low-loss fibre-chip coupling by laterally tapered InP/InGaAsP waveguide structure. *Electronics Letters*. 28:631–632.

Ziman, J.M. 1972. *Principles of the Theory of Solids*. 2d ed. Cambridge: Cambridge University Press.

Zory, P.S. 1993. *Quantum Well Lasers*. Boston: Academic Press.

Zucker, J.E., K.L. Jones, M.A. Newkirk, R.P. Gnall, B.I. Miller, M.G. Young, U. Koren, C.A. Burrus and B. Tell. 1992. Quantum well interferometric modulator monolithically integrated with 1.55 μm tunable distributed Bragg reflector laser. *Electronics Letters*. 28:1888–1889.

Index

Printing: Druckhaus Berlin-Mitte GmbH
Binding: Buchbinderei Stein & Lehmann, Berlin